Principles of Vibration and Sound

Second Edition

Springer
New York
Berlin
Heidelberg
Hong Kong
London
Milan
Paris
Tokyo

Thomas D. Rossing
Neville H. Fletcher

Principles of Vibration and Sound

Second Edition

With 182 Illustrations

Springer

Thomas D. Rossing
Physics Department
Northern Illinois University
DeKalb, IL 60015, USA

Neville H. Fletcher
Department of Physical Sciences
Research School of Physical
Sciences and Engineering
Australian National University
Canberra, ACT 0200 Australia

Library of Congress Cataloging-in-Publication Data
Rossing, Thomas, D., 1929–
Principles of vibration and sound / Thomas D. Rossing, Neville H.
Fletcher. — 2nd ed.
p. cm.
Includes bibliographical references and index.

1. Acoustical engineering. 2. Vibration. I. Fletcher, Neville H.
(Neville Horner) II. Title
TA365.R67 2003
534—dc21 2003054413

Printed in the United States of America. (ASC/EB)

ISBN 978-1-4419-2343-1

Springer-Verlag is a part of *Springer Science+Business Media*

springeronline.com

Preface to the Second Edition

The first edition of this book presented the principles of vibration and sound with only a little discussion of applications of these principles. During the past eight years, our own experience, as well as that of other teachers who used it as a textbook, has indicated that students would benefit from more discussion of applications. In this edition we have revised some of the material in the first nine chapters, but more importantly we have added four new chapters dealing with applications, including microphones, loudspeakers, and other transducers; acoustics of concert halls and studios; sound and noise outdoors; and underwater sound. Of course we could have selected many additional applications of vibration and sound, but that would have led to a book with too much material for the average acoustics course in physics and engineering departments. We think there is now ample material in the book so that instructors may select the applications of particular interest and omit the others without loss of continuity. We have continued to stress concepts over detailed theory, as seems most appropriate for an introductory course.

We appreciate the comments we have received from users, students, and teachers alike, and we continue to welcome feedback.

September 2003

Thomas D. Rossing
Neville H. Fletcher

Preface to the First Edition

Some years ago we set out to write a detailed book about the basic physics of musical instruments. There have been many admirable books published about the history of the development of musical instruments, about their construction as a master craft, and about their employment in musical performance; several excellent books have treated the acoustics of musical instruments in a semiquantitative way; but none to our knowledge had then attempted to assemble the hard acoustic information available in the research literature and to make it available to a wider readership. Our book *The Physics of Musical Instruments*, published by Springer-Verlag in 1991 and subsequently reprinted several times with only minor corrections, was the outcome of our labor.

Because it was our aim to make our discussion of musical instruments as complete and rigorous as possible, our book began with a careful introduction to vibrating and radiating systems important in that field. We treated simple linear oscillators, both in isolation and coupled together, and extended that to a discussion of some aspects of driven and autonomous nonlinear oscillators. Because musical instruments are necessarily extended structures, we then went on to discuss the vibrations of strings, bars, membranes, plates, and shells, paying particular attention to the mode structures and characteristic frequencies, for it is these that are musically important. The generation and propagation of acoustic waves in air is of obvious importance, and this too received fairly thorough discussion, at least in relation to those parts of the subject relevant to our major concern. Wind instruments, of course, consist of pipes and horns, and the propagation of waves in these structures, their normal modes, and their radiation properties were all carefully treated, again in the musical instrument context. The first third of our book thus presented a broad, but admittedly somewhat eclectic, treatment of the basic subject matter of vibrations and acoustics.

In response to several suggestions, the publishers have decided to issue this first section of *The Physics of Musical Instruments* as a separate book, suitable for use as a text in standard courses in vibrations and acoustics. We will not conceal the fact that, had we set out to write such a book in the first place, its content would probably have been rather different. But the subject matter of acoustics is so wide and the possible manners of approach so various that we believe the academic community may welcome this view of the subject. It is an unashamedly basic book, with emphasis on fundamental dynamical principles rather than on practical applications and with a moderately mathematical approach. It must therefore be left to supplementary reading to fill in fascinating and important material on such topics as physical acoustics, microphones, loudspeakers, architectural acoustics, and auditory physiology. Even for musical instruments the interested reader is referred to our complete book. The references in the text are similarly eclectic, with emphasis on those relating to musical applications.

To make the book more useful in general courses in acoustics and vibrations, we have added several new sections and one new chapter—on network analogs for acoustic systems. We have also included some problems at the end of each chapter to assist with the use of the book in a teaching environment.

January 1994

Neville H. Fletcher
Thomas D. Rossing

Contents

Part 1
Vibrating Systems

CHAPTER 1

Free and Forced Vibrations of Simple Systems

Mechanical, acoustical, or electrical vibrations are the sources of sound in musical instruments. Some familiar examples are the vibrations of strings (violin, guitar, piano, etc), bars or rods (xylophone, glockenspiel, chimes, clarinet reed), membranes (drums, banjo), plates or shells (cymbal, gong, bell), air in a tube (organ pipe, brass and woodwind instruments, marimba resonator), and air in an enclosed container (drum, violin, or guitar body).

In most instruments, sound production depends upon the collective behavior of several vibrators, which may be weakly or strongly coupled together. This coupling, along with nonlinear feedback, may cause the instrument as a whole to behave as a complex vibrating system, even though the individual elements are relatively simple vibrators.

In the first seven chapters, we will discuss the physics of mechanical and acoustical oscillators, the way in which they may be coupled together, and the way in which they radiate sound. Since we are not discussing electronic musical instruments, we will not deal with electrical oscillators except as they help us, by analogy, to understand mechanical and acoustical oscillators.

Many objects are capable of vibrating or oscillating. Mechanical vibrations require that the object possess two basic properties: a stiffness or springlike quality to provide a restoring force when displaced and inertia, which causes the resulting motion to overshoot the equilibrium position. From an energy standpoint, oscillators have a means for storing potential energy (spring), a means for storing kinetic energy (mass), and a means by which energy is gradually lost (damper). Vibratory motion involves the alternating transfer of energy between its kinetic and potential forms.

The inertial mass may be either concentrated in one location or distributed throughout the vibrating object. If it is distributed, it is usually the mass per unit length, area, or volume that is important. Vibrations in distributed mass systems may be viewed as standing waves.

The restoring forces depend upon the elasticity or the compressibility of some material. Most vibrating bodies obey Hooke's law; that is, the restoring force is proportional to the displacement from equilibrium, at least for small displacement.

1.1. Simple Harmonic Motion in One Dimension

The simplest kind of periodic motion is that experienced by a point mass moving along a straight line with an acceleration directed toward a fixed point and proportional to the distance from that point. This is called simple harmonic motion, and it can be described by a sinusoidal function of time t: $x(t) = A \sin 2\pi f t$, where the amplitude A describes the maximum extent of the motion, and the frequency f tells us how often it repeats.

The period of the motion is given by

$$T = \frac{1}{f}. \tag{1.1}$$

That is, each T seconds the motion repeats itself.

A simple example of a system that vibrates with simple harmonic motion is the mass–spring system shown in Fig. 1.1. We assume that the amount of stretch x is proportional to the restoring force F (which is true in most springs if they are not stretched too far), and that the mass slides freely without loss of energy. The equation of motion is easily obtained by combining Hooke's law, $F = -Kx$, with Newton's second law, $F = ma = m\ddot{x}$. Thus,

$$m\ddot{x} = -Kx$$

and

$$m\ddot{x} + Kx = 0,$$

where

$$\ddot{x} = \frac{d^2x}{dt^2}.$$

The constant K is called the spring constant or stiffness of the spring (expressed in newtons per meter). We define a constant $\omega_0 = \sqrt{K/m}$, so that the equation of motion becomes

$$\ddot{x} + \omega_0^2 x = 0. \tag{1.2}$$

This well-known equation has these solutions:

$$x = A \cos(\omega_0 t + \phi) \tag{1.3}$$

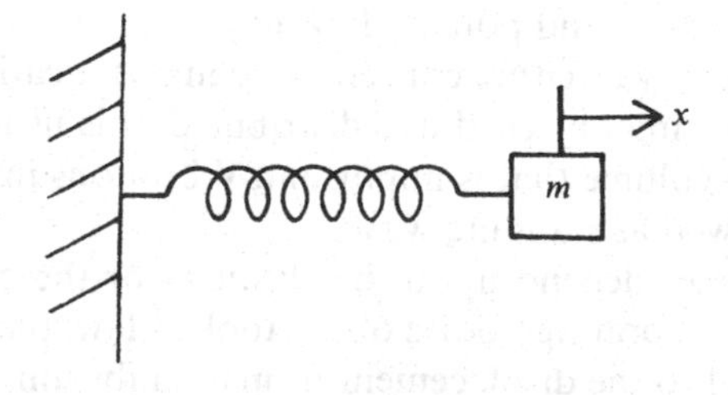

Fig. 1.1. Simple mass–spring vibrating system.

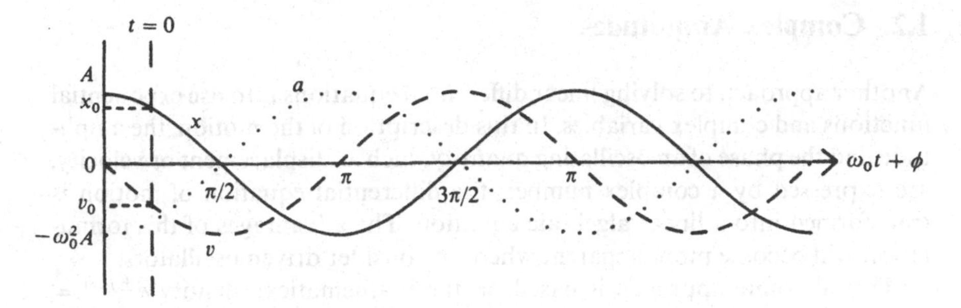

Fig. 1.2. Relative phase of displacement x, velocity v, and acceleration a of a simple vibrator.

or

$$x = B\cos\omega_0 t + C\sin\omega_0 t, \tag{1.4}$$

from which we recognize ω_0 as the natural angular frequency of the system.

The natural frequency f_0 of our simple oscillator is given by $f_0 = (1/2\pi)\sqrt{K/m}$, and the amplitude by $\sqrt{B^2 + C^2}$ or by A; ϕ is the initial phase of the motion. Differentiation of the displacement x with respect to time gives corresponding expressions for the velocity v and acceleration a:

$$v = \dot{x} = -\omega_0 A\sin(\omega_0 t + \phi), \tag{1.5}$$

and

$$a = \ddot{x} = -\omega_0^2 A\cos(\omega_0 t + \phi). \tag{1.6}$$

The displacement, velocity, and acceleration are shown in Fig. 1.2. Note that the velocity v leads the displacement by $\pi/2$ radians (90°), and the acceleration leads (or lags) by π radians (180°).

Solutions to second-order differential equations have two arbitrary constants. In Eq. (1.3) they are A and ϕ; in Eq. (1.4) they are B and C. Another alternative is to describe the motion in terms of constants x_0 and v_0, the displacement and velocity when $t = 0$. Setting $t = 0$ in Eq. (1.3) gives $x_0 = A\cos\phi$, and setting $t = 0$ in Eq. (1.5) gives $v_0 = -\omega_0 A\sin\phi$. From these we can obtain expressions for A and ϕ in terms of x_0 and v_0:

$$A = \sqrt{x_0^2 + \left(\frac{v_0}{\omega_0}\right)^2},$$

and (1.7)

$$\phi = \tan^{-1}\left(\frac{-v_0}{\omega_0 x_0}\right).$$

Alternatively, we could have set $t = 0$ in Eq. (1.4) and its derivative to obtain $B = x_0$ and $C = v_0/\omega_0$ from which

$$x = x_0\cos\omega_0 t + \frac{v_0}{\omega_0}\sin\omega_0 t. \tag{1.8}$$

1.2. Complex Amplitudes

Another approach to solving linear differential equations is to use exponential functions and complex variables. In this description of the motion, the amplitude and the phase of an oscillating quantity, such as displacement or velocity, are expressed by a complex number; the differential equation of motion is transformed into a linear algebraic equation. The advantages of this formulation will become more apparent when we consider driven oscillators.

This alternate approach is based on the mathematical identity $e^{\pm j\omega_0 t} = \cos\omega_0 t \pm j\sin\omega_0 t$, where $j = \sqrt{-1}$. In these terms, $\cos\omega_0 t = \mathrm{Re}(e^{\pm j\omega_0 t})$, where Re stands for the "real part of." Equation (1.3) can be written

$$\begin{aligned} x = A\cos(\omega_0 t + \phi) &= \mathrm{Re}[Ae^{j(\omega_0 t+\phi)}] = \mathrm{Re}(Ae^{j\phi}e^{j\omega_0 t}) \\ &= \mathrm{Re}(\tilde{A}e^{j\omega_0 t}). \end{aligned} \tag{1.9}$$

The quantity $\tilde{A} = Ae^{j\phi}$ is called the complex amplitude of the motion and represents the complex displacement at $t = 0$. The complex displacement $\tilde{x}$ is written

$$\tilde{x} = \tilde{A}e^{j\omega_0 t}. \tag{1.10}$$

The complex velocity $\tilde{v}$ and acceleration $\tilde{a}$ become

$$\tilde{v} = j\omega_0\tilde{A}e^{j\omega_0 t} = j\omega_0\tilde{x}, \tag{1.11}$$

and

$$\tilde{a} = -\omega_0^2\tilde{A}e^{j\omega_0 t} = -\omega_0^2\tilde{x}. \tag{1.12}$$

Each of these complex quantities can be thought of as a rotating vector or phasor rotating in the complex plane with angular velocity ω_0, as shown in Fig. 1.3. The real time dependence of each quantity can be obtained from the projection on the real axis of the corresponding complex quantities as they rotate with angular velocity ω_0.

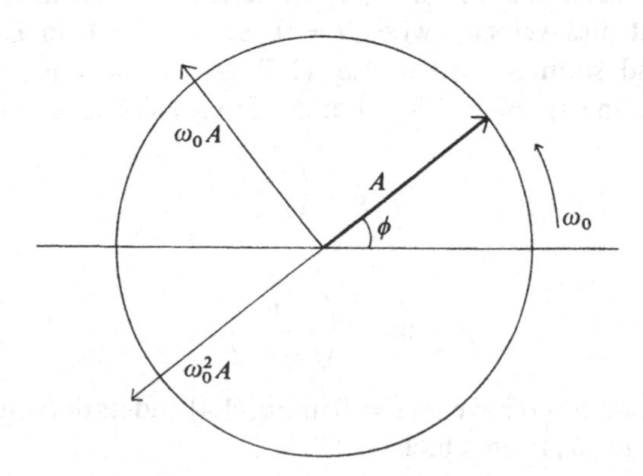

Fig. 1.3. Phasor representation of the complex displacement, velocity, and acceleration of a linear oscillator.

1.3. Superposition of Two Harmonic Motions in One Dimension

Frequently, the motion of a vibrating system can be described by a linear combination of the vibrations induced by two or more separate harmonic excitations. Provided we are dealing with a linear system, the displacement at any time is the sum of the individual displacements resulting from each of the harmonic excitations. This important principle is known as the principle of linear superposition. A linear system is one in which the presence of one vibration does not alter the response of the system to other vibrations, or one in which doubling the excitation doubles the response.

1.3.1. Two Harmonic Motions Having the Same Frequency

One case of interest is the superposition of two harmonic motions having the same frequency. If the two individual displacements are

$$\tilde{x}_1 = A_1 e^{j(\omega t+\phi_1)}$$

and

$$\tilde{x}_2 = A_2 e^{j(\omega t+\phi_2)},$$

their linear superposition results in a motion given by

$$\tilde{x}_1 + \tilde{x}_2 = (A_1 e^{j\phi_1} + A_2 e^{j\phi_2})e^{j\omega t} = Ae^{j(\omega t+\phi)}. \tag{1.13}$$

The phasor representation of this motion is shown in Fig. 1.4.

Expressions for A and ϕ can easily be obtained by adding the phasors $A_1 e^{j\omega\phi_1}$ and $A_2 e^{j\omega\phi_2}$ to obtain

$$A = \sqrt{(A_1 \cos\phi_1 + A_2 \cos\phi_2)^2 + (A_1 \sin\phi_1 + A_2 \sin\phi_2)^2}, \tag{1.14}$$

and

$$\tan\phi = \frac{A_1 \sin\phi_1 + A_2 \sin\phi_2}{A_1 \cos\phi_1 + A_2 \cos\phi_2}. \tag{1.15}$$

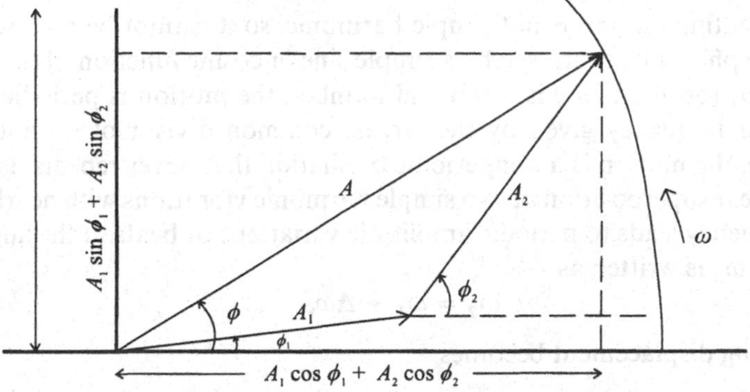

Fig. 1.4. Phasor representation of two simple harmonic motions having the same frequency.

What we have really done, of course, is to add the real and imaginary parts of $\tilde{x}_1$ and $\tilde{x}_2$ to obtain the resulting complex displacement $\tilde{x}$. The real displacement is

$$x = \mathrm{Re}(\tilde{x}) = A\cos(\omega t + \phi). \tag{1.16}$$

The linear combination of two simple harmonic vibrations with the same frequency leads to another simple harmonic vibration at this same frequency.

1.3.2. More Than Two Harmonic Motions Having the Same Frequency

The addition of more than two phasors is accomplished by drawing them in a chain, head to tail, to obtain a single phasor that rotates with angular velocity ω. This phasor has an amplitude given by

$$A = \sqrt{(\sum A_n \cos\phi_n)^2 + (\sum A_n \sin\phi_n)^2}, \tag{1.17}$$

and a phase angle ϕ obtained from

$$\tan\phi = \frac{\sum A_n \sin\phi_n}{\sum A_n \cos\phi_n} \tag{1.18}$$

The real displacement is the projection of the resultant phasor on the real axis, and this is equal to the sum of the real parts of all the component phasors:

$$x = A\cos(\omega t + \phi) = \sum A_n \cos(\omega t + \phi_n.) \tag{1.19}$$

1.3.3. Two Harmonic Motions with Different Frequencies: Beats

If two simple harmonic motions with frequencies f_1 and f_2 are combined, the resultant expression is

$$\tilde{x} = \tilde{x}_1 + \tilde{x}_2 = A_1 e^{j(\omega_1 t + \phi_1)} + A_2 e^{j(\omega_2 t + \phi_2)}, \tag{1.20}$$

where A, ω, and ϕ express the amplitude, the angular frequency, and the phase of each simple harmonic vibration.

The resulting motion is not simple harmonic, so it cannot be represented by a single phasor or expressed by a simple sine or cosine function. If the ratio of ω_2 to ω_1 (or ω_1 to ω_2) is a rational number, the motion is periodic with an angular frequency given by the largest common divisor of ω_2 and ω_1. Otherwise, the motion is a nonperiodic oscillation that never repeats itself.

The linear superposition of two simple harmonic vibrations with nearly the same frequency leads to periodic amplitude variations or beats. If the angular frequency ω_2 is written as

$$\omega_2 = \omega_1 + \Delta\omega, \tag{1.21}$$

the resulting displacement becomes

$$\begin{aligned} \tilde{x} &= A_1 e^{j(\omega_1 t + \phi_1)} + A_2 e^{j(\omega_1 t + \Delta\omega t + \phi_2)} \\ &= [A_1 e^{j\phi_1} + A_2 e^{j(\phi_2 + \Delta\omega t)}] e^{j\omega_1 t}. \end{aligned} \tag{1.22}$$

We can express this in terms of a time-dependent amplitude $A(t)$ and a time-dependent phase $\phi(t)$:

$$\tilde{x} = A(t)e^{j(\omega_1 t + \phi(t))}, \tag{1.23}$$

where

$$A(t) = \sqrt{A_1^2 + A_2^2 + 2A_1A_2\cos(\phi_1 - \phi_2 - \Delta\omega t)}, \tag{1.24}$$

and

$$\tan\phi(t) = \frac{A_1\sin\phi_1 + A_2\sin(\phi_2 + \Delta\omega t)}{A_1\cos\phi_1 + A_2\cos(\phi_2 + \Delta\omega t)}. \tag{1.25}$$

The resulting vibration could be regarded as approximately simple harmonic motion with angular frequency ω_1 and with both amplitude and phase varying slowly at frequency $\Delta\omega/2\pi$. The amplitude varies between the limits $A_1 + A_2$ and $|A_1 - A_2|$.

In the special case where the amplitudes A_1 and A_2 are equal and ϕ_1 and $\phi_2 = 0$, the amplitude equation [Eq. (1.24)] becomes

$$A(t) = A_1\sqrt{2 + 2\cos\Delta\omega_1 t} \tag{1.26}$$

and the phase equation [Eq. (1.25)] becomes

$$\tan\phi(t) = \frac{\sin\Delta\omega_1 t}{1 + \cos\Delta\omega_1 t}. \tag{1.27}$$

Thus, the amplitude varies between $2A_1$ and 0, and the beating becomes very pronounced.

The displacement waveform (the real part of $\tilde{x}$) is illustrated in Fig. 1.5. This waveform resembles the waveform obtained by modulating the amplitude of the vibration at a frequency $\Delta\omega/2\pi$, but they are not the same. Amplitude modulation results from nonlinear behavior in a system, which generates spectral components having frequencies ω_1 and $\omega_1 \pm \Delta\omega$. The spectrum of the waveform in Fig. 1.5. has spectral components ω_1 and $\omega_1 + \Delta\omega$ only.

Audible beats are heard whenever two sounds of nearly the same frequency reach the ear. The perception of combination tones and beats is discussed in Chapter 8 of Rossing (1982) and other introductory texts on musical acoustics.

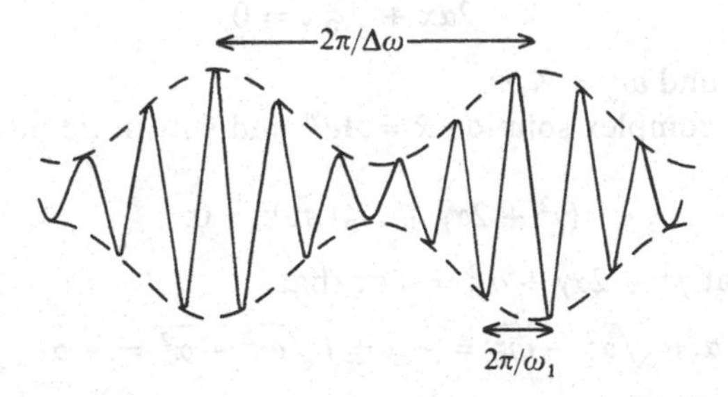

Fig. 1.5. Waveform resulting from linear superposition of simple harmonic motions with angular frequencies ω_1 and ω_2.

1.4. Energy

The potential energy E_p of our mass–spring system is equal to the work done in stretching or compressing the spring:

$$E_{\mathrm{p}} = -\int_0^x F\,dx = \int_0^x Kx\,dx = \frac{1}{2}Kx^2. \tag{1.28}$$

Using the expression for x in Eq. (1.3) gives

$$E_{\mathrm{p}} = \tfrac{1}{2}KA^2\cos^2(\omega_0 t + \phi). \tag{1.29}$$

The kinetic energy is $E_{\mathrm{k}} = \frac{1}{2}mv^2$, and using the expression for v in Eq. (1.5) gives

$$E_{\mathrm{k}} = \tfrac{1}{2}m\omega_0^2 A^2\sin^2(\omega_0 t + \phi) = \tfrac{1}{2}KA^2\sin^2(\omega_0 t + \phi). \tag{1.30}$$

The total energy E is then

$$E = E_{\mathrm{p}} + E_{\mathrm{k}} = \tfrac{1}{2}KA^2 = \tfrac{1}{2}m\omega_0^2 A^2 = \tfrac{1}{2}mU^2, \tag{1.31}$$

where U is the maximum velocity. The total energy in our loss-free system is constant and is equal either to the maximum potential energy (at maximum displacement) or the maximum kinetic energy (at the midpoint).

1.5. Damped Oscillations

There are many different mechanisms that can contribute to the damping of an oscillating system. Sliding friction is one example, and viscous drag in a fluid is another. In the latter case, the drag force F_r is proportional to the velocity:

$$F_r = -R\dot{x},$$

where R is the mechanical resistance. The drag force is added to the equation of motion:

$$m\ddot{x} + R\dot{x} + Kx = 0$$

or

$$\ddot{x} + 2\alpha\dot{x} + \omega_0^2 x = 0, \tag{1.32}$$

where $\alpha = R/2m$ and $\omega_0^2 = K/m$.

We assume a complex solution $\tilde{x} = \tilde{A}e^{\gamma t}$ and substitute into Eq. (1.32) to obtain

$$(\gamma^2 + 2\alpha\gamma + \omega_0^2)\tilde{A}e^{\gamma t} = 0. \tag{1.33}$$

This requires that $\gamma^2 + 2\alpha\gamma + \omega_0^2 = 0$ or that

$$\gamma = -\alpha \pm \sqrt{\alpha^2 - \omega_0^2} = -\alpha \pm j\sqrt{\omega_0^2 - \alpha^2} = -\alpha \pm j\omega_{\mathrm{d}}, \tag{1.34}$$

where $\omega_{\mathrm{d}} = \sqrt{\omega_0^2 - \alpha^2}$ is the natural angular frequency of the damped oscillator (which is less than that of the same oscillator without damping). The

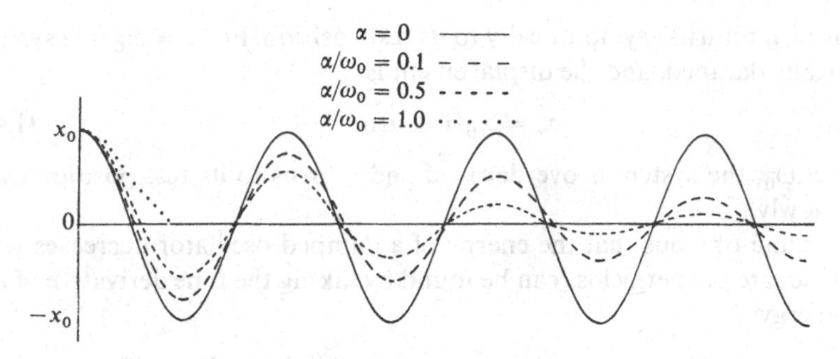

Fig. 1.6. Displacement of a harmonic oscillator with $v_0 = 0$ for different values of damping. The relaxation time is given by $1/\alpha$. Critical damping occurs when $\alpha = \omega_0$.

general solution is a sum of terms constructed by using each of the two values of γ:

$$\tilde{x} = e^{-\alpha t}(\tilde{A}_1 e^{j\omega_d t} + \tilde{A}_2 e^{-j\omega_d t}). \tag{1.35}$$

The real part of this solution, which gives the time history of the displacement, can be written in several different ways as in the loss-free case. The expressions that correspond to Eqs. (1.3) and (1.4) are

$$x = Ae^{-\alpha t}\cos(\omega_d t + \phi), \tag{1.36}$$

and

$$x = e^{-\alpha t}(B\cos\omega_d t + C\sin\omega_d t). \tag{1.37}$$

Setting $t = 0$ in Eq. (1.37) and its derivatives gives the displacement in terms of the initial displacement x_0 and initial velocity v_0:

$$x = e^{-\alpha t}\left(x_0\cos\omega_d t + \frac{v_0 + \alpha x_0}{\omega_d}\sin\omega_d t\right). \tag{1.38}$$

Figure 1.6 shows a few cycles of the displacement for different values of α when $v_0 = 0$.

The amplitude of the damped oscillator is given by $x_0 e^{-\alpha t}$, and its motion is not strictly periodic. Nevertheless, the time between zero crossings in the same direction remains constant and equal to $T_d = 1/f_d = 2\pi/\omega_d$, which is defined as the period of the oscillation. The time interval between successive maxima is also T_d, but the maxima and minima are not exactly halfway between the zeros.

One measure of the damping is the time required for the amplitude to decrease to $1/e$ of its initial value x_0. This time, τ, is called by various names, such as decay time, lifetime, relaxation time, and characteristic time; it is given by

$$\tau = \frac{1}{\alpha} = \frac{2m}{R}. \tag{1.39}$$

When $\alpha \geqq \omega_0$, the system is no longer oscillatory. When the mass is displaced, it returns asymptotically to its rest position. For $\alpha = \omega_0$, the system is critically damped, and the displacement is

$$x_c = x_0(1 + \alpha t)e^{-\alpha t}. \tag{1.40}$$

For $\alpha > \omega_0$, the system is overdamped and returns to its rest position even more slowly.

It is quite obvious that the energy of a damped oscillator decreases with time. The rate of energy loss can be found by taking the time derivative of the total energy:

$$\begin{aligned} d/dt(E_p + E_K) &= d/dt[\tfrac{1}{2}Kx^2 + \tfrac{1}{2}m\dot{x}^2] = Kx\dot{x} + m\dot{x}\ddot{x} \\ &= \dot{x}(Kx + m\ddot{x}) = \dot{x}(-R\dot{x}) = -2\alpha m\dot{x}^2, \end{aligned} \tag{1.41}$$

where use has been made of Eq. (1.32). Equation (1.41) tells us that the rate of energy loss is the friction force $-R\dot{x}$ times the velocity $\dot{x}$.

Often a Q factor or quality factor is used to compare the spring force to the damping force:

$$Q = \frac{Kx_0}{R\omega_0 x_0} = \frac{K}{R\omega_0} = \frac{\omega_0}{2\alpha}. \tag{1.42}$$

1.6. Other Simple Vibrating Systems

Besides the mass–spring system already described, the following are familiar examples of systems that vibrate in simple harmonic motion.

1.6.1. A Spring of Air

A piston of mass m, free to move in a cylinder of area-S and length-L [see Fig. 1.7(a)], vibrates in much the same manner as a mass attached to a spring. The spring constant of the confined air turns out to be $K = \gamma p_a S/L$, so the

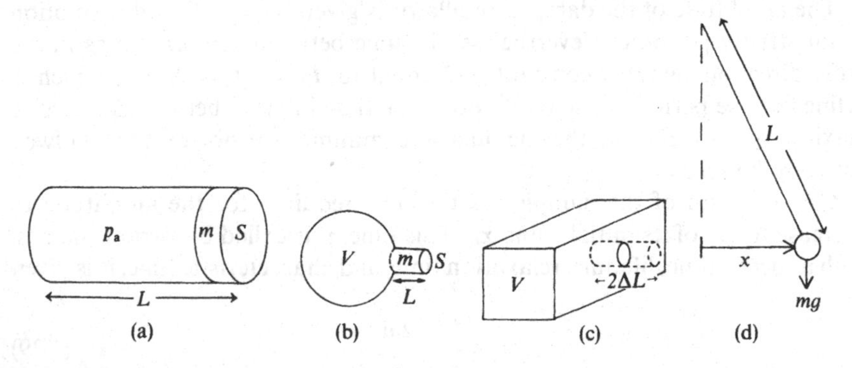

Fig. 1.7. Simple vibrating systems: (a) piston in a cylinder; (b) Helmholtz resonator with neck of length L; (c) Helmholtz resonator without a neck; and (d) simple pendulum.

natural frequency is

$$f_0 = \frac{1}{2\pi}\sqrt{\frac{\gamma p_a S}{mL}}, \tag{1.43}$$

where p_a is atmospheric pressure, m is the mass of the piston, and γ is a constant that is 1.4 for air.

1.6.2. Helmholtz Resonator

In the Helmholtz resonator shown in Fig. 1.7(b), the mass of air in the neck serves as the piston and the large volume of air V as the spring. The mass of air in the neck and the spring constant of the confined air are given by the expressions

$$m = \rho SL,$$

and (1.44)

$$K = \rho S^2 c^2 / V,$$

where ρ is the air density and c is the speed of sound.

The natural frequency of vibration is given by

$$f_0 = \frac{1}{2\pi}\sqrt{\frac{K}{m}} = \frac{c}{2\pi}\sqrt{\frac{S}{VL}}. \tag{1.45}$$

Note that the smaller the neck diameter, the lower the natural frequency of vibration, a result which may appear surprising at first glance.

The Helmholtz resonator in Fig. 1.7(c) has no neck to delineate the vibrating mass, but the effective length can be estimated by taking twice the "end correction" of a flanged tube (which is $8/3\pi \cong 0.85$ times the radius a). Thus,

$$m = \rho SL = \rho(\pi a^2)\left(\frac{16a}{3\pi}\right) = 5.33\rho a^3. \tag{1.46}$$

The natural frequency of a neckless Helmholtz resonator with a large face is thus expressed as

$$f_0 = \frac{c}{2\pi}\sqrt{\frac{1.85a}{V}}. \tag{1.47}$$

If the face of the resonator surrounding the hole is not large, the natural frequency will be slightly higher. The Helmholtz resonator is discussed in Section 6.5 and the end correction in Section 8.3.

1.6.3. Simple Pendulum

A simple pendulum, consisting of a mass m attached to a string of length l [Fig. 1.7(d)], oscillates in simple harmonic motion provided that $x \ll l$. Assuming that the mass of the string is much less than m, the natural frequency is given by

$$f_0 = \frac{1}{2\pi}\sqrt{\frac{g}{l}}, \tag{1.48}$$

where g is the acceleration due to gravity. Note that the frequency does not depend on the mass.

1.6.4. Electrical RLC Circuit

In the electrical circuit, shown in Fig. 1.8, the voltages across the inductor, the resistor, and the capacitor, respectively, should add to zero:

$$L\frac{di}{dt} + Ri + \frac{1}{C}\int i\,dt = 0.$$

Differentiating each term leads to an equation that is analogous to Eq. (1.32) for the simple mechanical oscillator:

$$L\ddot{i} + R\dot{i} + \frac{1}{C}i = 0$$

or

$$\ddot{i} + 2\alpha\dot{i} + \omega_0^2 i = 0, \tag{1.49}$$

where $\alpha = R/2L$ and $\omega_0^2 = 1/LC$.

The solution to Eq. (1.49) can be written as

$$i = I_0 e^{-\alpha t}\cos(\omega_d t + \phi), \tag{1.50}$$

which represents a current oscillating at a frequency $\sqrt{\omega_0^2 - \alpha^2}/2\pi$, with an amplitude that decays exponentially. If $\alpha \ll \omega_0$ (small damping), the frequency of oscillation is approximately

$$f_0 = \frac{\omega_0}{2\pi} = \frac{1}{2\pi\sqrt{LC}}, \tag{1.51}$$

and the current has a waveform similar to that shown in Fig. 1.6.

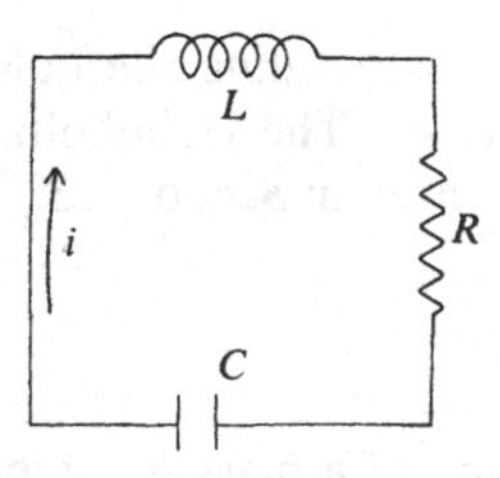

Fig. 1.8. Simple electrical oscillator with inductance L, resistance R, and capacitance C.

1.6.5. Combinations of Springs and Masses

Several combinations of masses and springs are shown in Fig. 1.9, along with their resonance frequencies. Note the effect of combining springs in series and parallel combinations. Two springs with spring constants K_1 and K_2 will have a combined spring constant $K_p = K_1 + K_2$ when connected in parallel but only $K_s = K_1 K_2/(K_1 + K_2)$ in series. When $K_1 = K_2$, the parallel and series values become $2K_1$ and $K_1/2$, respectively. The combinations in Fig. 1.9 all have a single degree of freedom. In Section 1.12, we discuss two-mass systems with two degrees of freedom; that is, the two masses move independently.

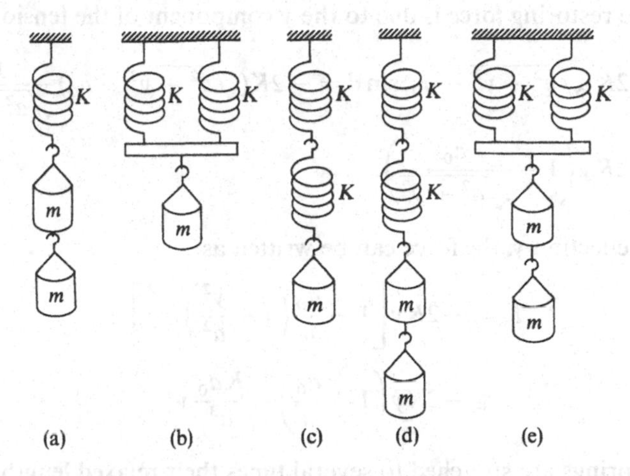

Fig. 1.9. Mass–spring combinations that vibrate at single frequencies: (a) $f_0 = (1/2\pi)\sqrt{K/2m}$; (b) $f_0 = (1/2\pi)\sqrt{2K/m}$; (c) $f_0 = (1/2\pi)\sqrt{K/2m}$; (d) $f_0 = (1/2\pi)\sqrt{K/4m}$; and (e) $f_0 = (1/2\pi)\sqrt{K/m}$.

1.6.6. Longitudinal and Transverse Oscillations of a Mass–Spring System

Consider the vibrating system shown in Fig. 1.10. Each spring has a spring constant K, a relaxed length a_0, and a stretched length a. Thus, each spring exerts a tension $K(a - a_0)$ on the mass when it is in its equilibrium position

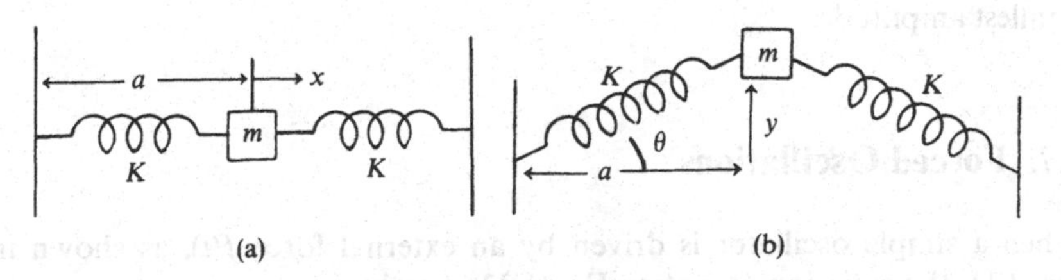

Fig. 1.10. Longitudinal (a) and transverse (b) oscillations of a mass–spring system.

($x = 0$). When the mass is displaced a distance x, the net restoring force is the difference between the two tensions:

$$F_x = K(a - x - a_0) - K(a + x - a_0) = -2Kx. \tag{1.52}$$

The natural frequency for longitudinal vibration is thus given by

$$f_l = \frac{1}{2\pi}\sqrt{\frac{2K}{m}}. \tag{1.53}$$

Now, consider transverse vibrations of the same systems, as shown in Fig. 1.10(b). When the mass is displaced a distance y from its equilibrium position, the restoring force is due to the y component of the tension:

$$F_y = -2K(\sqrt{a^2 + y^2} - a_0)\sin\theta = -2K(\sqrt{a^2 + y^2} - a_0)\frac{y}{\sqrt{a^2 + y^2}}$$

$$= -2Ky\left(1 - \frac{a_0}{\sqrt{a^2 + y^2}}\right). \tag{1.54}$$

For small deflection y, the force can be written as

$$F_y = -2Ky\left[1 - \frac{a_0}{a}\left(1 + \frac{y^2}{a^2}\right)^{-1/2}\right]$$

$$\cong -2Ky\left(1 - \frac{a_0}{a}\right) - \frac{Ka_0}{a^3}y^3. \tag{1.55}$$

When the springs are stretched to several times their relaxed length ($a \gg a_0$), the force is approximately $-2Ky$, and the natural frequency is practically the same as the frequency for longitudinal vibrations given in Eq. (1.53):

$$f_t \cong \frac{1}{2\pi}\sqrt{\frac{2K}{m}}. \tag{1.56}$$

When the springs are stretched only a small amount from their relaxed length ($a \cong a_0$), however, the first term in Eq. (1.55) becomes very small, so the vibration frequency is considerably smaller than that given in Eqs. (1.53) and (1.56). Furthermore, the contribution from the cubic term in Eq. (1.55) takes on increased importance, making the vibration nonsinusoidal for all but the smallest amplitude.

1.7. Forced Oscillations

When a simple oscillator is driven by an external force $f(t)$, as shown in Fig. 1.11, the equation of motion Eq. (1.32) then becomes

$$m\ddot{x} + R\dot{x} + Kx = f(t). \tag{1.57}$$

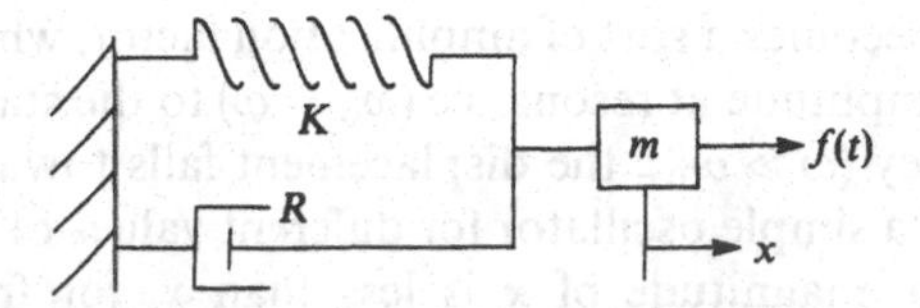

Fig. 1.11. A damped harmonic oscillator with driving force $f(t)$.

The driving force $f(t)$ may have harmonic time dependence, it may be impulsive, or it may even be a random function of time. For the case of a sinusoidal driving force $f(t) = F\cos\omega t$ turned on at some time, the solution to Eq. (1.57) consists of two parts: a transient term containing two arbitrary constants, and a steady-state term that depends only on F and ω.

To obtain the steady-state solution, it is advantageous to write the equation of motion in complex form:

$$m\ddot{\tilde{x}} + R\dot{\tilde{x}} + K\tilde{x} = Fe^{j\omega t}. \tag{1.58}$$

Since this equation is linear in x and the right-hand side is a harmonic function with angular frequency ω, in the steady state the left-hand side should be harmonic with the same frequency. Thus, we replace $\tilde{x}$ by $\tilde{A}e^{j\omega t}$ and obtain

$$\tilde{A}e^{j\omega t}(-\omega^2 m + j\omega R + K) = Fe^{j\omega t}. \tag{1.59}$$

The complex displacement is

$$\tilde{x} = \frac{Fe^{j\omega t}}{K - \omega^2 m + j\omega R} = \frac{\tilde{F}/m}{\omega_0^2 - \omega^2 + j\omega 2\alpha}, \tag{1.60}$$

where $\tilde{F} = Fe^{j\omega t}$, $\omega_0^2 = K/m$, and $\alpha = R/2m$.

Differentiation of $\tilde{x}$ gives the complex velocity $\tilde{v}$:

$$\tilde{v} = \frac{Fe^{j\omega t}}{R + j(\omega m - K/\omega)} = \frac{\tilde{F}\omega/m}{2\omega\alpha + j(\omega^2 - \omega_0^2)}. \tag{1.61}$$

The mechanical impedance $\tilde{Z}$ is defined as $\tilde{F}/\tilde{v}$:

$$\tilde{Z} = \tilde{F}/\tilde{v} = R + j(\omega m - K/\omega) = R + jX_m, \tag{1.62}$$

where $X_m = \omega m - K/\omega$ is the mechanical reactance. The actual steady-state displacement is given by the real part of Eq. (1.60):

$$x = \mathrm{Re}\frac{\tilde{F}}{j\omega\tilde{Z}} = \frac{F}{\omega Z}\sin(\omega t + \phi). \tag{1.63}$$

A quantity $x_s = F/K = F/m\omega_0^2$ can be defined as the static displacement of the oscillator produced by a constant force of magnitude F. At very low frequency, the displacement amplitude will approach F/K, and the oscillator is said to be stiffness dominated. When $\omega = \omega_d$, the amplitude becomes

$$x_0 = F/2\alpha m\omega_0 = Qx_s. \tag{1.64}$$

In other words, Q becomes a sort of amplification factor, which is the ratio of the displacement amplitude at resonance ($\omega_0 = \omega$) to the static displacement.

At high frequency ($\omega \gg \omega_0$), the displacement falls toward zero. The frequency response of a simple oscillator for different values of α (or Q) is shown in Fig. 1.12(a). The magnitude of x is less than x_s for frequencies above $\omega_0\sqrt{2-\delta^2}$ (where $\delta = 1/Q = 2\alpha/\omega_0$), which, for small values of α, is about $\sqrt{2}\omega_0$. If $\alpha > \omega_0/\sqrt{2}$, $x < x_s$ at all frequencies.

The phase angle between the displacement and the driving force is the phase

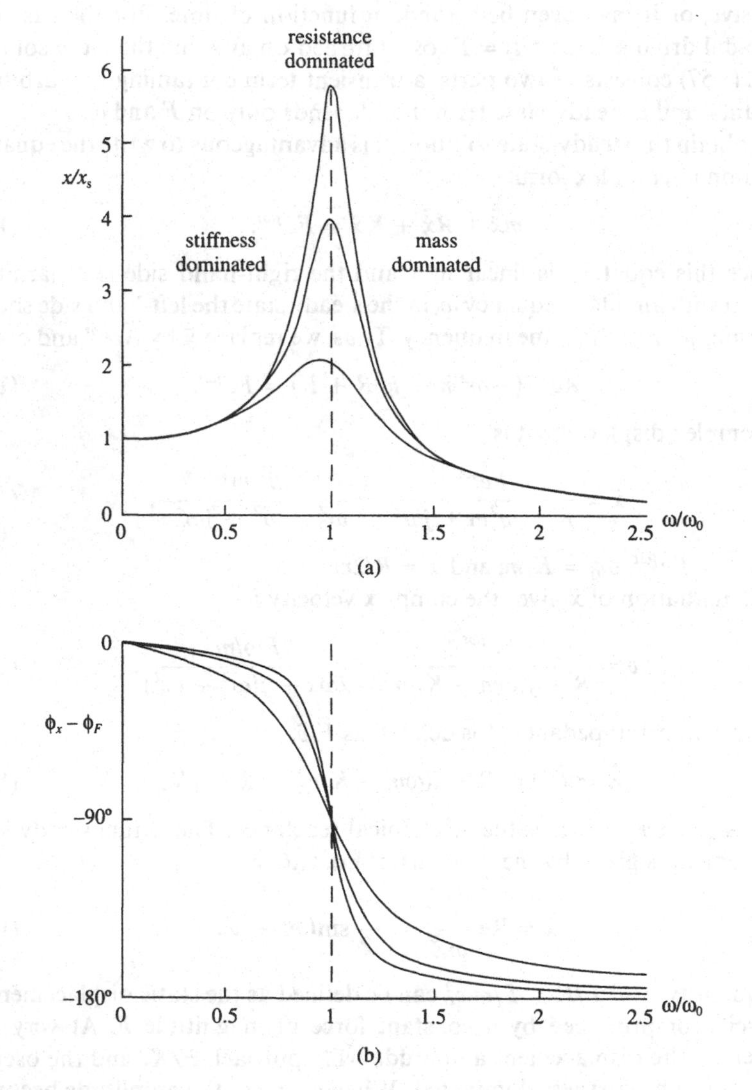

Fig. 1.12. Frequency dependence of the magnitude x and phase $(\phi_x - \phi_F)$ of the displacement of a linear harmonic oscillator.

angle of the denominator in Eq. (1.60):

$$\phi_x - \phi_F = \tan^{-1}\frac{2\alpha\omega}{\omega^2 - \omega_0^2}. \quad (1.65)$$

At low frequency ($\omega \cong 0$), $\phi_x - \phi_F = 0$. When $\omega = \omega_0$, $\phi_x - \phi_F = 90°$, and at high frequency ($\omega \gg \omega_0$), $\phi_x - \phi_F \cong 180°$, as shown in Fig. 1.12(b).

There are other convenient ways to represent the frequency response of a simple oscillator. One way is to show how the real and imaginary parts of the mechanical impedance $\tilde{Z}(=\tilde{F}/\tilde{v})$ or the mechanical admittance (mobility) $\tilde{Y} = 1/\tilde{Z}(=\tilde{v}/\tilde{F})$ vary with frequency. At resonance, the real part of the admittance has its maximum value, while that of the impedance remains equal to R at all frequencies. The imaginary parts of both quantities are zero at resonance. Figure 1.13 shows the real and imaginary parts of the mechanical impedance and admittance for an oscillator of the same type as in Fig. 1.12. The graph of imaginary part versus the real part in Fig. 1.13(c) is sometimes called a Nyquist plot.

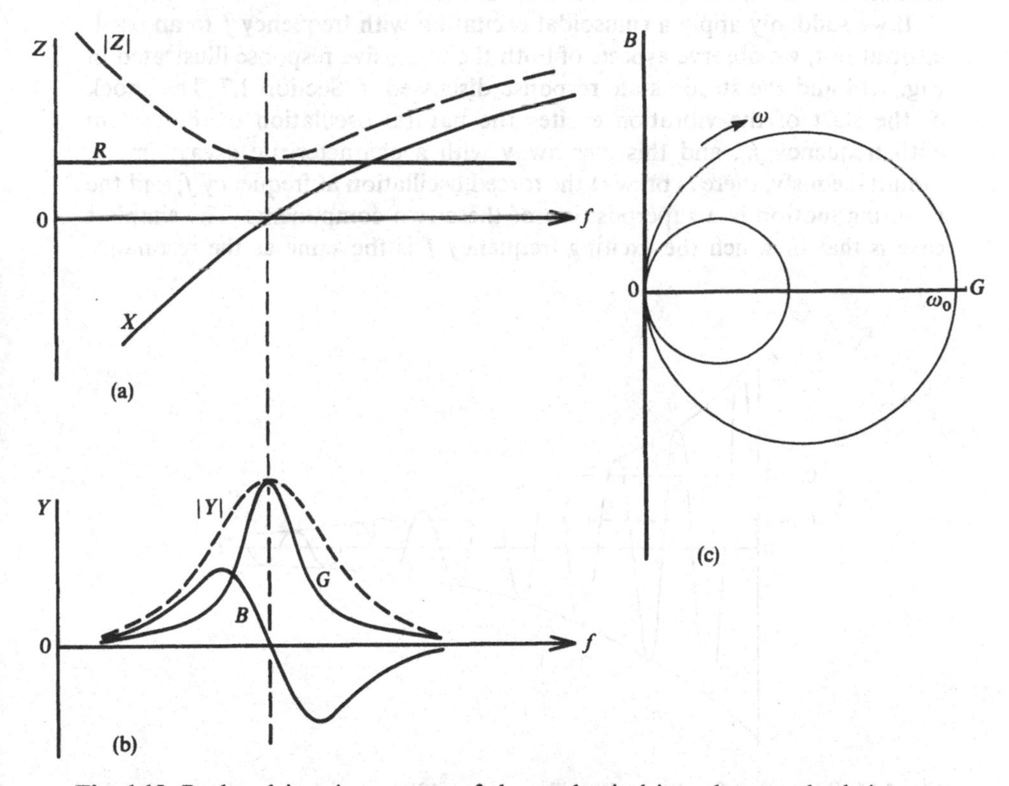

Fig. 1.13. Real and imaginary parts of the mechanical impedance and admittance for a harmonic oscillator of the same type as in Fig. 1.12: (a) mechanical impedance; (b) mechanical admittance or mobility; (c) Nyquist plot showing the imaginary part of admittance versus the real part, with frequency as a parameter.

1.8. Transient Response of an Oscillator

When a driving force is first applied to an oscillator, the motion can be quite complicated. We expect to find periodic motions at the natural frequency f_0 of the oscillator as well as the driving frequency f (or at all its component frequencies if the driving force is not harmonic). If the oscillator is heavily damped, the transient motion decays rapidly, and the oscillator quickly settles in to its steady-state motion. If the damping is small, however, the transient behavior may continue for many cycles of oscillation. If the driving frequency f is close to the natural frequency f_0, for example, strong beats may be observed.

In Section 1.5, the Q factor was defined by the equation $Q = \omega_0/2\alpha = \omega_0\tau/2$, where τ is the time required for the amplitude of a free damped oscillator to decrease to $1/e(=0.37)$ of its initial value. Thus, the decay time τ encompasses Q/π cycles of vibration. For $Q = 10$, for example, the amplitude falls to 37% of its initial value in just over three cycles, and it reaches 14% after six cycles, as shown in Fig. 1.14.

If we suddenly apply a sinusoidal excitation with frequency f to an oscillator at rest, we observe aspects of both the impulsive response illustrated in Fig. 1.14 and the steady-state response discussed in Section 1.7. The shock of the start of the vibration excites the natural oscillation of the system with frequency f_0, and this dies away with a characteristic decay time τ. Simultaneously, there is present the forced oscillation at frequency f, and the resulting motion is a superposition of these two components. The simplest case is that in which the exciting frequency f is the same as the resonance

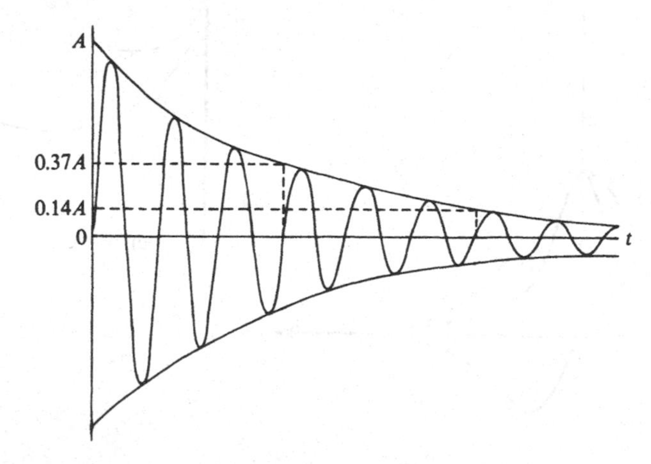

Fig. 1.14. Response of a damped oscillator ($Q = 10$) to impulsive excitation (by the application of a large force for a very short time, for example). The amplitude falls to 37% of its initial value in time τ, which corresponds to Q/π cycles.

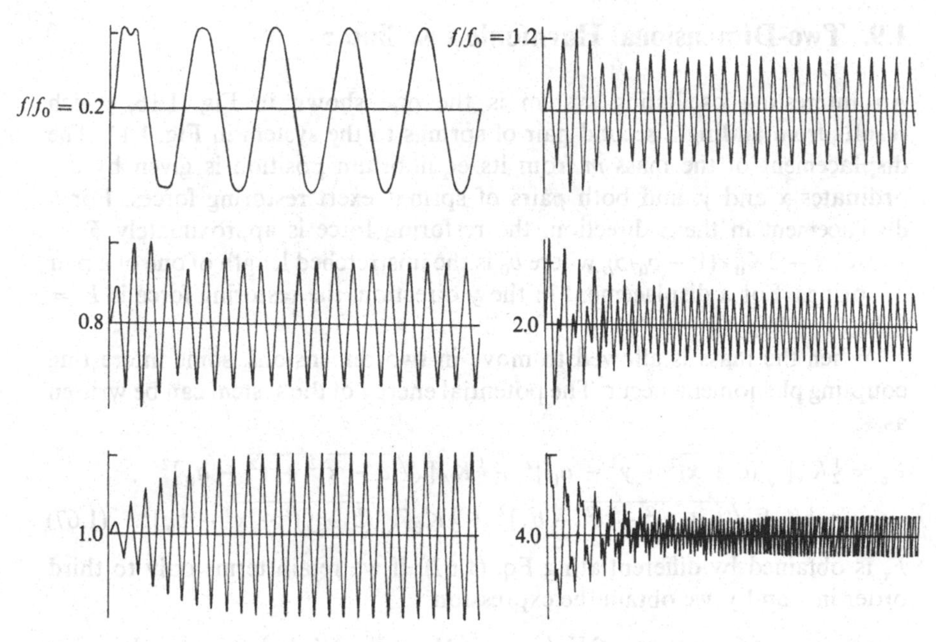

Fig. 1.15. Response of a simple oscillator to a sinusoidal force applied suddenly. The ratio f/f_0 varies from 0.2 to 4.0, and $Q = 10$ in each case. Note that the scale of amplitude is different in each case (from Fletcher, 1982).

frequency f_0, for the whole motion then builds steadily toward its final amplitude with time constant τ. More generally, however, we expect to see the presence of both frequencies f and f_0 during the duration τ of the attack transient and, if f is close to f_0, these may combine to produce beats at frequency $|f - f_0|$. These possibilities are illustrated in Fig. 1.15.

Mathematically, the problem is one of finding the appropriate general solution of Eq. (1.57). Because Eq. (1.57) is a linear equation, the general solution is a combination of the general solution of the homogeneous equation, Eq. (1.32), and a particular solution of Eq. (1.57), which we take to be the steady-state solution [Eq. (1.63)].

$$x = Ae^{-\alpha t}\cos(\omega_d t + \phi) + \frac{F}{\omega Z}\sin(\omega t + \theta), \tag{1.66}$$

where A and ϕ are arbitrary constants to be determined by the initial conditions. If the damping is small, ω_d can be replaced by ω_0.

When the driving frequency matches the natural frequency ($\omega = \omega_0$), the amplitude builds up exponentially to its final value without beats, as shown in Fig. 1.15(c). Note that irrespective of how the oscillator starts its motion, it eventually settles down to this steady-state motion.

1.9. Two-Dimensional Harmonic Oscillator

An interesting oscillating system is the one shown in Fig. 1.16, which results from adding a second pair of springs to the system in Fig. 1.11. The displacement of the mass m from its equilibrium position is given by coordinates x and y, and both pairs of springs exert restoring forces. For a displacement in the x direction, the restoring force is approximately $F_x = -2K_A x - 2K_B x(1 - b_0/b)$, where b_0 is the unstretched length of one of a pair of springs. For a displacement in the y direction, the restoring force is $F_y = -2K_A y - 2K_B y(1 - a_0/a)$.

When the mass is allowed to move in two dimensions, some interesting coupling phenomena occur. The potential energy of the system can be written as

$$E_p = \tfrac{1}{2}K_A[\sqrt{(a+x)^2 + y^2} - a_0]^2 + \tfrac{1}{2}K_A[\sqrt{(a-x)^2 + y^2} - a_0]^2 + \tfrac{1}{2}K_B[\sqrt{(b+y)^2 + x^2} - b_0]^2 + \tfrac{1}{2}K_B[\sqrt{(b-y)^2 + x^2} - b_0]^2. \quad (1.67)$$

F_x is obtained by differentiating Eq. (1.67). If we retain terms only to third order in x and y, we obtain the expression

$$F_x \cong -2(K_A + K_B)x + \frac{2K_B b_0}{b}x + \left(\frac{2K_A a_0}{a^3} + \frac{2K_B b_0}{b^3}\right)xy^2 - \frac{K_B b_0}{b^3}x^3. \quad (1.68)$$

Note that the third term, which is of third order, couples the x and y motions. For small amplitudes of vibration, however, the x and y motions are independent. Thus, we can solve the independent equations of motion,

$$\ddot{x} + \frac{2K_A + 2K_B(1 - b_0/b)}{m}x = 0,$$

and (1.69)

$$\ddot{y} + \frac{2K_A(1 - a_0/a) + 2K_B}{m}y = 0,$$

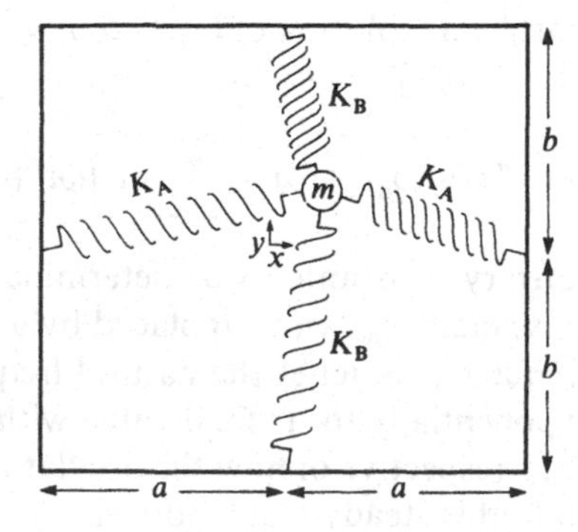

Fig. 1.16. Two-dimensional oscillator consisting of a mass m and two pairs of springs.

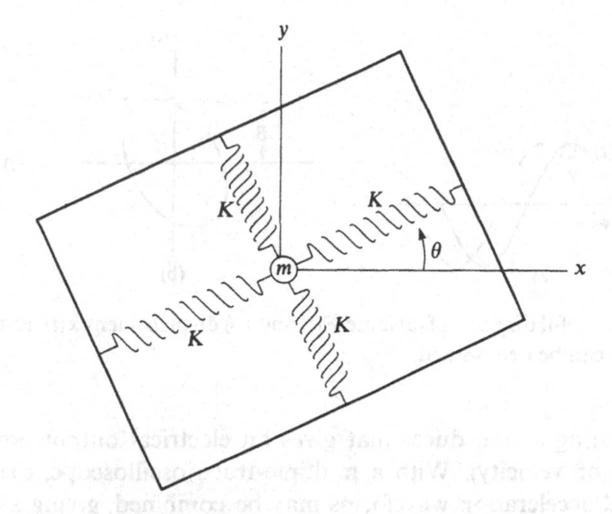

Fig. 1.17. Two-dimensional oscillator of Fig. 1.16 rotated through an angle θ. The normal modes remain unchanged, but the normal coordinates are no longer x and y.

to obtain two independent or normal modes of vibration with natural frequencies:

$$f_1 = \frac{1}{2\pi}\sqrt{\frac{2K_A + 2K_B(1 - b_0/b)}{m}},$$

and (1.70)

$$f_2 = \frac{1}{2\pi}\sqrt{\frac{2K_A(1 - a_0/a) + 2K_B}{m}}.$$

When a_0 and b_0 are much smaller than a and b, f_1 and f_2 differ only slightly. When this oscillating system is set up on an air table and the mass is initially set into motion at 45° to the x and y axes, a slowly changing Lissajous figure is observed as the x and y components of motion change their relative phases.

Since each of the normal modes corresponds to motion along one coordinate only, we call the x and y coordinates the normal coordinates of the motion. In general, the normal coordinates of a two-dimensional oscillator will not be the x and y axes. If the springs were oriented at angles θ to the axes, for example, the normal coordinates (which still lie in the directions of the springs) would be $\psi_1 = x\cos\theta + y\sin\theta$ and $\psi_2 = y\cos\theta - x\sin\theta$ as illustrated in Fig. 1.17.

1.10. Graphical Representations of Vibrations: Lissajous Figures

There are several useful ways to represent a vibrating object with a graphic display device, such as a cathode-ray oscilloscope or an X–Y plotter. Perhaps the most common way is to make a plot of position (or velocity) versus time

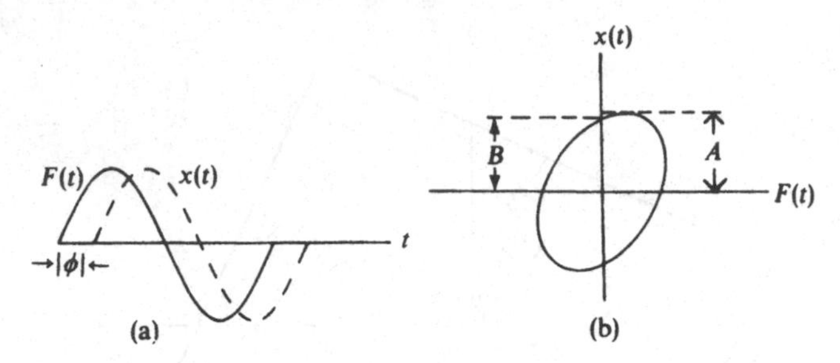

Fig. 1.18. Two useful displays of (a) force $F(t)$ and (b) displacement $x(t)$ from which the phase angle ϕ can be determined.

by incorporating a transducer that gives an electrical output proportional to position (or velocity). With a multiple-trace oscilloscope, the position, velocity, and acceleration waveforms may be combined, giving a display of the type shown in Fig. 1.2.

Another useful display combines force and displacement [Eq. (1.63)] or force and velocity. This can be done by displaying force and displacement as functions of time, as in Fig. 1.18(a), or by making a plot of displacement as a function of force, as in Fig. 1.18(b). In Fig. 1.18(a), the phase angle ϕ would be determined as a fraction of the total period (multiplied by 360° to obtain the phase angle in degrees). In Fig. 1.18(b), the phase angle is obtained from the relationship

$$\phi = \sin^{-1}\frac{B}{A}. \tag{1.71}$$

Note that the display must be centered when measuring A and B.

Two related harmonic motions with different frequencies are often represented in a display like that of Fig. 1.18(b). If $\omega_2 = \omega_1 + \Delta\omega$, as in Eq. (1.19), the display will cycle between a straight line ($\phi = 0$, 180°), a horizontal or vertical ellipse ($\phi = 90°, 270°$), and ellipses of other orientations, as ϕ advances with a frequency $\Delta\omega/2\pi$.

When ω_2 and ω_1 are related by the relationship $m\omega_1 = n\omega_2$, where m and n are integers, stable patterns result. These patterns are called Lissajous figures in honor of Jules Antoine Lissajous. Examples of such figures are shown in Fig. 1.19.

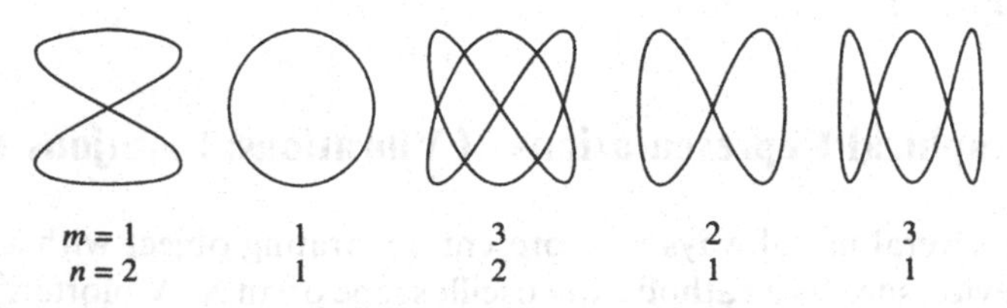

Fig. 1.19. Lissajous figures obtained by displaying $\cos\omega_2 t$ versus $\cos\omega_1 t$, where $m\omega_1 = n\omega_2$ for different integers m, n.

1.11. Normal Modes of Two-Mass Systems

Further understanding of normal modes and normal coordinates of oscillating systems comes from considering the two-mass system in Fig. 1.20(a). The analysis is simplified by letting all three spring constants and both masses be the same. Letting x_1 and x_2 be the displacements of the two masses, we write the equations of motion:

$$m\ddot{x}_1 + Kx_1 + K(x_1 - x_2) = 0,$$

and (1.72)

$$m\ddot{x}_2 + Kx_2 + K(x_2 - x_1) = 0.$$

In order to find the normal modes, we assume harmonic solutions $x_1 = A_1 \cos \omega t$ and $x_2 = A_2 \cos \omega t$, and substitute them into Eq. (1.72) to obtain

$$-\omega^2 A_1 + \frac{2K}{m} A_1 - \frac{K}{m} A_2 = 0,$$

and

$$-\omega^2 A_2 + \frac{2K}{m} A_2 - \frac{K}{m} A_1 = 0.$$

Letting $K/m = \omega_0^2$ as before, these equations can be written as

$$(\omega^2 - 2\omega_0^2)A_1 + \omega_0^2 A_2 = 0,$$

and (1.73)

$$\omega_0^2 A_1 + (\omega^2 - 2\omega_0^2)A_2 = 0.$$

The normal mode frequencies are obtained by setting the determinant of the coefficients equal to zero:

$$\begin{vmatrix} \omega^2 - 2\omega_0^2 & \omega_0^2 \\ \omega_0^2 & \omega^2 - 2\omega_0^2 \end{vmatrix} = 0 = \omega^4 - 4\omega_0^2\omega^2 + 4\omega_0^4 - \omega_0^4, \qquad (1.74)$$

from which $\omega^2 = 2\omega_0^2 \pm \omega_0^2$ and $\omega = \omega_0, \sqrt{3}\omega_0$.

It is easy to deduce the nature of these normal modes. The one with angular frequency $\omega_0(=K/m)$ describes the two masses moving together in the same

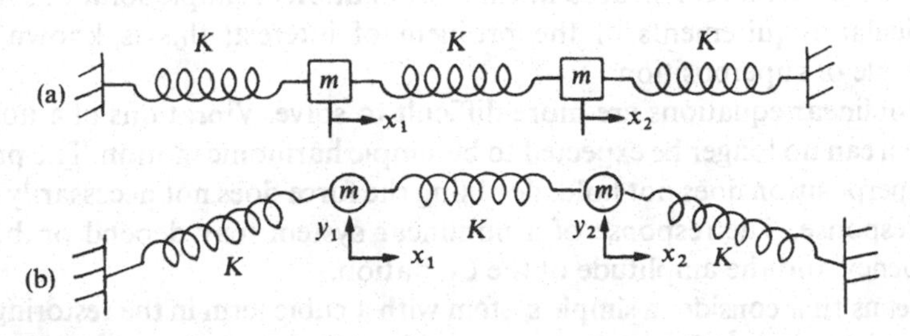

Fig. 1.20. Oscillating systems consisting of two masses and three springs. In (a) the masses move in a line; in (b) they move in a plane, so transverse oscillations are possible as well as longitudinal oscillations.

direction ($x_1 = x_2$) so that the center spring is not stretched. Thus, each mass is acted on by one spring, and the frequency is the same as the one-mass system in Fig. 1.1. The normal mode of higher frequency $\sqrt{3}\omega_0$ consists of the masses moving in opposite directions ($x_1 = -x_2$), so that the center spring is stretched twice as much as either of the end springs.

This result can be obtained in a more formal way by substituting each value of ω into the Eq. (1.73), in turn, and solving for A_1 and A_2:

$$\omega = \omega_0: \quad (\omega_0^2 - 2\omega_0^2)A_1 + \omega_0^2 A_2 = 0 \quad \text{from which} \quad A_1 = A_2,$$

and (1.75)

$$\omega = \sqrt{3}\omega_0: \quad (3\omega_0^2 - 2\omega_0^2)A_1 + \omega_0^2 A_2 = 0 \quad \text{from which} \quad A_1 = -A_2.$$

The normal coordinates are thus written as

$$\psi_1 = x_1 + x_2 \quad \text{and} \quad \psi_2 = x_1 - x_2.$$

Now, consider the two-mass system in Fig. 1.20(b), where each mass is free to move in two directions. We define four coordinates x_1, x_2, y_1, and y_2. By analogy with the one-dimensional oscillator, we can see that there are now four normal modes: two transverse modes (motion in the y directions) and two longitudinal modes (motion in the x directions). Each longitudinal mode will be higher in frequency than the corresponding transverse mode, as in the one-mass system discussed in Section 1.9. Each mode can be described as motion along a normal coordinate. A system given an initial excitation along a single normal coordinate (or vibrating in a single normal mode) would ideally remain in that same normal mode of vibration until it runs out of energy. We will return to this subject in a later chapter.

1.12. Nonlinear Vibrations of a Simple System

Thus far, we have dealt almost exclusively with linear systems in which the restoring force is proportional to the displacement. Vibrations of such a system are harmonic; the equations of motion are linear differential equations. The sum of any two solutions to a linear differential equation is itself a solution. Thus, we construct a linear combination of simple solutions to fit the particular requirements of the problem of interest; this is known as the principle of superposition.

Nonlinear equations are more difficult to solve. Vibrations of a nonlinear system can no longer be expected to be simple harmonic motion. The principle of superposition does not hold; doubling the force does not necessarily double the response. The response of a nonlinear system may depend on both the frequency and the amplitude of the excitation.

Let us first consider a simple system with a cubic term in the restoring force: $F_x = -kx - bx^3$. One such system could be the mass-spring system in Section 1.6.6, where b is found to be Ka_0/a^3 for transverse oscillations. If the amplitude is small, the ratio x^3/a^3 is so small that the cubic force term can be ignored; this is an example of a *linear approximation*. For increasing amplitude, however, a point is reached where this is no longer advisable.

The nonlinear equation of motion is

$$m\ddot{x} = -Kx - bx^3 - F_0 \cos \omega t. \tag{1.76}$$

As a first approximation to x, we select $x_1 = A \cos \omega t$ and substitute into the right hand side of Eq. (1.76) to obtain a second approximation

$$m\ddot{x}_2 = -KA \cos \omega t - bA^3 \cos^3 \omega t - F_0 \cos \omega t, \tag{1.77}$$

where x_2 is the second approximation to the exact value of x.

Using the identity $\cos^3 x = 3/4 \cos x + 1/4 \cos 3x$ gives

$$\ddot{x}_2 = -\left(\frac{KA}{m} + \frac{3bA^3}{4m} + \frac{F_0}{m}\right) \cos \omega t - \frac{bA^3}{4m} \cos 3\omega t. \tag{1.78}$$

Integration of Eq. (1.78) gives

$$x_2 = \left(\frac{KA}{m\omega^2} + \frac{3bA^3}{4m\omega^2} + \frac{F_0}{m\omega^2}\right) \cos \omega t + \frac{bA^3}{36m\omega^2} \cos 3\omega t. \tag{1.79}$$

This process of successive approximation, sometimes called Duffing's method, works well if b, A, and F_0 are sufficiently small. Note that the term bx^3 in the force is responsible for the generation of a third harmonic ($\cos 3\omega t$ term in the expression for x_2).

If we equate the $\cos \omega t$ term in Eq. (1.79) to $A \cos \omega t$, which was our first approximation for x, we obtain

$$A\omega^2 = \frac{K}{m} A + \frac{3bA^3}{4m} + \frac{F_0}{m}$$

or (1.80)

$$A^2 = \frac{4m}{3b}\left(\omega^2 - \frac{K}{m} - \frac{F_0}{mA}\right).$$

The relationship of amplitude to frequency (for a given driving force F_0) from Eq. (1.80) is shown in Fig. 1.21. Note that the curve is double valued for a certain range of frequency. The dotted curve, obtained by setting $F_0 = 0$ in Eq. (1.80), describes free oscillations of the system.

When b is positive, the effective spring constant increases with amplitude; we call this a *hardening spring* system. In a hardening spring system, the free-oscillation frequency increases with amplitude. When b is negative, the effective spring constant and the frequency decrease with increasing amplitude; this is called a *softening spring* system. Response curves for softening and hardening spring systems are shown in Fig. 1.22. Damping has been added in order to limit the amplitude.

Many other types of nonlinear oscillators exist in physics, and their equations of motion are, in general, more difficult to solve than the one we have

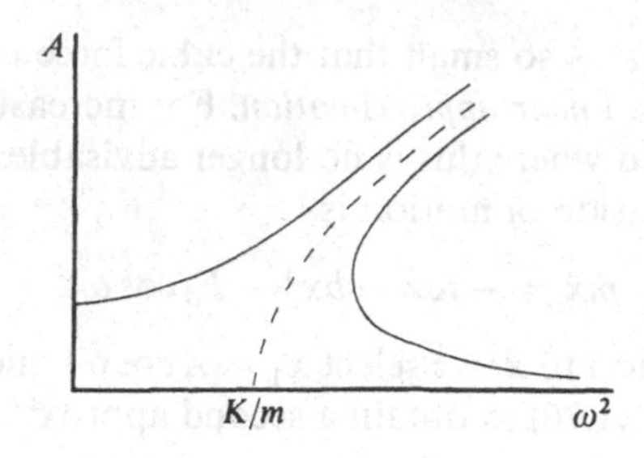

Fig. 1.21. Amplitude versus frequency for a nonlinear oscillating system. The dashed curve represents free oscillations.

briefly discussed. In an oscillator with several normal modes (and normal coordinates $\psi_1, \psi_2, \ldots, \psi_j$), the spring constant K_i used in the equation describing the normal mode may include a function of other normal coordinates: $K_i = K^0 + K^1(\psi_j)$. In general, this leads to coupling between the normal modes at a finite amplitude of oscillation. Forces resulting from this nonlinear coupling may be added to the driving force in the equation of motion:

$$m_i\ddot{\psi}_i + R\dot{\psi}_i + K^{\circ}\psi_i = f(t) + \sum F_{ij}(\psi_j). \tag{1.81}$$

In musical instruments, we encounter quite a number of cases where the forcing function $f(t)$ depends upon the vibration amplitude of the system being driven. The force between a violin bow and string, for example, depends upon their relative velocities, and the air flow through a clarinet reed depends upon the pressure difference across it. Nonlinearities of these types, however, are quite different from the nonlinear vibrations of simple systems discussed in this section.

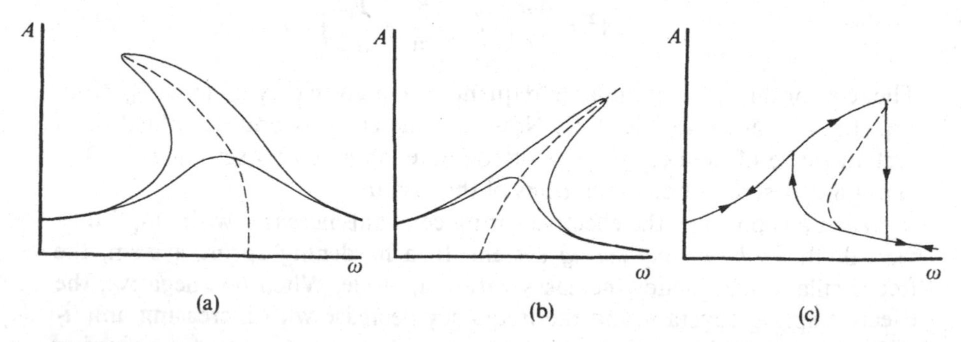

Fig. 1.22. (a) Response curves for oscillating system with softening spring behavior at small and large amplitudes. (b) Response curve for hardening spring behavior. (c) Response curve illustrating amplitude jumps at certain frequencies that lead to hysteresis.

APPENDIX

A.1. Alternative Ways of Expressing Harmonic Motion

We have written the solutions to the equation $\ddot{x} + \omega_0^2 x = 0$ in three ways:

$$x = A\cos(\omega_0 t + \phi), \tag{1.3}$$

$$x = B\cos\omega_0 t + C\sin\omega_0 t, \tag{1.4}$$

and

$$x = \mathrm{Re}(\tilde{A}e^{j\omega_0 t}), \tag{A1.1}$$

where

$$\tilde{A} = Ae^{j\phi} = A\cos\phi + jA\sin\phi. \tag{A1.2}$$

In order to establish a relationship between these constants, we can expand the cosine in Eq. (1.3):

$$x = A\cos\phi\cos\omega_0 t - A\sin\phi\sin\omega_0 t. \tag{A1.3}$$

Comparison with Eq. (1.4) gives the relationships

$$B = A\cos\phi,$$

$$C = -A\sin\phi,$$

and

$$\phi = \arctan(-C/B). \tag{A1.4}$$

Comparing Eq. (1.4) with Eq. (A1.2), it is clear that

$$\mathrm{Re}(\tilde{A}) = A\cos\phi = B,$$

and

$$\mathrm{Im}(\tilde{A}) = A\sin\phi = -C.$$

Yet, a fourth useful form is obtained by writing $x = De^{pt}$ and noting that this will be a solution to the differential equation when $p^2 = -\omega_0^2$ or $p = \pm j\omega_0$. The general solution can then be written as

$$x = \tilde{D}e^{j\omega_0 t} + \tilde{D}^* e^{-j\omega_0 t}. \tag{A1.5}$$

$\tilde{D}$ and $\tilde{D}^*$ are complex conjugates, as are $e^{j\omega_0 t}$ and $e^{-j\omega_0 t}$, of course; thus, the general solution is *real* (since any number added to its complex conjugate is real).

Expanding the exponentials in Eq. (A1.5) gives

$$x = \tilde{D}\cos\omega_0 t + j\tilde{D}\sin\omega_0 t + \tilde{D}^*\cos\omega_0 t - j\tilde{D}^*\sin\omega_0 t.$$

Comparison with Eq. (A1.1) gives

$$\tilde{D} + \tilde{D}^* = 2\,\mathrm{Re}(\tilde{D}) = A\cos\phi,$$

and

$$j(\tilde{D} - \tilde{D}^*) = -2\,\mathrm{Im}(\tilde{D}) = -A\sin\phi. \tag{A1.6}$$

To summarize, we have four forms of the solution given by Eqs. (1.3), (1.4), (A1.1), and (A1.5). Each form includes two arbitrary constants. Although in Eqs. (A1.1) and (A1.5) the constants are complex, x is real in each case. In Eq. (A1.1), the real displacement x is obtained by taking the real part of a complex displacement x; in Eq. (A1.5), however, the real displacement is obtained by adding two terms that are complex conjugates.

A.2. Equivalent Electrical Circuit for a Simple Oscillator

Many mechanical systems are mathematically equivalent to corresponding electrical systems. It is often helpful to represent a mechanical oscillating system by an equivalent electrical circuit, so that electrical network theory can be applied. The simple mechanical oscillator in Fig. 1.11 [and in Fig. A1.1(a)], for example, can be represented by the equivalent electrical circuit in Fig. A1.1(c). In the two electrical circuits, we identify velocity $\dot{x}$ with current i, displacement x with charge q, and force $f(t)$ with voltage $v(t)$. Mass m is then analogous to inductance L and stiffness to reciprocal capacitance $1/C$; resistance R appears in both circuits.

The mechanical and electrical impedances are

$$Z_m = R + jX_m = R + j(\omega m - K/\omega), \tag{A1.7a}$$

and

$$Z_e = R + jX_e = R + j(\omega L - 1/\omega C). \tag{A1.7b}$$

The resonance frequencies are

$$f_0 = \frac{1}{2\pi}\sqrt{\frac{K}{m}},$$

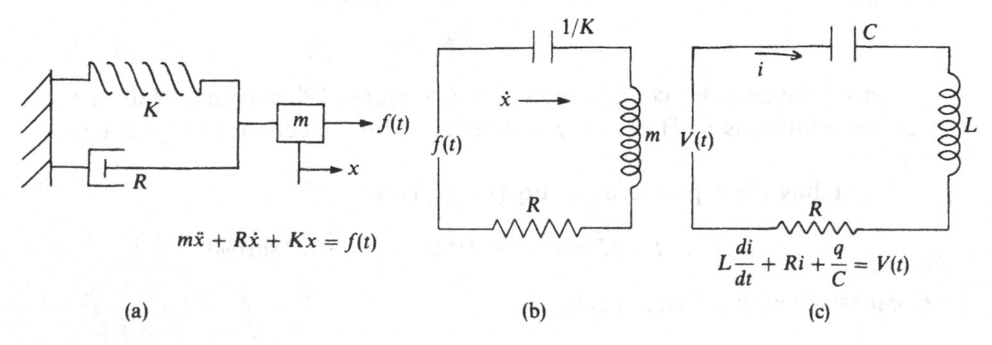

Fig. A1.1. (a) A simple mechanical oscillator. (b) Its equivalent electrical circuit. (c) Circuit of electrical analogs.

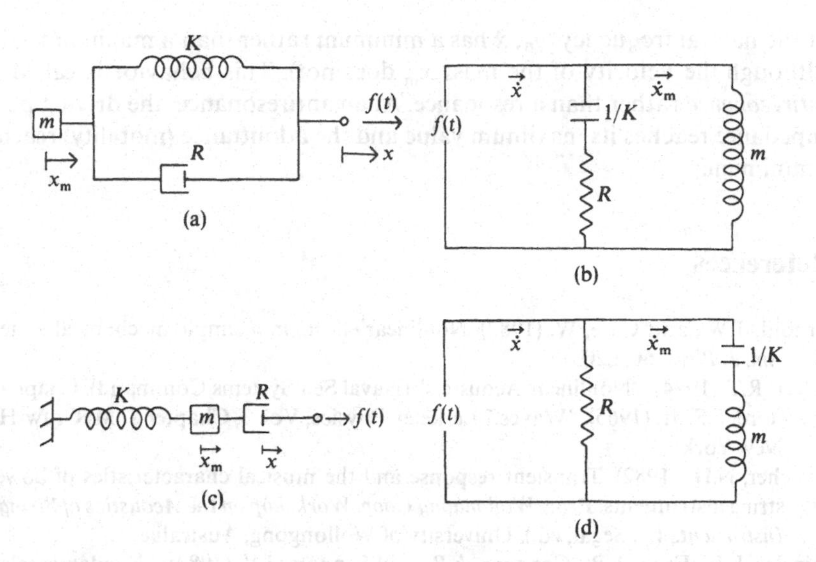

Fig. A1.2. Mechanical oscillating systems and their equivalent electrical circuits.

and (A1.8)

$$f_0 = \frac{1}{2\pi\sqrt{LC}}.$$

The oscillator in Fig. A1.1(a) is represented by a series circuit, because all the elements experience the same displacement x. If the force were applied to the end of the spring opposite the mass, as in Fig. A1.2(a), the system would be represented by the parallel circuit shown in Fig. A1.2(b). Similarly, the system in Fig. A1.2(c) has the equivalent circuit shown in Fig. A1.2(d).

The reciprocal of electrical impedance is electrical admittance. In mechanical systems, the reciprocal of impedance is called *mechanical admittance* or *mobility*. Mobility Y is velocity divided by force.

Note that the oscillating system in Fig. A1.2(a) is represented by a circuit [Fig. A1.2(b)] in which the two reactive elements ($1/K$ and m) are in parallel.

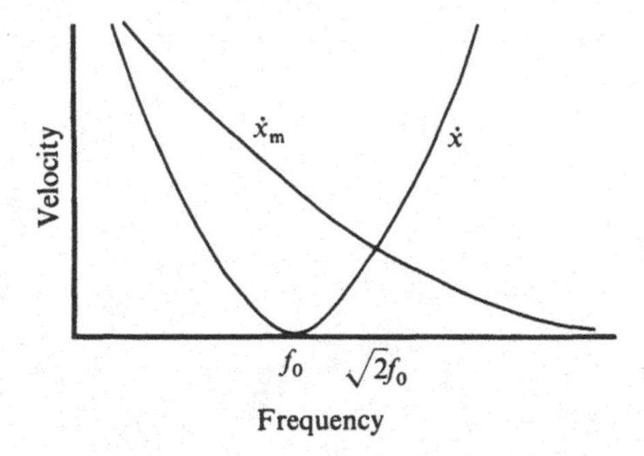

Fig. A1.3. Frequency response of the oscillating system shown in Fig. A1.2(a) as represented by the equivalent circuit in Fig. A1.2(b).

At the natural frequency ω_0, $\dot{x}$ has a minimum rather than a maximum value (although the velocity of the mass $\dot{x}_m$ does not). This behavior is called an *antiresonance* rather than a resonance. At an antiresonance, the driving point impedance reaches its maximum value and the admittance (mobility) reaches a minimum.

References

Arnold, T.W., and Case, W. (1982). Nonlinear effects in a simple mechanical system. *Am. J. Phys.* **50**, 220.

Beyer, R.T. (1974). "Nonlinear Acoustics" (Naval Sea Systems Command), Chapter 2.

Crawford, F.S. Jr. (1965). "Waves." Berkeley Physics, Vol. 3, Chapter 1. McGraw-Hill, New York.

Fletcher, N.H. (1982). Transient response and the musical characteristics of bowed-string instruments. *Proc. Wollongong Coop. Workshop on the Acoustics of Stringed Instruments* (A. Segal, ed.). University of Wollongong, Australia.

Kinsler, L.E., Frey, A.R., Coppens, A.B., and Sanders, J.V. (1982). "Fundamentals of Acoustics," 3rd ed., Chapter 1. Wiley, New York.

Lee, E.W. (1960). Non-linear vibrations. *Contemp. Phys.* **2**, 143.

Main, I.G. (1978). "Vibrations and Waves in Physics." Cambridge Univ. Press, London and New York.

Morse, P.M. (1948). "Vibration and Sound," 2nd ed., Chapter 2. McGraw-Hill, New York. Reprinted 1976, Acoustical Soc. Am., Woodbury, New York.

Rossing, T.D. (1982). "The Science of Sound," Addison-Wesley, Reading, Massachusetts.

Skudrzyk, E. (1968). "Simple and Complex Vibrating Systems," Chapters 1 and 2. Pennsylvania State University, University Park, Pennsylvania.

CHAPTER 2

Continuous Systems in One Dimension: Strings and Bars

In the last chapter, we considered vibrating systems consisting of one or more masses, springs, and dampers. In this chapter, we will focus on systems in which these elements are distributed continuously throughout the system rather than appearing as discrete elements. We begin with a system composed of several discrete elements, then allow the number of elements to grow larger, eventually leading to a continuum.

2.1. Linear Array of Oscillators

The oscillating system with two masses in Fig. 1.20 was shown to have two transverse vibrational modes and two longitudinal modes. In both the longitudinal and transverse pairs, there is a mode of low frequency in which the masses move in the same direction and a mode of higher frequency in which they move in opposite directions.

The normal modes of a three-mass oscillator are shown in Fig. 2.1. The masses are constrained to move in a plane, and so there are six normal modes of vibration, three longitudinal and three transverse. Each longitudinal mode will be higher in frequency than the corresponding transverse mode. If the masses were free to move in three dimensions, there would be $3 \times 3 = 9$ normal modes, three longitudinal and six transverse.

Increasing the number of masses and springs in our linear array increases the number of normal modes. Each new mass adds one longitudinal mode and (provided the masses move in a plane) one transverse mode. The modes of transverse vibration for mass/spring systems with $N = 1$ to 24 masses are shown in Fig. 2.2; note that as the number of masses increases, the system takes on a wavelike appearance. A similar diagram could be drawn for the longitudinal modes.

As the number of masses in our linear system increases, we take less and less notice of the individual elements, and our system begins to resemble a vibrating string with mass distributed uniformly along its length. Presumably, we could describe the vibrations of a vibrating string by writing N equations

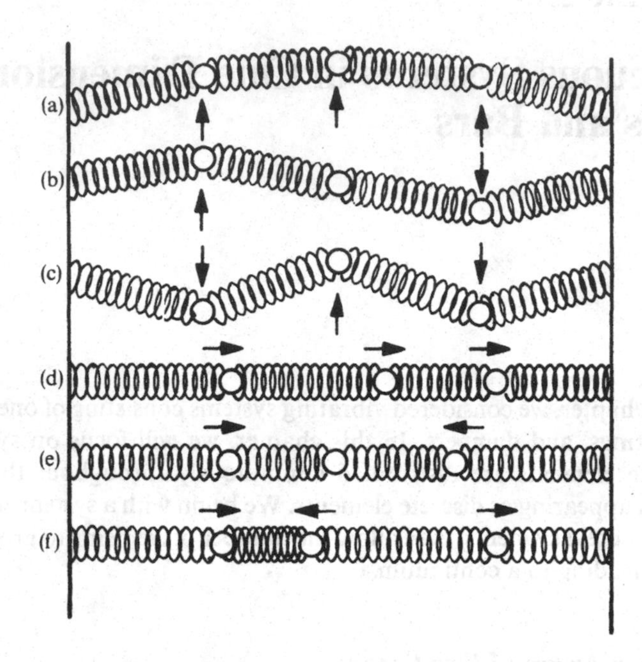

Fig. 2.1. Normal modes of a three–mass oscillator. Transverse mode (a) has the lowest frequency and longitudinal mode (f) the highest.

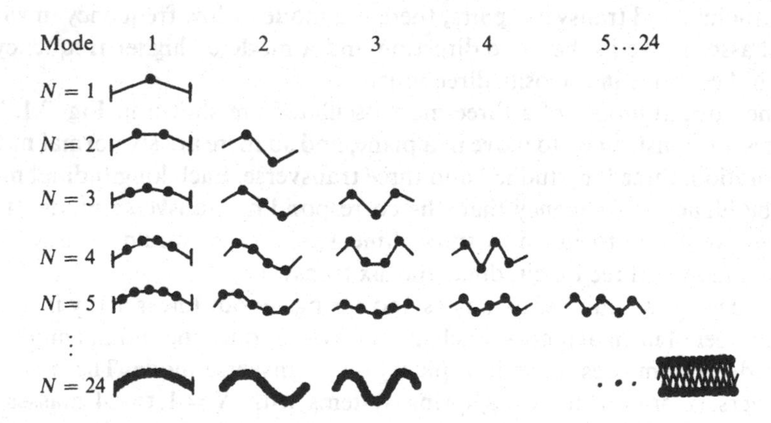

Fig. 2.2. Modes of transverse vibration for mass/spring systems with different numbers of masses. A system with N masses has N modes.

of motion for N equally spaced masses and letting N go to infinity, but it is much simpler to consider the shape of the string as a whole.

2.2. Transverse Wave Equation for a String

The study of vibrating strings has a long history. Pythagoras is said to have observed how the division of a stretched string into two segments gave pleasing sounds when the lengths of these two segments had a simple ratio (2 : 1, 3 : 1, 3 : 2, etc.). These are examples of normal modes of a string fixed at its ends. Closer examination of the motion of a string reveals that the normal modes depend upon the mass of the string, its length, the tension applied, and the end conditions.

Consider a uniform string (Fig. 2.3) with linear density μ (kg/m) stretched to a tension T (newtons). The net force dF, restoring segments ds to its equilibrium position, is the difference between the y components of T at the two ends of the segment:

$$dF_y = (T \sin\theta)_{x+dx} - (T\sin\theta)_x.$$

Applying the Taylor's series expansion $f(x + dx) = f(x) + \dfrac{\partial f(x)}{\partial x} dx + \cdots$ to $T \sin\theta$ and keeping first-order terms gives

$$dF_y = \left[(T\sin\theta)_x + \frac{\partial(T\sin\theta)}{\partial x} dx\right] - (T\sin\theta)_x = \frac{\partial(T\sin\theta)}{\partial x} dx. \tag{2.1}$$

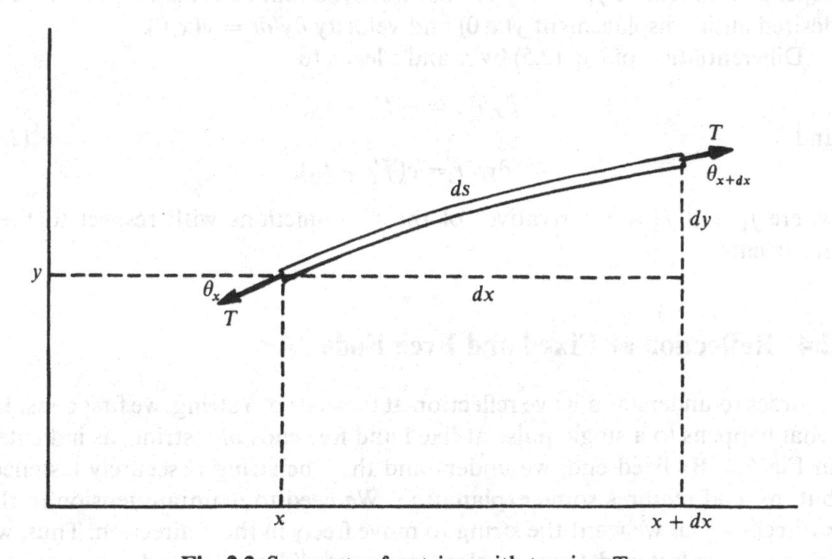

Fig. 2.3. Segments of a string with tension T.

For small displacement y, $\sin\theta$ can be replaced by $\tan\theta$, which is also $\partial y/\partial x$:

$$dF_y = \frac{\partial(T\,\partial y/\partial x)}{\partial x}dx = T\frac{\partial^2 y}{\partial x^2}dx. \tag{2.2}$$

The mass of the segment ds is μds, so Newton's second law of motion becomes

$$T\frac{\partial^2 y}{\partial x^2}dx = (\mu ds)\frac{\partial^2 y}{\partial t^2}. \tag{2.3}$$

Since dy is small, $ds \cong dx$. Also, we write $c^2 = T/\mu$ and obtain

$$\frac{\partial^2 y}{\partial t^2} = \frac{T}{\mu}\frac{\partial^2 y}{\partial x^2} = c^2\frac{\partial^2 y}{\partial x^2}. \tag{2.4}$$

This is the well-known equation for transverse waves in a vibrating string.

2.3. General Solution of the Wave Equation: Traveling Waves

The general solution of Eq. (2.4) can be written in a form credited to d'Alembert (1717–1783):

$$y = f_1(ct - x) + f_2(ct + x). \tag{2.5}$$

The function $f_1(ct - x)$ represents a wave traveling to the right with a velocity c; similarly, $f_2(ct + x)$ represents a wave traveling to the left with the same velocity. The nature of functions f_1 and f_2 is arbitrary; they could be sinusoidal or they could describe impulsive waves, for example. In fact, the two independent functions f_1 and f_2 can be chosen so that their sum represents any desired initial displacement $y(x, 0)$ and velocity $\partial y/\partial t = \dot{y}(x, 0)$.

Differentiation of Eq. (2.5) by x and t leads to

$$\partial y/\partial x = -f_1' + f_2',$$

and

$$\partial y/\partial t = c(f_1' + f_2'), \tag{2.6}$$

where f_1' and f_2' are derivatives of the two functions with respect to their arguments.

2.4. Reflection at Fixed and Free Ends

In order to understand wave reflection at the ends of a string, we first consider what happens to a single pulse at fixed and free ends of a string, as indicated in Fig. 2.4. By fixed end, we understand that the string is securely fastened, but free end requires some explanation. We need to maintain tension in the x direction, but we want the string to move freely in the y direction. Thus, we imagine it is fastened to a massless ring that slides up and down on a rod without friction.

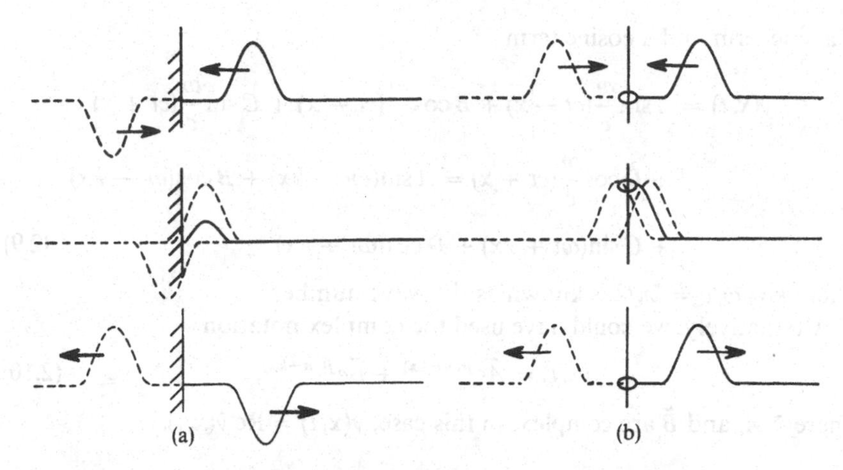

Fig. 2.4. Reflection of a pulse at a fixed end (a) and at a free end (b). In (a) the appropriate boundary condition can be met by having an imaginary pulse of opposite phase meet the real pulse at $x = 0$. In (b) the imaginary pulse has the same phase.

1. At a *fixed end*, $y = 0$. Assuming that the string is fixed at $x = 0$, the general solution [Eq. (2.5)] becomes

$$y = 0 = f_1(ct - 0) + f_2(ct + 0),$$

from which (2.7)

$$f_1(ct) = -f_2(ct).$$

Thus, an up pulse reflects as a down pulse, as shown in Fig. 2.4(a).

2. At a *free end*, $\partial y/\partial x = 0$ because no transverse force is possible. Thus, from Eq. (2.6),

$$f'_1(ct) = f'_2(ct). \tag{2.8a}$$

Integration of Eq. (2.8a) gives

$$f_1(ct) = f_2(ct). \tag{2.8b}$$

An up pulse now reflects as an up pulse, as shown in Fig. 2.4(b).

Of couse, many other end conditions are possible. For example, the string may be attached to a string with a different linear density μ, to a spring, or to a mass. A particularly important case is that of an end support that is nearly fixed but yields slightly, such as the bridge of a piano or violin.

2.5. Simple Harmonic Solutions to the Wave Equation

In order to see how simple harmonic motions are propagated along a string, we let the functions f_1 and f_2 in the general solution [Eq. (2.5)] each consist of a sine term and a cosine term

$$y(x,t) = A\sin\frac{\omega}{c}(ct - x) + B\cos\frac{\omega}{c}(ct - x) + C\sin\frac{\omega}{c}(ct + x)$$

$$+ D\cos\frac{\omega}{c}(ct + x) = A\sin(\omega t - kx) + B\cos(\omega t - kx)$$

$$+ C\sin(\omega t + kx) + D\cos(\omega t + kx), \tag{2.9}$$

where $k = \omega/c = 2\pi/\lambda$ is known as the wave number.

Alternatively, we could have used the complex notation

$$\tilde{y}(x,t) = \tilde{A}e^{j(\omega t - kx)} + \tilde{B}e^{j(\omega t + kx)}, \tag{2.10}$$

where $\tilde{y}$, $\tilde{A}$, and $\tilde{B}$ are complex. In this case, $y(x,t) = \operatorname{Re}\tilde{y}(x,t)$.

2.6. Standing Waves

Consider a string of length L fixed at $x = 0$ and $x = L$. The first condition $y(0,t) = 0$ requires that $A = -C$ and $B = -D$ in Eq. (2.9), so

$$y = A[\sin(\omega t - kx) - \sin(\omega t + kx)] + B[\cos(\omega t - kx) - \cos(\omega t + kx)]. \tag{2.11}$$

Using the sum and difference formulas, $\sin(x \pm y) = \sin x \cos y \pm \cos x \sin y$ and $\cos(x \pm y) = \cos x \cos y \mp \sin x \sin y$,

$$y = 2A\sin kx\cos\omega t - 2B\sin kx\sin\omega t$$

$$= 2[A\cos\omega t - B\sin\omega t]\sin kx. \tag{2.12}$$

The second condition $y(L,t) = 0$ requires that $\sin kL = 0$ or $\omega L/c = n\pi$. This restricts ω to values $\omega_n = n\pi c/L$ or $f_n = n(c/2L)$. Thus, the string has normal modes of vibration:

$$y_n(x,t) = (A_n\sin\omega_n t + B_n\cos\omega_n t)\sin\frac{\omega_n x}{c}. \tag{2.13}$$

These modes are harmonic, because each f_n is n times $f_1 = c/2L$.

The general solution of a vibrating string with fixed ends can be written as a sum of the normal modes:

$$y = \sum_n (A_n\sin\omega_n t + B_n\cos\omega_n t)\sin k_n x, \tag{2.14}$$

and the amplitude of the nth mode is $C_n = \sqrt{A_n^2 + B_n^2}$. At any point $y(x,t) = \sum_n y_n(x,t)$.

Alternatively, the general solution could be written as

$$y = \sum_n C_n\sin(\omega_n t + \phi_n)\sin k_n x, \tag{2.15}$$

where C_n is the amplitude of the nth mode and ϕ_n is its phase.

2.7. Energy of a Vibrating String

When a string vibrates in one of its normal modes, the kinetic and potential energies alternately take on their maximum value, which is equal to the total energy, just as in the simple mass-spring system discussed in Section 1.3. Thus, the energy of a mode can be calculated by considering either the kinetic or the potential energy. The maximum kinetic energy of a segment vibrating in its nth mode is

$$dE_n = \frac{\omega_n^2 \mu}{2}(A_n^2 + B_n^2)\sin^2 \frac{n\pi x}{L} dx.$$

Integrating over the entire length gives

$$E_n = \frac{\omega_n^2 \mu L}{4}(A_n^2 + B_n^2) = \frac{\omega_n^2 \mu L}{4} C_n^2. \tag{2.16}$$

The potential and kinetic energies of each mode have a time average value that is $E_n/2$. The total energy of the string can be found by summing up the energy in each normal mode:

$$E = \sum_n E_n.$$

2.8. Plucked String: Time and Frequency Analyses

When a string is excited by bowing, plucking, or striking, the resulting vibration can be considered to be a combination of several modes of vibration. For example, if the string is plucked at its center, the resulting vibration will consist of the fundamental plus the odd-numbered harmonics. Fig. 2.5 illustrates how the modes associated with the odd-numbered harmonics, when each is present in the right proportion, add up at one instant in time to give the initial shape of the center-plucked string. Modes 3, 7, 11, etc., must be opposite in phase from modes, 1, 5, and 9 in order to give maximum displacement at the center, as shown at the top. Finding the normal mode spectrum of a string given its initial displacement calls for *frequency analysis* or *Fourier analysis*.

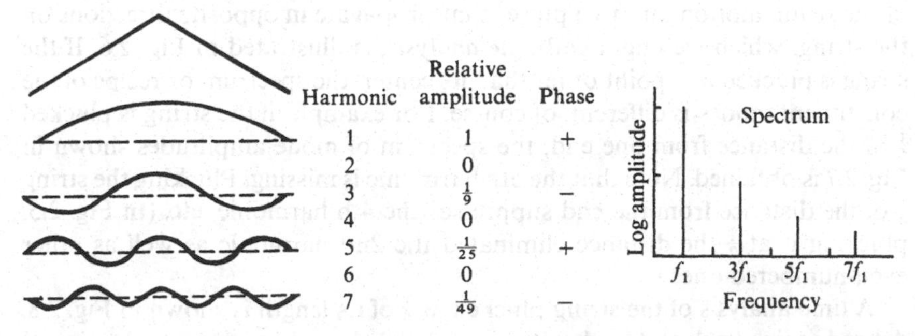

Fig. 2.5. Frequency analysis of a string plucked at its center. Odd-numbered modes of vibration add up in appropriate amplitude and phase to give the shape of the string.

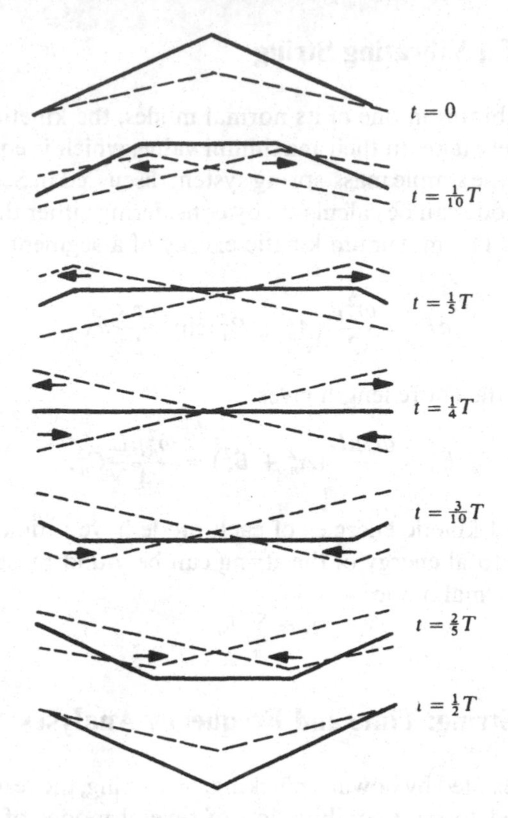

Fig. 2.6. Time analysis of the motion of a string plucked at its midpoint through one half cycle. Motion can be thought of as due to two pulses traveling in opposite directions.

Since all the modes shown in Fig. 2.5 have different frequencies of vibration, they quickly get out of phase, and the shape of the string changes rapidly after plucking. The shape of the string at each moment can be obtained by adding the normal modes at that particular time, but it is more difficult to do so because each of the modes will be at a different point in its cycle. The resolution of the string motion into two pulses that propagate in opposite directions on the string, which we might call *time analysis*, is illustrated in Fig. 2.6. If the string is plucked at a point other than its center, the spectrum or recipe of the constituent modes is different, of course. For example, if the string is plucked $\frac{1}{5}$ of the distance from one end, the spectrum of mode amplitudes shown in Fig. 2.7 is obtained. Note that the 5th harmonic is missing. Plucking the string $\frac{1}{4}$ of the distance from the end suppresses the 4th harmonic, etc. (In Fig. 2.5, plucking it at $\frac{1}{2}$ the distance eliminated the 2nd harmonic as well as other even-numbered ones.)

A time analysis of the string plucked at $\frac{1}{5}$ of its length is shown in Fig. 2.8. A bend racing back and forth within a parallelogram boundary can be viewed as the resultant of two pulses (dashed lines) traveling in opposite directions.

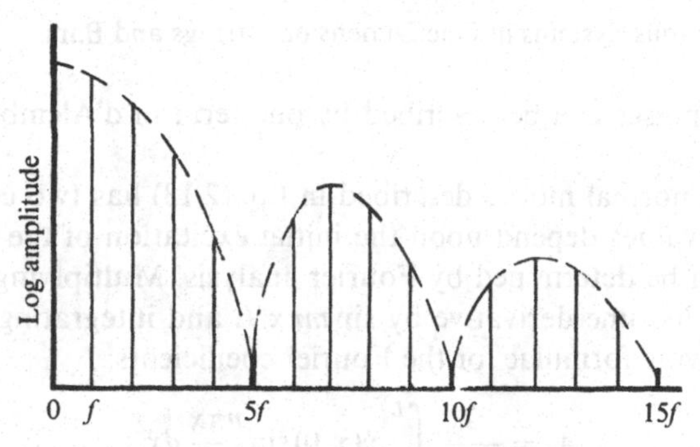

Fig. 2.7. Spectrum of a string plucked one-fifth of the distance from one end.

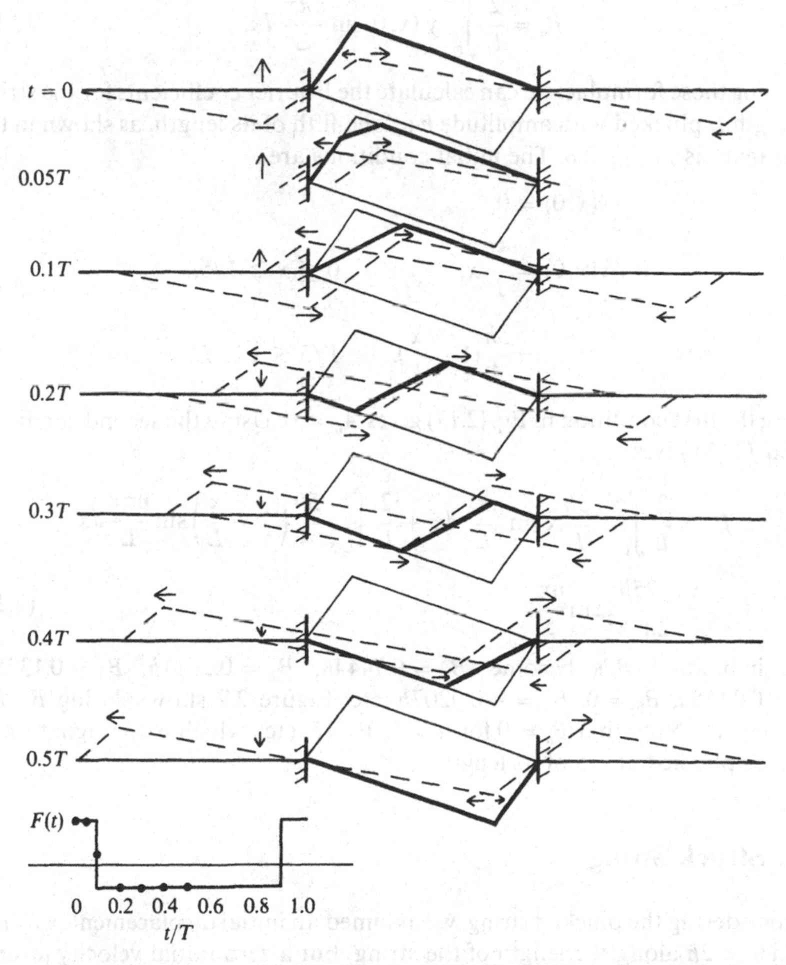

Fig. 2.8. Time analysis through one half cycle of the motion of a string plucked one-fifth of the distance from one end. The motion can be thought of as due to two pulses [representing the two terms in Eq. (2.5)] moving in opposite directions (dashed curves). The resultant motion consists of two bends, one moving clockwise and the other counterclockwise around a parallelogram. The normal force on the end support, as a function of time, is shown at the bottom.

Each of these pulses can be described by one term in d'Alembert's solution [Eq. (2.5)].

Each of the normal modes described in Eq. (2.13) has two coefficients A_n and B_n whose values depend upon the initial excitation of the string. These coefficients can be determined by Fourier analysis. Multiplying each side of Eq. (2.14) and its time derivative by $\sin m\pi x/L$ and integrating from 0 to L gives the following formulae for the Fourier coefficients:

$$A_n = \frac{2}{\omega_n L}\int_0^L \dot{y}(x,0)\sin\frac{n\pi x}{L}\,dx, \tag{2.17}$$

$$B_n = \frac{2}{L}\int_0^L y(x,0)\sin\frac{n\pi x}{L}\,dx. \tag{2.18}$$

Using these formulae, we can calculate the Fourier coefficients for the string of length L plucked with amplitude h at one-fifth of its length, as shown in the time analysis in Fig. 2.8. The initial conditions are

$$\begin{aligned} \dot{y}(x,0) &= 0, \\ y(x,0) &= \frac{5h}{L}x, \qquad 0 \leqq x \geqq L/5, \\ &= \frac{5h}{4}\left(1 - \frac{x}{L}\right), \qquad L/5 \leqq x \geqq L. \end{aligned} \tag{2.19}$$

Using the first condition in Eq. (2.17) gives $A_n = 0$. Using the second condition in Eq. (2.18) gives

$$\begin{aligned} B_n &= \frac{2}{L}\int_0^{L/5} \frac{5h}{L} x\sin\frac{n\pi x}{L}\,dx + \frac{2}{L}\int_{L/5}^{L} \frac{5h}{4}\left(1 - \frac{x}{L}\right)\sin\frac{n\pi x}{L}\,dx \\ &= \frac{25h}{2n^2\pi^2}\sin\frac{n\pi}{5}. \end{aligned} \tag{2.20}$$

The individual B_n's become: $B_1 = 0.7444h$, $B_2 = 0.3011h$, $B_3 = 0.1338h$, $B_4 = 0.0465h$, $B_5 = 0$, $B_6 = -0.0207h$, etc. Figure 2.7 shows $20\log|B_n|$ for $n = 0$ to 15. Note that $B_n = 0$ for $n = 5$, 10, 15, etc., which is the signature of a string plucked at 1/5 of its length.

2.9. Struck String

In considering the plucked string, we assumed an initial displacement (varying from 0 to $2h$ along the length of the string) but a zero initial velocity (everywhere) at $t = 0$. Now, we consider the opposite set of conditions: zero initial displacement with a specified initial velocity. This velocity could be imparted by a hard hammer that strikes the string at $t = 0$, for example. Of course, a blow by a real hammer does not instantly impart a velocity to the string; in fact, the nature of the initial velocity depends in a complicated way on a number of factors, such as the compliance of the hammer. Various models of hammer–string interaction are discussed in a series of papers by Hall (1986, 1987a,b).

Suppose that the string is struck by a hard, narrow hammer having a velocity V. After a short time t, a portion of the string with length $2ct$ and mass $2\mu ct$ is set into motion. As this mass increases and becomes comparable to the hammer mass M, the hammer is slowed down and would eventually be stopped. With a string of finite length, however, reflected impulses return while the hammer still has appreciable velocity, and these reflected impulses interact with the hammer in a rather complicated way, causing it to be thrown back from the string.

At the point of contact, the string and hammer together satisfy the equation

$$M\frac{\partial^2 y}{\partial t^2} = T\,\Delta\left(\frac{\partial y}{\partial x}\right), \tag{2.21}$$

while elsewhere the string continues to satisfy Eq. (2.4). The discontinuity in the string slope $\Delta\left(\frac{\partial y}{\partial x}\right)$, according to Eq. (2.21), is responsible for the force that slows down the hammer. Equation (2.21) is satisfied at the contact point by a velocity

$$v(t) = Ve^{-t/\tau}, \tag{2.22}$$

where $\tau = Mc/2T$ may be termed the deceleration time (Hall, 1986). The corresponding displacement is

$$y(t) = V\tau(1 - e^{-t/\tau}). \tag{2.23}$$

The displacement at the contact point approaches $VMc/2T$, and the velocity approaches zero. If the string were very long, the displacement and velocity elsewhere on the string could be found by substituting $t - [(x - x_0)/c]$ for t in Eqs. (2.24) and (2.25), as shown in Fig. 2.9.

Only when the string is very long or the hammer is very light does the hammer stop, as in Fig. 2.9. In a string of finite length, reflected pulses return from both ends of the strings and interact with the moving hammer in a fairly complicated way. Eventually, the hammer is thrown clear of the string, and the string vibrates freely in its normal modes.

In general, the harmonic amplitudes in the vibration spectrum of a struck string fall off less rapidly with frequency than those of the plucked strings shown in Figs. 2.5 and 2.7. For a very light hammer whose mass M is much less than the mass of the string M_s, the spectrum dips to zero for harmonic numbers that are multiples of $1/\beta$ (where the string is struck at a fraction β of its length), but otherwise does not fall off with frequency, as shown in Fig. 2.10(a). (The spectrum of sound radiated by a piano, which may be quite different from the vibration spectrum of the string, will be discussed in Chapter 12).

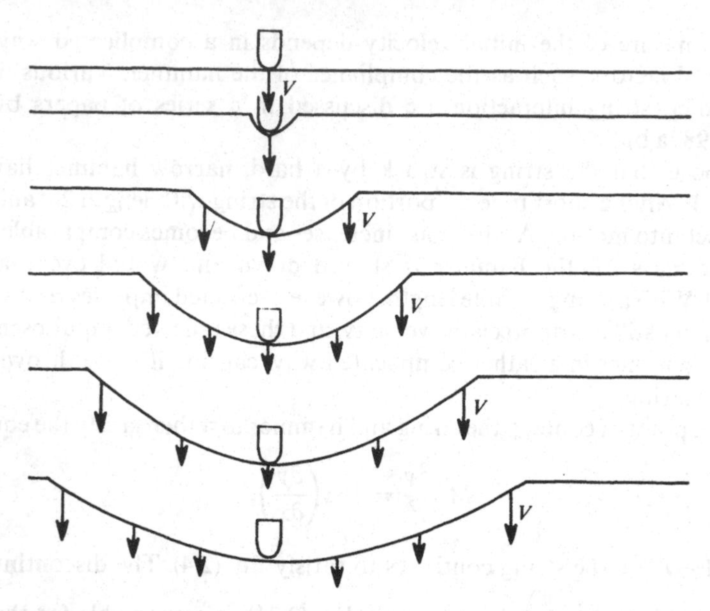

Fig. 2.9. Displacement and velocity of a long string at successive times after being struck by a hard narrow hammer having a velocity V.

If the hammer mass is small but not negligible compared to the mass of the string, the spectrum envelope falls off as $1/n$ (6 dB/octave) above a mode number given by $n_m = 0.73\ M_s/M$, as shown in Fig. 2.10(b). Note that for high harmonic (mode) numbers, there are missing modes between those in Fig. 2.10(a).

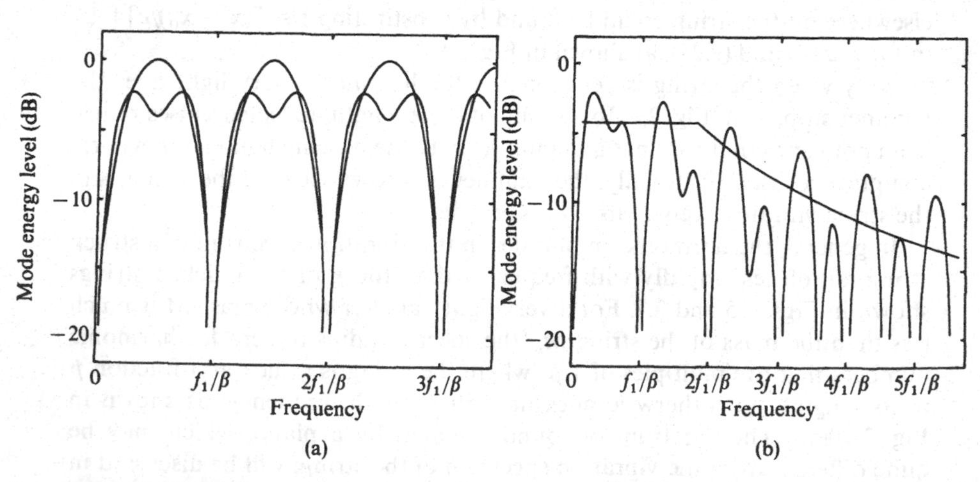

Fig. 2.10. Spectrum envelopes for a string struck at a fraction β of its length: (a) hammer mass $M \ll$ string mass M_s; (b) $M = 0.4\ \beta M_s$ (from Hall, 1986).

2.10. Bowed String

The motion of a bowed string has interested physicists for many years, and much has been written on the subject. In this chapter, we give only a brief description of some of the important features.

As the bow is drawn across the string of a violin, the string appears to vibrate back and forth smoothly between two curved boundaries, much like a string vibrating in its fundamental mode. However, this appearance of simplicity is deceiving. Over a hundred years ago, Helmholtz (1877) showed that the string more nearly forms two straight lines with a sharp bend at the point of intersection. This bend races around the curved path that we see, making one round trip each period of the vibration.

To observe the string motion, Helmholtz constructed a vibration microscope, consisting of an eyepiece attached to a tuning fork. This was driven in sinusoidal motion parallel to the string, and the eyepiece was focused on a bright-colored spot on the string. When Helmholtz bowed the string, he saw a Lissajous figure (see Section 1.10). The figure was stationary when the tuning fork frequency was an integral fraction of the string frequency. Helmholtz noted that the displacement of the string followed a triangular pattern at whatever point he observed it, as shown in Fig. 2.11. The velocity waveform at each point alternates between two values.

Other early work on the subject was published by Krigar–Menzel and Raps (1891) and by Nobel laureate C.V. Raman (1918). More recent experiments by Schelleng (1973), McIntyre, et al. (1981), Lawergren (1980), Kondo and Kubata (1983), and by others have verified these early findings and have greatly added to our understanding of bowed strings. An excellent discussion of the bowed string is given by Cremer (1981).

The motion of a bowed string is shown in Fig. 2.12. A time analysis in Fig. 2.12(A) shows the Helmholtz-type motion of the string; as the bow moves ahead at a constant speed, the bend races around a curved path. Fig. 2.12(B) shows the position of the point of contact at successive times; the letters correspond to the frames in Fig. 2.12(A). Note that there is a single bend in the bowed string, whereas in the plucked string (Fig. 2.8), we had a double bend.

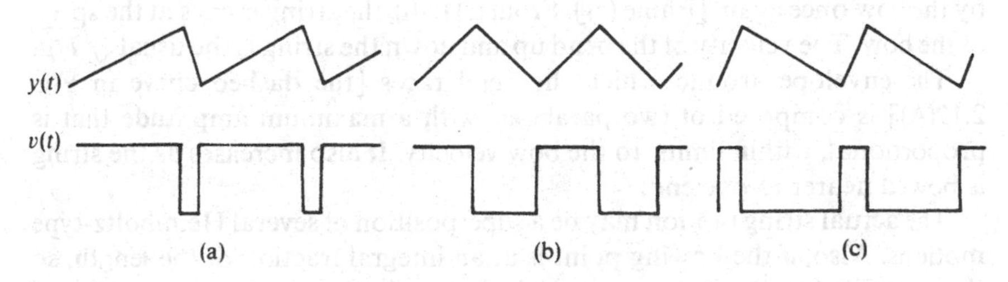

Fig. 2.11. Displacement and velocity of a bowed string at three positions along its length: (a) at $x = L/4$, (b) at the center, and (c) at $x = 3L/4$.

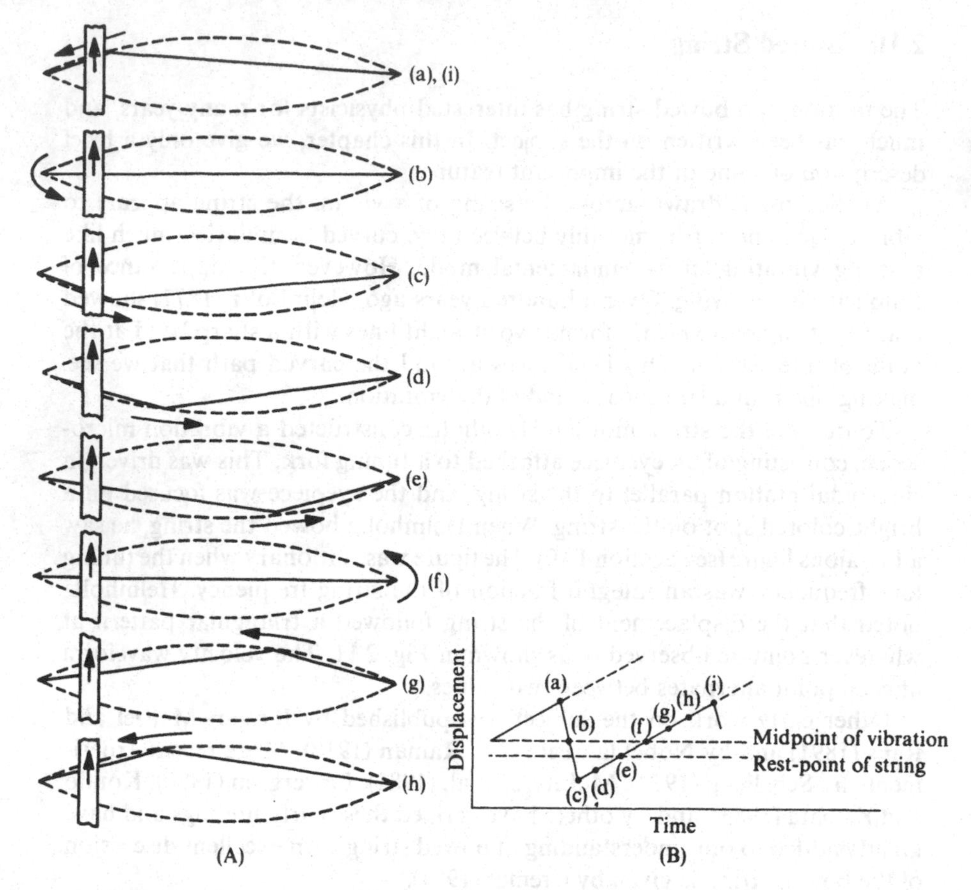

Fig. 2.12. Motion of a bowed string. (A) Time analysis of the motion, showing the shape of the string at eight successive times during the cycle. (B) Displacement of the bow (dashed line) and the string at the point of contact (solid line) at successive times. The letters correspond to the letters in (A).

The action of the bow on the string is often described as a stick and slip action. The bow drags the string along until the bend arrives [from (a) in Fig. 2.12(A)] and triggers the slipping action of the string until it is picked up by the bow once again [frame (c)]. From (c) to (i), the string moves at the speed of the bow. The velocity of the bend up and down the string is the usual $\sqrt{T/\mu}$.

The envelope around which the bend races [the dashed curve in Fig. 2.12(A)] is composed of two parabolas with a maximum amplitude that is proportional, within limits, to the bow velocity. It also increases as the string is bowed nearer to one end.

The actual string motion may be a superposition of several Helmholtz-type motions. Also, if the bowing point is at an integral fraction of the length, so that certain harmonics are not excited, the displacement curves take on ripples. These and many other details of bowed string motion are treated elegantly by Cremer (1981).

2.11. Driven String: Impedance

One way to excite a string is to apply a transverse force $f(t)$ to one end. We first consider an infinite string with tension T and a transverse force $\tilde{f}(t) = Fe^{j\omega t}$ as shown in Fig. 2.13. Since the string is infinitely long and the force is applied at the left end, the solution consists only of waves moving to the right.

$$y(x, t) = \tilde{A}e^{j(\omega t - kx)},$$

where $\tilde{A}$ is a complex constant giving the amplitude and the phase with respect to the driving force and $k = \omega/c = 2\pi/\lambda$.

Since there is no mass concentrated at $x = 0$, the driving force should balance the transverse component of the tension:

$$\tilde{F} = -T\sin\theta \cong -T(\partial\tilde{y}/\partial x) \qquad x = 0. \tag{2.24}$$

Substitution of $\tilde{f}(t) = Fe^{j\omega t}$ and $\tilde{y}(x, t) = \tilde{A}e^{j(\omega t - kx)}$ gives

$$Fe^{j\omega t} = jkT\tilde{A}e^{j\omega t}$$

or

$$\tilde{A} = F/jkT, \tag{2.25}$$

so

$$\tilde{y}(x, t) = \frac{-jF}{kT}e^{j(\omega t - kx)}.$$

The velocity $\tilde{u} = \partial\tilde{y}/\partial t$ becomes

$$\tilde{u}(x, t) = \frac{F\omega}{kT}e^{j(\omega t - kx)} = \frac{Fc}{T}e^{j(\omega t - kx)}. \tag{2.26}$$

We define the mechanical input impedance Z_{in} as the ratio of force to velocity at the driving point, so

$$Z_{in} = \frac{\tilde{f}(t)}{\tilde{u}(0, t)}. \tag{2.27}$$

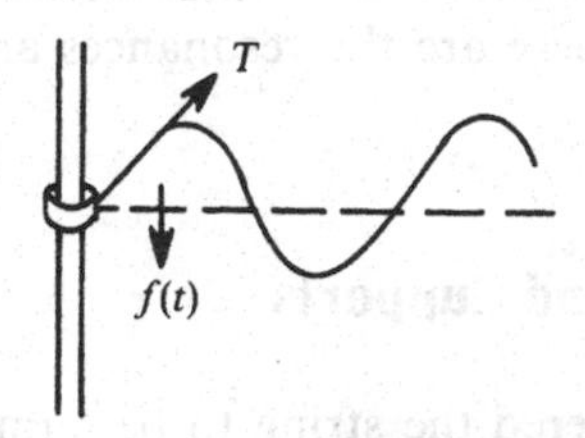

Fig. 2.13. Forces at $x = 0$ on a string free to move in the y direction.

In a string of infinite length (or a string terminated so that no reflections occur), $\tilde{Z}_{in}$ equals *the characteristic impedance* Z_0, which is a real quantity; the input impedance is purely resistive in an infinite string.

$$Z_0 = \frac{T}{c} = \sqrt{T\mu} = \mu c. \tag{2.28}$$

The behavior of a string of finite length is more complicated because $\partial\tilde{y}/\partial x$ at $x = 0$ depends upon the reflected wave as well.

$$\tilde{y}(x, t) = \tilde{A}e^{j(\omega t - kx)} + \tilde{B}e^{j(\omega t + kx)}. \tag{2.29}$$

Assume that the string is fixed at $x = L$ and driven at $x = 0$ as before. Substitution of Eq. (2.29) into Eq. (2.24) gives

$$Fe^{j\omega t} = T(jk\tilde{A} - jk\tilde{B})e^{j\omega t}. \tag{2.30}$$

The boundary condition at $x = L$ gives

$$0 = \tilde{A}e^{-jkL} + \tilde{B}e^{jkL}. \tag{2.31}$$

Solving Eqs. (2.30) and (2.31) together gives

$$\tilde{A} = \frac{Fe^{jkL}}{2jkT\cos kL},$$

and

$$\tilde{B} = \frac{Fe^{-jkL}}{-2jkT\cos kL},$$

from which (2.32)

$$\tilde{y}(x, t) = \frac{F}{kT}\frac{\sin k(L - x)}{\cos kL}e^{j\omega t},$$

and

$$\tilde{u}(x, t) = \frac{j\omega F}{kT}\frac{\sin k(L - x)}{\cos kL}e^{j\omega t}. \tag{2.33}$$

The input impedance at $x = 0$ is

$$\tilde{Z}_{in} = f(t)/\tilde{u}(x, t) = \frac{-jkT}{\omega}\cot kL = -jZ_0 \cot kL. \tag{2.34}$$

This impedance is purely reactive and varies from 0 ($kL = \pi/2$, $3\pi/2$, etc.) to $\pm j\infty$ ($kL = 0$, π, etc.). These are the resonances and antiresonances of the string, respectively.

2.12. Motion of the End Supports

In Section 2.6, we considered the string to be terminated by two rigid end supports ($y = 0$ at $x = 0$ and $x = L$). We will now consider what happens when one of the end supports is not completely rigid.

We can generally describe the termination by writing its complex impedance. If the imaginary part of the complex impedance is masslike, the resonances of the string will be raised slightly above those given by Eq. (2.13); if it is springlike, on the other hand, the resonance frequencies will be lowered. The real part of the complex impedance is indicative of the rate of energy transfer from the spring to the support (the bridge and soundboard of a guitar, for example).

Let us consider a string fixed at $x = 0$ and terminated at $x = L$ by a support that can be characterized by a mass m. The transverse force exerted on the mass by the string is $-T(\partial\tilde{y}/\partial x)_{x=L}$. By Newton's second law,

$$-T(\partial\tilde{y}/\partial x)_L = m(\partial^2\tilde{y}/\partial t^2)_L. \tag{2.35}$$

Applying the boundary condition at $x = 0$ to Eq. (2.11) gives

$$0 = \tilde{A}e^{j\omega t} + \tilde{B}e^{j\omega t}, \tag{2.36}$$

so $\tilde{A} = -\tilde{B}$, and the harmonic solution becomes

$$\tilde{y}(x,t) = \tilde{A}(-e^{-kx} + e^{kx})e^{j\omega t} = \tilde{A}\sin kx e^{j\omega t}. \tag{2.37}$$

Substituting Eq. (2.37) into Eq. (2.35) gives

$$-kT\tilde{A}\cos kL\, e^{j\omega t} = -\omega^2 m\tilde{A}\sin kL\, e^{j\omega t},$$

and (2.38)

$$\cot kL = \frac{\omega^2 m}{kT} = \frac{ky^2 m}{T}\frac{km}{\mu} = \frac{m}{M}kL,$$

where $M = \mu L$ is the total mass of the string. The transcendental equation, Eq. (2.38), can be solved graphically, as shown in Fig. 2.14, for two values of m/M. As $m \gg M$, the roots approach the values $k = n\pi$ for the string fixed at $x = L$ as well as at $x = 0$. Note that the normal mode frequencies obtained

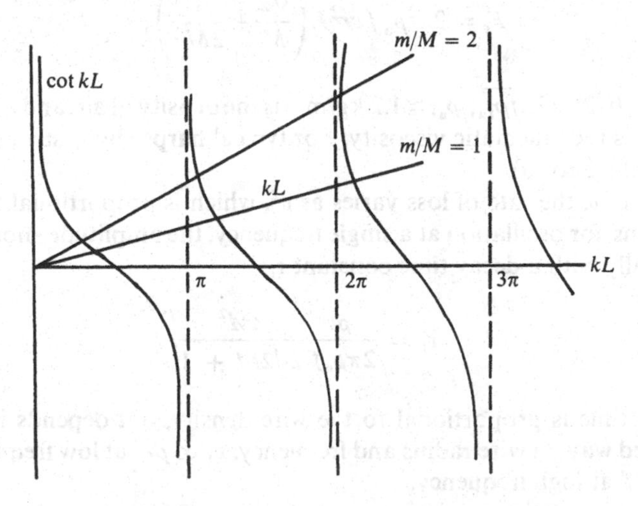

Fig. 2.14. Graphical solution of $\cot kL = (m/M)kL$.

from Eq. (2.38) are slightly compressed from the harmonic relationship; the frequency of the lowest mode is raised slightly more than the second.

2.13. Damping

Damping of vibrating strings can generally be attributed to three different loss mechanisms: (1) air damping, (2) internal damping, and (3) transfer of energy to other vibrating systems. The damping due to these mechanisms will vary with frequency, and their contributions will be comparable in size in many systems. (Fletcher, 1976, 1977).

2.13.1. Air Damping

A vibrating string is not a good sound radiator. The reason for this is that the string acts as a dipole source, producing a compression in front and a rarefaction behind as it moves; its radius is so small that these effectively cancel each other. This does not mean, however, that the string has little interaction with the air. Viscous flow of air around the moving string may be the major cause of damping of its vibrations under some conditions.

The complex problem of viscous drag on a vibrating string was solved long ago by Stokes, who showed that the force on the string has two components. One is an additional masslike load that lowers the mode frequencies very slightly; the other produces exponential decay of amplitude.

Over a range of wire diameters and frequencies encountered in musical instruments, the retarding force experienced by a cylinder of length L and radius r moving with a velocity v and frequency f is

$$F_{\mathrm{r}} = 2\pi^2 \rho_{\mathrm{a}} f v r^2 L\left(\frac{\sqrt{2}}{M} + \frac{1}{2M^2}\right), \tag{2.39}$$

where $M = (r/2)\sqrt{2\pi f/\eta_{\mathrm{a}}}$, $\rho_{\mathrm{a}}(\cong 1.2\ \mathrm{kg/m^3})$ is the density of air, and $\eta_{\mathrm{a}}(\cong 1.5 \times 10^{-5}\ \mathrm{m^2/s})$ is the kinematic viscosity. For typical harpsichord strings, M is in the range of 0.3 to 1.0.

Since $F_{\mathrm{r}} \propto v$, the rate of loss varies as v^2, which is proportional to kinetic energy. Thus, for oscillation at a single frequency, the amplitude should decay exponentially with a decay time constant τ.

$$\tau_1 = \frac{\rho}{2\pi\rho_{\mathrm{a}} f} \frac{2M^2}{2\sqrt{2}M + 1}. \tag{2.40}$$

The decay time is proportional to the wire density, but depends in a more complicated way on wire radius and frequency. $\tau_1 \propto \rho r^2$ at low frequency and $\tau_1 \propto \rho r/\sqrt{f}$ at high frequency.

2.13.2. Internal Damping

String material has so far been characterized by its radius, its density, and its Young's modulus, but more can be said than this. All real materials show an elastic behavior in which, when a stress is applied, an instantaneous strain occurs and then, over some characteristic time τ, the strain increases slightly. This second elongation may be moderately large or extremely small, and the time τ may be anything from less than a millisecond to many seconds. In viscoelastic materials, the second elongation increases slowly but without limit.

Such behavior can be represented by making the Young's modulus for the material complex:

$$E = E_1 + jE_2. \tag{2.41}$$

According to a relaxation formula attributable to Debye, E_2 has a peak at the relaxation frequency $\omega = 1/\tau$. Equation (2.41) can, however, be used in the more general case where many relaxation times contribute, both E_1 and E_2 varying with frequency. This behavior is simple to understand, E_1 being contributed by normal elastic bond distortions and E_2 by relaxation processes such as dislocation motion or the movement of kinks in polymer chains. Typically E_2/E_1 may be less than 10^{-4} in hard crystals, rather larger in metals, and perhaps as large as 10^{-1} in some polymer materials, though in such cases it may also depend on temperature. One elastic constant is really inadequate to describe even isotropic materials, but we shall neglect this added complication here.

By substituting Eq. (2.41) into the equation of motion, the decay time for this internal damping can be found to be

$$\tau_2 = \frac{1}{\pi f}\frac{E_1}{E_2}. \tag{2.42}$$

Clearly, internal damping is a material property independent of string radius, length, or tension. It is generally negligible for solid metal strings but may become the prime damping mechanism for gut or nylon strings or, more particularly, for strings of nylon overspun with metal. The decay time due to this mechanism is clearly shortest at high frequencies if, as is often the case, E_1 is nearly independent of frequency.

2.13.3. Energy Loss Through the Supports

In considering energy transfer to a movable support (and through it to other vibrating systems), it is easier to consider admittance than impedance. *Admittance* (the reciprocal of impedance) is the ratio of velocity to force, and its real part G is called *conductance*.

For a given string mode n, the velocity imparted to the support can be written

$$v_n = \alpha G F_n, \tag{2.43}$$

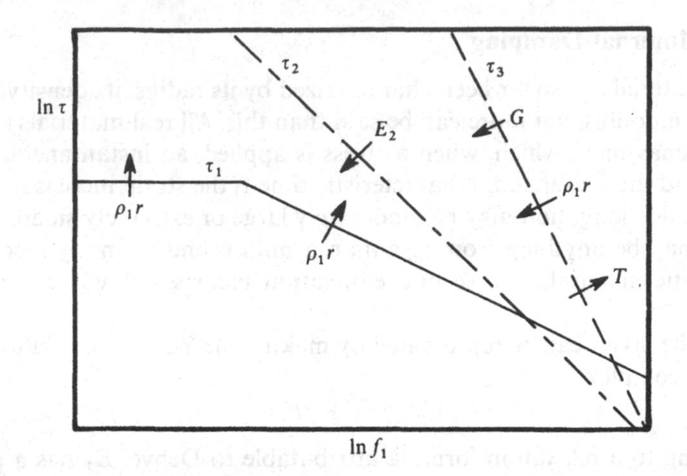

Fig. 2.15. Schematic behavior of decay times τ_i caused by various mechanisms as functions of the fundamental frequency f_1 of the string, which is assumed to be varied by changing only the string length. τ_1 is determined by air damping, τ_2 by internal damping, and τ_3 by loss to the support. Arrows indicate the directions in which the curves would be shifted by an increase in the string radius r, the density tension T, and the imaginary part of the Young's modulus E_2, and by the mechanical conductivity G of the bridge (Fletcher, 1976).

where F_n is the vertical component of the force and α is a constant. An analysis of the energy loss process again leads to an exponential energy decay with a time constant given by (Fletcher, 1977)

$$\tau_3 = (8\mu L f^2 G)^{-1}. \tag{2.44}$$

When all three mechanisms contribute to damping, the decay time τ is obtained by adding reciprocals:

$$1/\tau = 1/\tau_1 + 1/\tau_2 + 1/\tau_3. \tag{2.45}$$

This relationship is shown schematically in Fig. 2.15 on the assumption that G and E_2 are independent of frequency. The curves show the various contributions to the decay time as functions of frequency on the assumption that we are dealing with a single string whose frequency is raised by reducing its length. Also indicated are the directions in which the various curves would move in response to increases in various string parameters. The curve for the resultant decay time is a smoothed lower envelope to the individual decay time curves.

In most musical instruments, the rate at which energy is transferred from the string to the bridge and soundboard is quite small. For thin metal strings, the decay time is determined mostly by air viscosity, and so the decay time for the upper partials varies as $1/\sqrt{f}$. For instruments with gut or nylon strings, internal damping becomes dominant for most modes, and the decay time for the upper partials varies as $1/f$. Such strings therefore have a much less brilliant sound than do metal strings. If a finger tip is used to stop the string or if the bridge is so light that end losses predominate, the decay time for the upper partials varies more nearly as $1/f^2$.

2.14. Longitudinal Vibrations of a String or Thin Bar

Longitudinal waves in a string are much less common than transverse waves. Nevertheless, they do occur, and they may give rise to standing waves or longitudinal modes of vibration. Unlike transverse waves, their velocity (and hence their frequency) does not change with tension (except for possible changes in the physical properties of the string). Longitudinal waves in a thin bar travel at the same velocity as do longitudinal waves in a string of the same material.

When a bar is strained, elastic forces are produced. Consider a short segment of length dx of a bar having a cross section area S, as shown in Fig. 2.16. The plane at x moves a distance w to the right while the plane at $x + dx$ moves a distance $w + dw$. The stress is given by F/S and the strain (change in length per unit of original length) by $\partial\omega/\partial x$. If E is Young's modulus, Hooke's law can be written

$$\frac{F}{S} = E\frac{\partial w}{\partial x} \tag{2.46}$$

Expanding $F + dF$ in a Taylor's series, and differentiating Eq. (2.46) gives

$$dF = F(x+dx) - F(x) = \frac{\partial F}{\partial x}\,dx = SE\frac{\partial^2 w}{\partial x^2}\,dx. \tag{2.47}$$

The mass segment under consideration is $\rho S\,dx$, and thus, the equation of motion becomes

$$\begin{aligned} \rho S\,dx\frac{\partial^2 w}{\partial t^2} &= SE\frac{\partial^2 w}{\partial x^2}dx, \\ \frac{\partial^2 w}{\partial t^2} = \frac{E}{\rho}\frac{\partial^2 w}{\partial x^2} &= c_L^2\frac{\partial^2 w}{\partial x^2}. \end{aligned} \tag{2.48}$$

This is a one-dimensional wave equation for waves with a velocity $c_L = \sqrt{E/\rho}$.

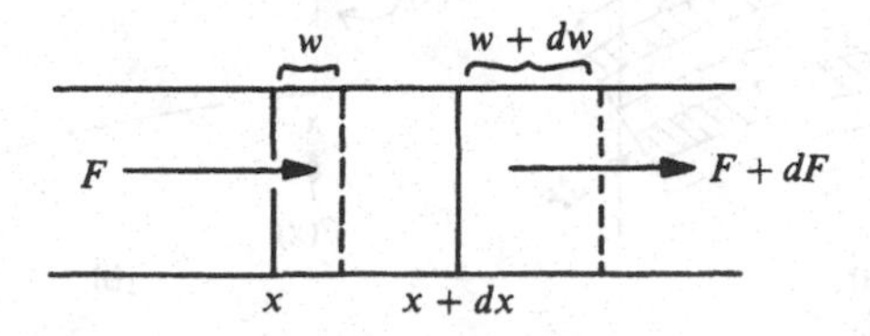

Fig. 2.16. Forces and strains in a short segment of a bar (or string).

The general solution of Eq. (2.48) has the same form as the equation (2.5) for transverse waves in a string:

$$w = w_1(c_L t - x) + w_2(c_L t + x). \tag{2.49}$$

The normal modes of vibration depend upon the end conditions. If both ends are fixed, or if both ends are free, the mode frequencies are given by

$$f_n = n\frac{c_L}{2L} \qquad n = 1, 2, 3, \ldots, \tag{2.50}$$

and for a bar fixed at one end and free at the other,

$$f_m = m\frac{c_L}{4L} \qquad m = 1, 3, 5, \ldots. \tag{2.51}$$

If the bar (or string) is terminated by a movable support, the modal frequencies are found by methods similar to that described in Section 2.12.

2.15. Bending Waves in a Bar

A bar or rod is capable of transverse vibrations in somewhat the same manner as a string. The dependence of the frequency on tension is more complicated than it is in a string, however. In fact, a bar vibrates quite nicely under zero tension, the elastic forces within the bar supplying the necessary restoring force in this case.

When a bar is bent, the outer part is stretched and the inner part is compressed. Somewhere in between is a neutral axis whose length remains unchanged, as shown in Fig. 2.17. A filament located at a distance z below the neutral axis is compressed by an amount $d\phi$. The strain is $z\,d\phi/dx$, and the amount of force required to produce the strain is

$$E\,dS\,z\frac{d\phi}{dx}, \tag{2.52}$$

where dS is the cross sectional area of the filament and E is Young's modulus.

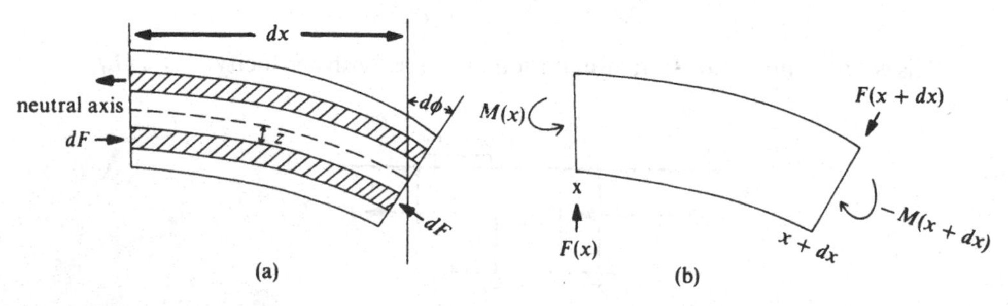

Fig. 2.17. (a) Bending strains in a bar. (b) Bending moments and shear forces in a bar.

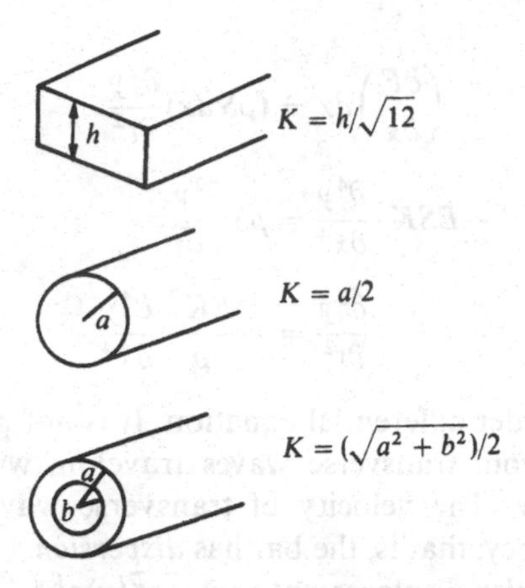

Fig. 2.18. Radii of gyration for some simple shapes.

The moment of this force about the center line is $dM = [E\,dS\,z(d\phi/dx)]z$, and so the total moment to compress all the filaments is

$$M = \int dM = E\frac{d\phi}{dx}\int z^2\,dS. \tag{2.53}$$

It is customary to define a constant K called the *radius of gyration* of the cross section such that

$$K^2 = \frac{1}{S}\int z^2\,dS, \tag{2.54}$$

where $S = \int dS$ is the total cross section. The radius of gyration for a few familiar shapes is shown in Fig. 2.18. The bending moment is thus

$$M = E\frac{d\phi}{dx}SK^2 \cong -ESK^2\frac{\partial^2 y}{\partial x^2}, \tag{2.55}$$

since $d\phi \cong -\left(\frac{\partial^2 y}{\partial x^2}\right)dx$ for small $d\phi$.

The bending moment is not the same for every part of the bar. In order to keep the bar in equilibrium, there must be a shearing force F with a moment $F\,dx$, as shown in Fig. 2.17(b).

$$F\,dx = (M + dM) - M = dM,$$

and

$$F = \frac{\partial M}{\partial x} = -ESK^2\frac{\partial^3 y}{\partial x^3}.$$

But the shearing force F is not constant, either; the net force $dF = (\partial F/\partial x)\,dx$ produces an acceleration perpendicular to the axis of the bar. The equation of motion is

$$\left(\frac{\partial F}{\partial x}\right)dx = (\rho S\,dx)\frac{\partial^2 y}{\partial t^2}$$

$$-ESK^2\frac{\partial^4 y}{\partial x^4} = \rho S\frac{\partial^2 y}{\partial t^2} \tag{2.56}$$

$$\frac{\partial^2 y}{\partial t^2} = -\frac{EK^2}{\rho}\frac{\partial^4 y}{\partial x^4}.$$

This is a fourth-order differential equation. It is not possible to construct a general solution from transverse waves traveling with velocity v, as in the longitudinal case. The velocity of transverse waves is, in fact, quite dependent on frequency; that is, the bar has *dispersion*.

We write the complex displacement as $\tilde{y} = \tilde{Y}(x)e^{j\omega t}$;

$$\frac{\partial^2 \tilde{y}}{\partial t^2} = -\omega^2 \tilde{Y}e^{j\omega t} \quad \text{and} \quad \frac{\partial^4 \tilde{y}}{\partial x^4} = \frac{d^4 \tilde{Y}}{dx^4}e^{j\omega t}, \quad \text{so} \quad -\omega^2 \tilde{Y} = -\frac{EK^2}{\rho}\frac{d^4 \tilde{Y}}{dx^4}$$

or

$$\frac{d^4 \tilde{Y}}{dx^4} = \frac{\rho\omega^2}{EK^4}\tilde{Y} = \frac{\omega^4}{v^4}\tilde{Y}$$

where

$$v^2 = \omega K\sqrt{E/\rho} = \omega K c_L.$$

Note that the wave velocity $v(f)$ is proportional to $\sqrt{\omega}$.

We now write $\tilde{Y}(x) = \tilde{A}e^{\gamma x}$ and substitute

$$\gamma^4 A e^{\gamma x} = \frac{\omega^4}{v^4}Ae^{\gamma x},$$

$$\gamma^2 = \pm\frac{\omega^2}{v^2},$$

or

$$\gamma = \pm\frac{\omega}{v} \quad \text{or} \quad \pm j\frac{\omega}{v}. \tag{2.57}$$

The complete solution is a sum of four terms, each corresponding to one of the roots of Eq. (2.57):

$$\tilde{y}(x,t) = e^{j\omega t}(\tilde{A}e^{\omega x/v} + \tilde{B}e^{-\omega x/v} + \tilde{C}e^{j\omega x/v} + \tilde{D}e^{-j\omega x/v}). \tag{2.58}$$

Since $e^{\pm x} = \cosh x \pm \sinh x$ and $e^{\pm jx} = \cos x \pm j\sin x$, another way of writing Eq. (2.58) is

$$y = \cos(\omega t + \phi)\left[A\cosh\frac{\omega x}{v} + B\sinh\frac{\omega x}{v} + C\cos\frac{\omega x}{v} + D\sin\frac{\omega x}{v}\right], \tag{2.59}$$

where A, B, C, and D are now real constants (Kinsler et al., 1982).

Since the equation of motion is a fourth-order equation, we have four arbitrary constants. We thus need four boundary conditions (two at each end) to determine them.

2.16. Bars with Fixed and Free Ends

We will consider three different end conditions for a bar: free, supported (hinged), and clamped. For each of these, we can write a pair of boundary conditions. At a free end, there is no torque and no shearing force, so the second and third derivatives are both zero, as given in Fig. 2.19. At a simply supported (or hinged) end, there is no displacement and no torque, so y and its second derivative are zero. At a clamped end, y and its first derivative are zero.

1. Example I: A bar of length L free at both ends. The boundary conditions at $x = 0$ become

$$\frac{\partial^2 y}{\partial x^2} = 0 = \cos(\omega t + \phi)\left(\frac{\omega}{v}\right)^2 (A - C),$$

and

$$\frac{\partial^3 y}{\partial x^3} = 0 = \cos(\omega t + \phi)\left(\frac{\omega}{v}\right)^3 (B - D),$$

from which $A = C$ and $B = D$, so the general solution [Eq. (2.59)] becomes

$$y(x, t) = \cos(\omega t + \phi)\left[A\left(\cosh\frac{\omega x}{v} + \cos\frac{\omega x}{v}\right) + B\left(\sinh\frac{\omega x}{v} + \sin\frac{\omega x}{v}\right)\right]. \tag{2.60}$$

At $x = L$, the boundary conditions become

$$\frac{\partial^2 y}{\partial x^2} = 0 = \cos(\omega t + \phi)\left(\frac{\omega}{v}\right)^2\left[A\left(\cosh\frac{\omega L}{v} - \cos\frac{\omega L}{v}\right) + B\left(\sinh\frac{\omega L}{v} - \sin\frac{\omega L}{v}\right)\right],$$

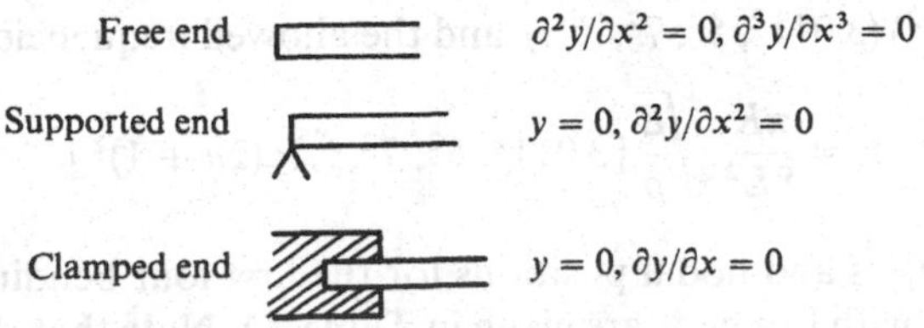

Free end		$\partial^2 y/\partial x^2 = 0, \partial^3 y/\partial x^3 = 0$
Supported end		$y = 0, \partial^2 y/\partial x^2 = 0$
Clamped end		$y = 0, \partial y/\partial x = 0$

Fig. 2.19. Three different end conditions for a bar.

$$\frac{\partial^3 y}{\partial x^3} = 0 = \cos(\omega t + \phi)\left(\frac{\omega}{v}\right)^3 \left[A\left(\sinh\frac{\omega L}{v} + \sin\frac{\omega L}{v}\right)\right.$$

$$\left. + B\left(\cosh\frac{\omega L}{v} - \cos\frac{\omega L}{v}\right)\right].$$

These equations can have a common solution only for certain values of ω. Setting the expressions in brackets equal to zero, and dividing the first by the second gives

$$\frac{\cosh \omega L/v - \cos \omega L/v}{\sinh \omega L/v + \sin \omega L/v} = \frac{\sinh \omega L/v - \sin \omega L/v}{\cosh \omega L/v + \sin \omega L/v}.$$

Cross multiply and note that $\sin^2 x + \cos^2 x = \cosh^2 x - \sinh^2 x = 1$:

$$\cosh^2 \omega L/v - 2\cosh \omega L/v \cos \omega L/v + \cos^2 \omega L/v = \sinh^2 \omega L/v - \sin^2 \omega L/v,$$

$$2 - 2\cosh\frac{\omega L}{v}\cos\frac{\omega L}{v} = 0,$$

or

$$\cosh\frac{\omega L}{v} = \frac{1}{\cos \omega L/v}. \tag{2.61}$$

This equation could be solved by graphing the two functions, but this is not very practical since the hyperbolic cosine increases exponentially. An alternative is to make use of the indentities:

$$\tan\frac{x}{2} = \sqrt{\frac{1 - \cos x}{1 + \cos x}}$$

and

$$\tanh\frac{x}{2} = \sqrt{\frac{\cosh x - 1}{\cosh x + 1}},$$

so that Eq. (2.61) becomes

$$\tan\frac{\omega L}{2v} = \pm\tanh\frac{\omega L}{2v}. \tag{2.62}$$

A graph of these two functions is shown in Fig. 2.20. The intersections of these curves give roots $\omega L/2v = \pi/4$ (3.011, 5, ...). But $v^2 = \omega K\sqrt{E/\rho}$, so $\omega^2 = (v^2\pi^2/4L^2)(3.011^2, 5^2, 7^2, \ldots)$, and the allowed frequencies are given by

$$f_n = \frac{\pi K}{8L^2}\sqrt{\frac{E}{\rho}}[3.011^2, 5^2, 7^2, \ldots, (2n + 1)^2]. \tag{2.63}$$

The frequencies and nodal positions for the first four bending vibrational modes of a bar with free ends are given in Table 2.1. Note that the frequencies are not harmonically related as they were for longitudinal modes.

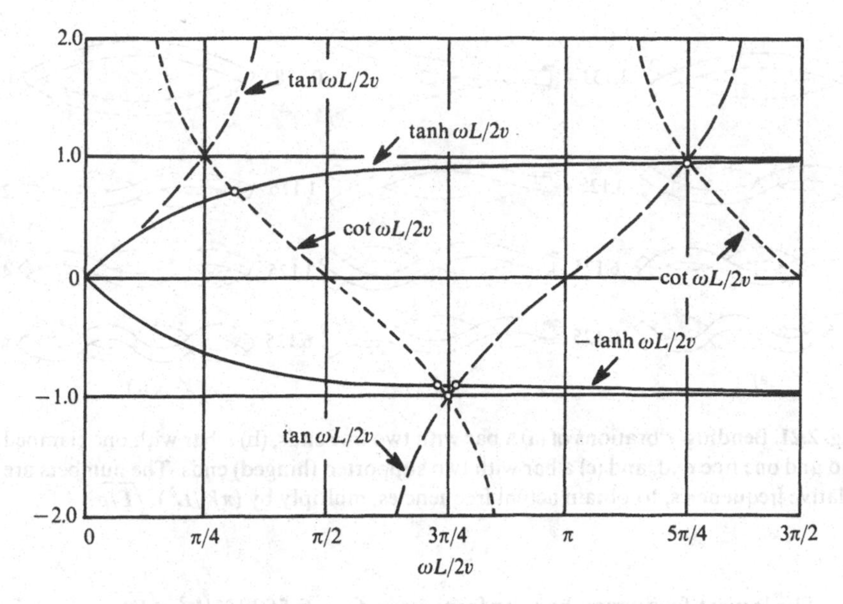

Fig. 2.20. Curves showing tangent, cotangent, and hyperbolic tangent functions (from Kinsler et al., "Fundamentals of Acoustics," 3rd ed., copyright © 1982, John Wiley & Sons, Inc. Reprinted by permission).

2. Example II: A bar of length of L clamped at $x = 0$ and free at $x = L$. The boundary conditions at $x = 0$ now lead to the result that $A + C = 0 = B + D$. The transcendental equation, which gives the allowed values of frequency, is now [see equation 3.55 in Kinsler et al. (1982)]

$$\cot\frac{\omega L}{2v} = \pm\tanh\frac{\omega L}{2v}.$$

Again, we obtain the frequencies by using Fig. 2.20:

$$f_n = \frac{\pi K}{8L^2}\sqrt{\frac{E}{\rho}}[1.194^2, 2.988^2, 5^2, \ldots, (2n-1)^2]. \tag{2.64}$$

These frequencies are in the ratios $f_2 = 6.267f_1$, $f_3 = 17.55f_1$, $f_4 = 35.39f_1$,

Table 2.1. Characteristics of transverse vibrations in a bar with free ends.

Frequency (Hz)	Wavelength (m)	Nodal positions (m from end of 1-m bar)
$f_1 = 3.5607K/L^2\sqrt{E/\rho}$	$1.330L$	0.224, 0.776
$2.756f_1$	$0.800L$	0.132, 0.500, 0.868
$5.404f_1$	$0.572L$	0.094, 0.356, 0.644, 0.906
$8.933f_1$	$0.445L$	0.073, 0.277, 0.500, 0.723, 0.927

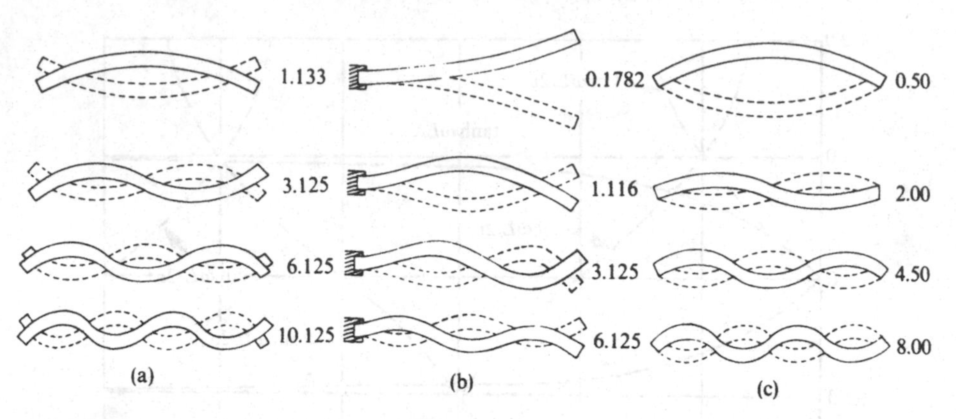

Fig. 2.21. Bending vibrations of (a) a bar with two free ends, (b) a bar with one clamped end and one free end, and (c) a bar with two supported (hinged) ends. The numbers are relative frequencies; to obtain actual frequencies, multiply by $(\pi K/L^2)\sqrt{E/\rho}$.

etc. The lowest frequency has the frequency $f_1 = 0.5598K/L^2\sqrt{E/\rho}$, which is only about $\frac{1}{6}$ of the lowest frequency of the same bar with two free ends.

3. Example III: A bar of length L with simple supports (hinges) at the ends. The frequencies are given by

$$f_n = \frac{\pi K}{2L^2}\sqrt{\frac{E}{\rho}}m^2 \qquad m = 1, 2, 3, \ldots. \tag{2.65}$$

These frequencies are considerably lower than those given by Eq. (2.63), since the bending wavelengths are longer than the corresponding modes of the free bar, as shown in Fig. 2.21.

2.17. Vibrations of Thick Bars: Rotary Inertia and Shear Deformation

Thus far, we have considered transverse motion of the bar due to the bending moment only. Such a simplified model is often called the Euler–Bernoulli beam. It is essentially correct for a long, thin bar or rod. The Timoshenko beam, a model that considers rotary inertia and shear stress, is prefered in considering thick bars.

As a beam bends, the various elements rotate through some small angle. The rotary inertia is thus equivalent to an increase in mass and results in a slight lowering of vibrational frequencies, especially the higher ones.

Shear forces, which we considered in deriving the equation of motion [Eq. (2.56)], tend to deform the bar; in particular they cause rectangular elements to become parallelograms and thus decrease the transverse deflection slightly. Therefore, the frequencies of the higher modes are decreased slightly in a thick bar as compared with a thin one.

2.18. Vibrations of a Stiff String

In real strings, the restoring force is partly due to the applied tension and partly due to the stiffness of the string (although the former usually dominates). Thus, the equation of motion of a flexible string [Eq. (2.4)] can be modified by adding a term appropriate to bending stiffness:

$$\mu \frac{\partial^2 y}{\partial t^2} = T \frac{\partial^2 y}{\partial x^2} - ESK^2 \frac{\partial^4 y}{\partial x^4}. \tag{2.66}$$

In this equation, μ is mass per unit length, T is tension, E is Young's modulus, S is the cross-sectional areas and K is the radius of gyration, as before.

Solving this equation is difficult, but if the stiffness of the string is small, the mode frequencies can be written (Morse, 1948) as

$$f_n = nf_1^\circ [1 + \beta + \beta^2 + \frac{n^2\pi^2}{8} \beta^2], \tag{2.67}$$

where f_1° is the fundamental frequency of the same string without stiffness and $\beta = (2K/L)\sqrt{ES/T}$. For a string with a circular cross section, K is half the radius a, so $\beta = (a^2/L)\sqrt{\pi E/T}$. The second two terms in Eq. (2.67) raise the frequency of all the modes, but the fourth term depends upon n^2 and thus stretches the intervals between the higher modes.

String stiffness is of considerable importance in the tuning of piano strings. To minimize beating between the upper strings and the inharmonic overtones of the lower strings, octaves are stretched to ratios that are greater than 2 : 1. In a 108-cm (42-inch) upright piano, for example, the fourth harmonic of C_4 (middle C) is about 4 cents (0.2%) sharp, but this increases to about 18 cents (1.1%) for C_5 (Kent, 1982). In a large grand piano with long strings, inharmonicity is considerably less, but in small spinets it is substantially greater.

The stiffness of a violin string is of considerable importance when it is excited by bowing. In our discussion of Helmholtz-type motion in Section 2.10, we envisioned a very sharp bend propagating back and forth on an ideal string with great flexibility. On a real string, however, the bend is rounded appreciably by the stiffness of the string (and to a lesser extent by damping of high-frequency components). The rounded Helmholtz bend is sharpened each time it passes the bow, however, and so the resultant motion represents an equilibrium between the rounding and sharpening process. Two effects that depend upon rounding of the Helmholtz corner are noise due to small variations in period (jitter) and the note flattening with increased bow pressure (Cremer, 1981; McIntyre and Woodhouse, 1982).

2.19. Dispersion in Stiff and Loaded Strings: Cutoff Frequency

Waves on an ideal string travel without dispersion; that is, the wave velocity is independent of frequency. Thus, a pulse does not change its shape as it propagates back and forth on an ideal, lossless string. Two sources of dispersion in real strings are stiffness and mass loading. In addition, loss mechanisms in strings are frequency dependent.

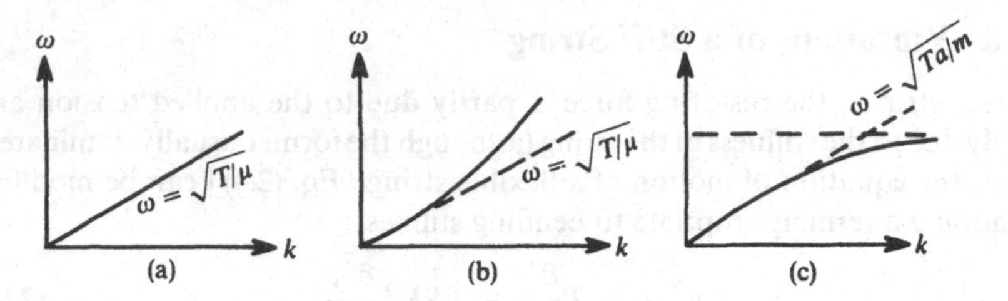

Fig. 2.22. Graphs of ω versus k for (a) an ideal string, (b) a stiff string, and (c) a string loaded with masses spaced a distance a apart.

In an ideal string, ω and k are related by the simple expression $\omega = ck$, so the wave velocity equals the slope of the line obtained by plotting ω versus k, as in Fig. 2.22(a). In order to draw a graph of ω versus k for a stiff string, we write Eq. (2.67) as

$$\omega_n = 2\pi \frac{n}{2L}\sqrt{\frac{T}{\mu}}\left(1 + \beta + \beta^2 + \frac{4K^2}{8L^2}\frac{ES}{T}\pi^2 n^2\right)$$

$$= k\sqrt{\frac{T}{\mu}}\left(1 + \beta + \beta^2 + \frac{K^2 ES}{2T}k^2\right)$$

$$= k\sqrt{\frac{T}{\mu}}\left(1 + \beta + \beta^2 + \alpha k^2\right). \tag{2.68}$$

The graph of Eq. (2.68) is shown in Fig. 2.22(b). When the graph of ω versus k is a curved line, we observe two different wave velocities: a phase velocity v_ϕ and a group velocity v_g. These are given by

$$v_\phi = \omega/k \quad \text{and} \quad v_g = d\omega/dk. \tag{2.69}$$

The phase velocity is the velocity of a wave crest or a given phase angle, whereas the group velocity is the velocity of the wave envelope of a given amplitude of the wave packet.

For a string loaded with equally spaced masses, dispersion of a different type is observed. The dispersion relationship can be written (see p. 76 in Crawford, 1965) as

$$\omega(k) = 2\sqrt{\frac{T}{ma}}\sin ka, \tag{2.70}$$

where a is the spacing between beads of mass m. The maximum value of ω is $2\sqrt{T/ma}$, which occurs when $k = \pi/a$. The allowed values of k between 0 and π/a equal the number of normal modes of the system, which is also equal to the number of equally spaced masses.

The maximum value of ω divided by 2π is called the cutoff frequency f_c:

$$f_c = \frac{1}{\pi}\sqrt{\frac{T}{ma}}. \tag{2.71}$$

It represents the highest frequency of wave disturbance that can propagate on the loaded string. Note that when $k = \pi/a$, the group velocity $v_g = d\omega/dk = 0$.

2.20. Torsional Vibrations of a Bar

Torsional waves are a third type of wave motion possible in a bar or rod. The equation of motion for torsional waves is derived by equating the net torque acting on an element of the bar to the product of moment of inertia and angular acceleration. Young's modulus is replaced by the shear modulus. The resulting wave equation is quite similar to the equation for longitudinal waves.

Torsional waves in a bar, like compressional waves (but unlike bending waves), are nondispersive; that is, they have a wave velocity that is independent of frequency

$$c_T = \sqrt{\frac{GK_T}{\rho I}}, \tag{2.72}$$

where GK_T is the torsional stiffness factor that relates a twist to the shearing strain produced, ρI is the polar moment of inertia per unit length, ρ is density, and G is the shear modulus. For a circular rod, $K_T \cong I$, so the velocity is $\sqrt{G/\rho}$; for square and rectangular bars, it is slightly less, as shown in Fig. 2.23.

In many materials the shear modulus G is related to the Young's modulus E and Poisson's ratio ν by the equation

$$G = \frac{E}{2(1+\nu)}. \tag{2.73}$$

In aluminum, for example, $\nu = 0.33$, so $G = 0.376E$, and the ratio of torsional to longitudinal wave velocity in a circular aluminum rod is 0.61.

The torsional modes of vibration of a bar with free ends have frequencies that equal the torsional wave velocity times $n/2L$ in direct analogy to the longitudinal modes. (If one end of the bar is clamped, the frequencies become $m/4L$ ($m = 1, 3, 5, \ldots$).

Bowing a violin string excites torsional waves as well as transverse waves. For a steel E string tuned to 660 Hz, the torsional wave speed v_T is about 7.5 times the transverse wave speed c, but for a gut E string, $v_T/c \cong 2$ (Schelleng, 1973). Gillan and Elliott (1989) found values of v_T/c from 2.6 to 7.6 in violin strings and 5.7 for a steel cello string. Furthermore, they found damping factors from 1% to 7.7%, which suggests that torsional damping is dominated by internal damping in the string rather than reflection losses at the bridge and nut. Torsional waves change the effective compliance of the string and affect the mechanics of the bow/string interaction in other ways as well (Cremer, 1981).

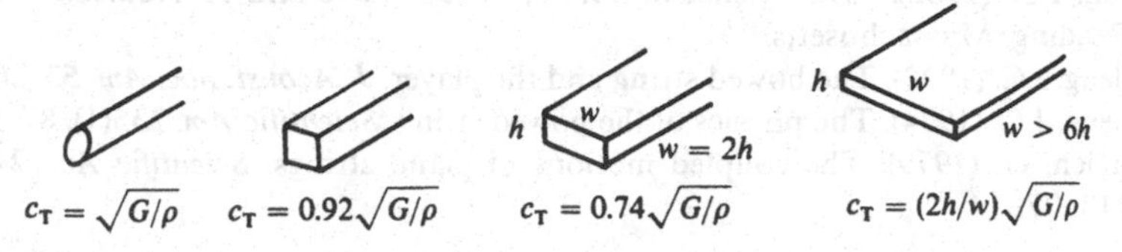

Fig. 2.23. Torsional wave velocities for bars with different cross sections.

References

Crawford, F.S., Jr. (1965). "Waves," Chapter 2. McGraw-Hill, New York.

Cremer, L. (1981). "Physik der Geige." Hirtzel Verlag, Stuttgart. English translation by J.S. Allen, MIT Press, Cambridge, Massachusetts, 1984.

Fletcher, N.H. (1976). Plucked strings—a review. *Catgut Acoust. Soc. Newsletter*, No. 26, 13–17.

Fletcher, N.H. (1977). Analysis of the design and performance of harpsichords. *Acustica* **37**, 139–147.

Gillan, F.S., and Elliott, S.J. (1989). Measurement of the torsional modes of vibration of strings on instruments of the violin family. *J. Sound Vib.* **130**, 347–351.

Hall, D.E. (1986). Piano string excitation in the case of small hammer mass. *J. Acoust. Soc. Am.* **79**, 141–147.

Hall, D.E. (1987a). Piano string excitation II: General solution for a hard narrow hammer. *J. Acoust. Soc. Am.* **81**, 535–546.

Hall, D.E. (1987b). Piano string excitation III: General solution for a soft narrow hammer. *J. Acoust. Soc. Am.* **81**, 547–555.

Helmholtz, H.L.F. (1877). "On the Sensations of Tone," 4th ed., trans. A.J. Ellis (Dover, New York, 1954).

Kent, E.L. (1982). Influence of irregular patterns in the inharmonicity of piano-tone partials upon piano tuning practice. *Das Musikinstrument* **31**, 1008–1013.

Kinsler, L.E., Frey, A.R., Coppens, A.B., and Sanders, J.V. (1982). "Fundamentals of Acoustics," 3rd ed., Chapters 2 and 3. New York.

Kondo, M., and Kubota, H. (1983). A new identification expression of Helmholtz waves. *SMAC 83*. (A. Askenfelt, S. Felicetti, E. Jansson, and J. Sundberg, eds.), pp. 245–261. Royal Swedish Acad. Music, Stockholm.

Krigar-Menzel, O. and Raps, A. (1981), Aus der Sitzungberichten, *Ann. Phys. Chem.* **44**, 613–641.

Lawergren B. (1980). On the motion of bowed violin strings. *Acustica* **44**, 194–206.

McIntyre, M.E., Schumacher, R.T., and Woodhouse, J. (1981). Aperiodicity of bowed string motion. *Acustica* **49**, 13–32.

McIntyre, M.E., and Woodhouse, J. (1982). The acoustics of stringed musical instruments. *Interdisc. Sci. Rev.* **3**, 157–173. Reprinted in "Musical Acoustics: Selected Reprints" (T.D. Rossing, ed.), Am. Assn. Physics Teach., College Park, Maryland, 1988.

Morse, P.M. (1948). "Vibration and Sound," 2nd ed., Chapters 3 and 4. McGraw-Hill, New York. Reprinted by Acoust. Soc. Am., Woodbury, New York, 1982.

Raman, C.V. (1918). On the mechanical theory of the vibrations of bowed strings and of musical instruments of the violin family, with experimental verification of the results: Part I. *Indian Assoc. Cultivation Sci. Bull.* **15**, 1–158. Excerpted in "Musical Acoustics, Part I" (C.M. Hutchins, ed.) Dowden, Hutchinson, and Ross, Stroudsburg, Pennsylvania, 1975.

Rossing, T.D. (1982). "The Science of Sound," Chapters 10 and 13. Addison-Wesley, Reading, Massachusetts.

Schelleng, J.C. (1973). The bowed string and the player. *J. Acoust. Soc. Am.* **53**, 26–41.

Schelleng, J.C. (1974). The physics of the bowed string. *Scientific Am.* **235**(1), 87–95.

Weinreich, G. (1979). The coupled motions of piano strings. *Scientific Am.* **240**(1), 118–27.

CHAPTER 3

Two-Dimensional Systems: Membranes and Plates

In this chapter, we will consider two-dimensional, continuous vibrating systems, with and without stiffness. An ideal membrane, like an ideal string, has no stiffness of its own, and thus, its oscillations depend upon the restoring force supplied by an externally applied tension. A plate, on the other hand, like a bar, can vibrate with fixed or free ends and with or without external tension.

3.1. Wave Equation for a Rectangular Membrane

The simplest two-dimensional system we can consider is a rectangular membrane with dimensions L_x and L_y, with fixed edges, and with a surface tension T that is constant throughout.

Consider an element with area density σ, as shown in Fig. 3.1. It has been displaced a small distance dz, and the surface tension T acts to restore it to equilibrium. The forces acting on the edges dx have the magnitude $T\,dx$, and their vertical components are $-T\sin\alpha\,dx$ and $-T\sin\beta\,dx$. For small α and β,

$$\sin\alpha \simeq \tan\alpha = \left(\frac{\partial z}{\partial y}\right)_{y+dy}$$

and

$$\sin\beta \simeq \tan\beta = \left(\frac{\partial z}{\partial y}\right)_{y}.$$

Therefore,

$$F_y = T\,dx\left[\left(\frac{\partial z}{\partial y}\right)_{y+dy} - \left(\frac{\partial z}{\partial y}\right)_{y}\right] = T\,dx\frac{\partial^2 z}{\partial y^2}\,dy.$$

Similarly, the vertical component of the forces acting on the edges dy is

$$F_x = T\,dy\frac{\partial^2 z}{\partial y^2}\,dx.$$

The total force on element $dx\,dy$ is $F = F_x + F_y$, so the equation of motion

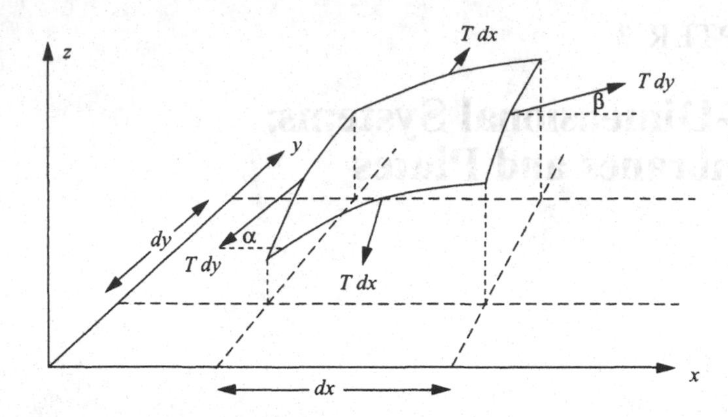

Fig. 3.1. Forces on a rectangular membrane element.

$(F = ma)$ becomes

$$T\,dx\,dy\left(\frac{\partial^2 z}{\partial x^2} + \frac{\partial^2 z}{\partial y^2}\right) = \sigma\,dx\,dy\frac{\partial^2 z}{\partial t^2}$$

or

$$\frac{\partial^2 z}{\partial t^2} = \frac{T}{\sigma}\left(\frac{\partial^2 z}{\partial x^2} + \frac{\partial^2 z}{\partial y^2}\right) = c^2\nabla^2 z. \tag{3.1}$$

This is a wave equation for transverse waves with a velocity $c = \sqrt{T/\sigma}$. It is easily solved by writing the deflection $z(x, y, t)$ as a product of three functions, each of a single variable: $z(x, y, t) = X(x)Y(y)T(t)$. The second derivatives are

$$\frac{\partial^2 z}{\partial x^2} = \frac{d^2 X}{dx^2}YT, \qquad \frac{\partial^2 z}{\partial y^2} = \frac{d^2 Y}{dy^2}XT, \qquad \text{and} \qquad \frac{\partial^2 z}{\partial t^2} = \frac{d^2 T}{dt^2}XY,$$

so that the equation becomes

$$\frac{1}{T}\frac{d^2 T}{dt^2} = \frac{c^2}{X}\frac{d^2 X}{dx^2} + \frac{c^2}{Y}\frac{d^2 Y}{dy^2}. \tag{3.2}$$

This equation can only be true if each side of the equation is a constant, which we denote as $-\omega^2$. This gives two equations:

$$\frac{d^2 T}{dt^2} + \omega^2 T = 0,$$

with solutions $T(t) = E\sin\omega t + F\cos\omega t$, and

$$\frac{1}{X}\frac{d^2 X}{dx^2} + \frac{\omega^2}{c^2} = -\frac{1}{Y}\frac{d^2 Y}{dy^2}.$$

Again, each side must equal a constant, which we will call k^2. This gives

$$\frac{d^2X}{dx^2} + \left(\frac{\omega^2}{c^2} - k^2\right)X = 0,$$

with solutions $X(x) = A\sin\sqrt{(\omega^2/c^2) - k^2}\,x + B\cos\sqrt{(\omega^2/c^2) - k^2}\,x$, and

$$\frac{d^2Y}{dy^2} + k^2Y = 0,$$

with solutions $Y(y) = C\sin ky + D\cos ky$. For a rectangular membrane of dimensions L_x by L_y, fixed at all four sides, the boundary conditions require that $z = 0$ for $x = 0$, $x = L_x$, $y = 0$, and $y = L_y$. From the first condition, we see that $B = 0$; from the second,

$$A\sin\sqrt{\frac{\omega^2}{c^2} - k^2}\,L_x = 0, \qquad \text{so} \qquad \sqrt{\frac{\omega^2}{c^2} - k^2}\,L_x = m\pi, \qquad \text{and}$$

$$X(x) = A\sin\frac{m\pi x}{L_x},$$

with $m = 1, 2, \ldots$. From the third, $D = 0$; and from the fourth, $C\sin kL_y = 0$, so $kL_y = n\pi$ and $Y(y) = C\sin(n\pi/L_y)y$, with $n = 1, 2, \ldots$. Therefore,

$$z_{mn} = A\sin\frac{m\pi x}{L_x}C\sin\frac{n\pi y}{L_y}(E\sin\omega t + F\cos\omega t)$$

$$= \sin\frac{m\pi x}{L_x}\sin\frac{n\pi y}{L_y}(M\sin\omega t + N\cos\omega t), \qquad m = 1, 2, \ldots. \tag{3.3}$$

To determine the modal frequencies, solve $\sqrt{(\omega^2/c^2) - k^2} = m\pi/L_x$ for ω:

$$\omega^2 = \left(\frac{m\pi}{L_x}\right)^2 c^2 + k^2c^2 = \left(\frac{m\pi}{L_x}\right)^2 c^2 + \left(\frac{n\pi}{L_y}\right)^2 c^2,$$

and

$$f_{mn} = \frac{1}{2}\sqrt{\frac{T}{\sigma}}\sqrt{\frac{m^2}{L_x^2} + \frac{n^2}{L_y^2}}, \qquad m, n = 1, 2, \ldots. \tag{3.4}$$

Comparison of Eqs. (3.3) and (3.4) with Eqs. (2.13) and (2.14), which describe the modes in a string, suggests that the normal modes of a rectangular membrane might be called two-dimensional string modes. Standing waves in the x direction appear to be independent of standing waves in the y direction. Some of the modes are illustrated in Fig. 3.2.

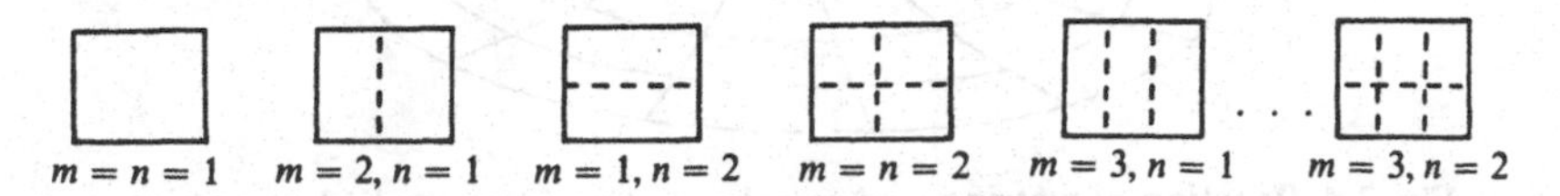

Fig. 3.2. Some normal modes of a rectangular membrane.

3.2. Square Membranes: Degeneracy

In a square membrane ($L_x = L_y$), $f_{mn} = f_{nm}$; the mn and nm modes are said to be *degenerate*, since they have the same frequency. Although there are now fewer allowed frequencies, there are just as many characteristic functions as in the rectangular case. In fact, the membrane can vibrate with simple harmonic motion at a frequency f_{mn} with any of an infinite number of different shapes corresponding to different values of a and b in the equation.

$$z(x, y, t) = (az_{mn} + bz_{nm}) \cos \omega_{mnt} \qquad \text{where} \quad a^2 + b^2 = 1. \tag{3.5}$$

Various combinations of z_{13} and z_{31} are shown in Fig. 3.3.

An important difference between a string and a membrane is in the reaction to a force applied at a single point, as shown in Fig. 3.4. A string pulled aside by a force F applied a distance x from one end will deflect a distance h so that $T(h/x)$ and $T[h/(L - x)]$ add up to F. An ideal membrane, on the other hand, cannot support a point force F, and the displacement theoretically becomes infinite no matter how small the force! If a force F is applied to a small circle

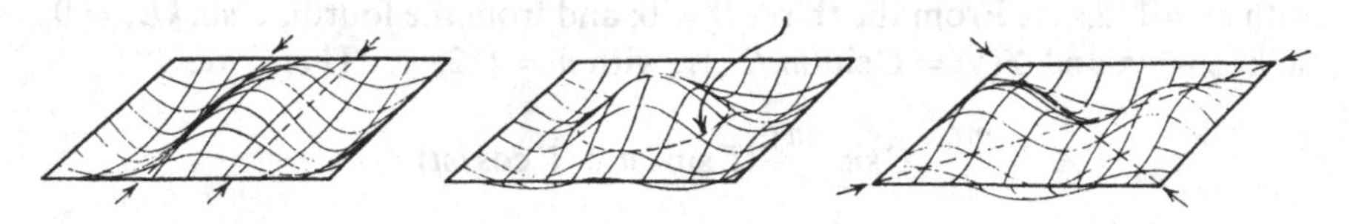

Fig. 3.3. Degenerate modes of vibration in a square membrane corresponding to different values of a and b in Eq. (3.5). Arrows point to the nodal lines (Morse, 1948).

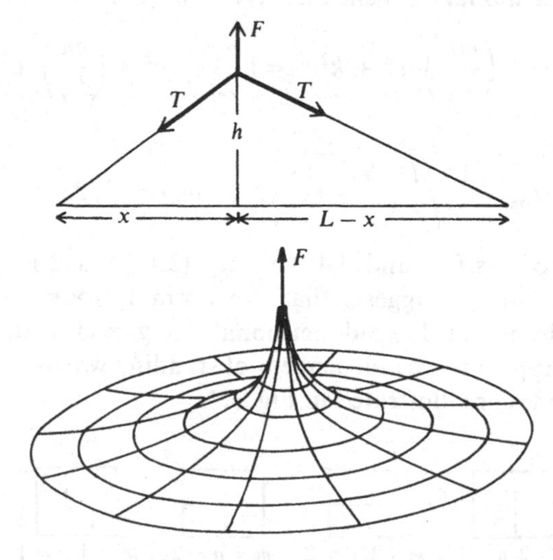

Fig. 3.4. Reaction of a string and membrane to a force applied at a point.

of radius r at the center of a membrane of radius a, the displacement becomes

$$z = \frac{2F}{T} \ln \frac{a}{r}, \tag{3.6}$$

which goes to infinity as $r \to 0$ (Morse, 1948, p. 176).

3.3. Circular Membranes

For a circular membrane, the wave equation [Eq. (3.1)] should be written in polar coordinates by letting $x = r\cos\phi$ and $y = r\sin\phi$.

$$\frac{\partial^2 z}{\partial t^2} = c^2\left(\frac{\partial^2 z}{\partial r^2} + \frac{1}{r}\frac{\partial z}{\partial r} + \frac{1}{r^2}\frac{\partial^2 z}{\partial \phi^2}\right). \tag{3.7}$$

We write solutions of the form $z(r, \phi, t) = R(r)\Phi(\phi)e^{j\omega t}$ leading to the equations:

$$\frac{d^2 R}{dr^2} + \frac{1}{r}\frac{dR}{dr} + \left(\frac{\omega^2}{c^2} - \frac{m^2}{r^2}\right)R = 0,$$

and (3.8)

$$\frac{d^2\Phi}{d\phi^2} + m^2\Phi = 0.$$

The solution to the second equation is $\Phi(\phi) = Ae^{\pm jm\phi}$. The first equation is a form of Bessel's equation $(d^2y/dx^2) + (1/x)(dy/dx) + [1 - (m^2/x^2)]y = 0$ with $y = R$ and $x = kr = \omega r/c$. The solutions are Bessel functions of order m. Each of these functions $J_0(x), J_1(x), \ldots, J_m(x)$ goes to zero for several values of x as shown in Fig. 3.5.

$J_0(x) = 0$ when $x = 2.405, 5.520, 8.654, \ldots$.

$J_1(x) = 0$ when $x = 0, 3.83, 7.02, 10.17, \ldots$.

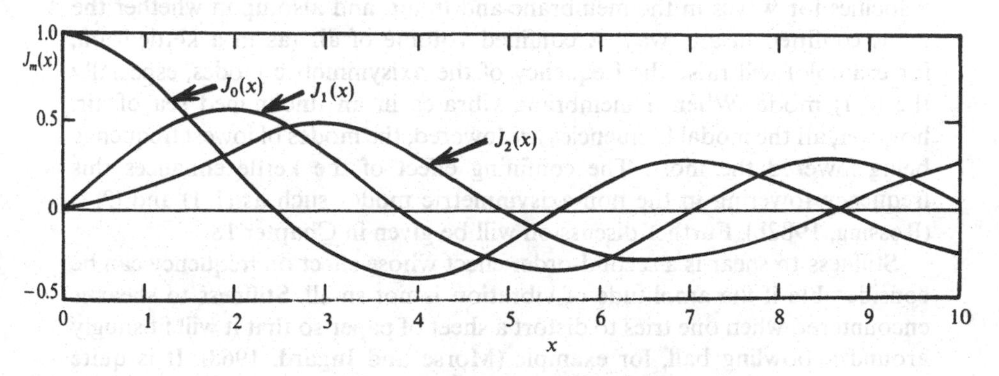

Fig. 3.5. First three Bessel functions.

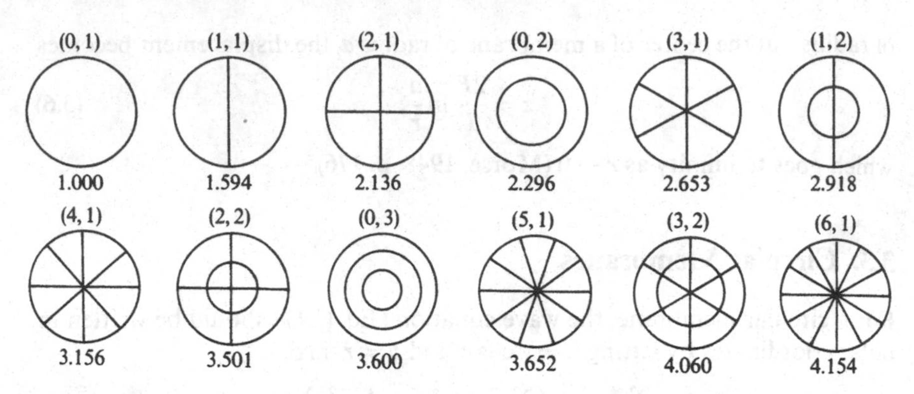

Fig. 3.6. First 12 modes of an ideal membrane. The mode designation (m, n) is given above each figure and the relative frequency below. To convert these to actual frequencies, multiply by $(2.405/2\pi a)\sqrt{T/\sigma}$, where a is the membrane radius.

The nth zero of $J_m(kr)$ gives the frequency of the (m, n) mode, which has m nodal diameters and n nodal circles (including one at the boundary). In the fundamental $(0, 1)$ mode, the entire membrane moves in phase. The first 14 modes of an ideal membrane and their relative frequencies are given in Fig. 3.6.

3.4. Real Membranes: Stiffness and Air Loading

The normal mode frequencies of real membranes may be quite different from those of an ideal membrane given in Fig. 3.6. The principal effects in the membrane acting to change the mode frequencies are air loading, bending stiffness, and stiffness to shear. In general, air loading lowers the modal frequencies, while the other two effects tend to raise them. In thin membranes, air loading is usually the dominant effect.

The effect on frequency of the air loading depends upon the comparative velocities for waves in the membrane and in air, and also upon whether the air is confined in any way. A confined volume of air (as in a kettledrum, for example) will raise the frequency of the axisymmetric modes, especially the $(0, 1)$ mode. When a membrane vibrates in an unconfined sea of air, however, all the modal frequencies are lowered, the modes of lowest frequency being lowered the most. The confining effect of the kettle enhances this frequency lowering in the non-axisymmetric modes such as $(1, 1)$ and $(2, 1)$ (Rossing, 1982b). Further discussion will be given in Chapter 18.

Stiffness to shear is a second-order effect whose effect on frequency can be considerable if the amplitude of vibration is not small. Stiffness to shear is encountered when one tries to distort a sheet of paper so that it will fit snugly around a bowling ball, for example (Morse and Ingard, 1968). It is quite different from bending stiffness, which is encountered when the paper is rolled up. Bending stiffness will be discussed in Section 3.12.

3.5. Waves in a Thin Plate

A plate may be likened to a two-dimensional bar or a membrane with stiffness. Like a bar, it can transmit compressional waves, shear waves, torsional waves, or bending waves; and it can have three different boundary conditions: free, clamped, or simply supported (hinged).

A plate might be expected to transmit longitudinal (compressional) waves at the same velocity as a bar: $c_L = \sqrt{E/\rho}$. This is not quite the case, however, since the slight lateral expansion that accompanies a longitudinal compression is constrained in the plane of the plate, thus adding a little additional stiffness. The correct expression for the velocity of *longitudinal* waves in an infinite plate is

$$c_L = \sqrt{\frac{E}{\rho(1 - \nu^2)}}, \tag{3.9}$$

where ν is Poisson's ratio ($\nu \simeq 0.3$ for most materials).

Actually, pure longitudinal waves [Fig. 3.7(a)] occur only in solids whose dimensions in all directions are greater than a wavelength. These waves travel at a speed c'_L, which is slightly less than the quasi-longitudinal waves [Fig. 3.7(b)] that propagate in a bar or a plate (see Cremer et al., 1973).

$$c'_L = \sqrt{\frac{E(1 - \nu)}{\rho(1 + \nu)(1 - 2\nu)}}. \tag{3.10}$$

Transverse waves in a solid involve mainly shear deformations, although both shear stresses and normal stresses may be involved. Solids not only

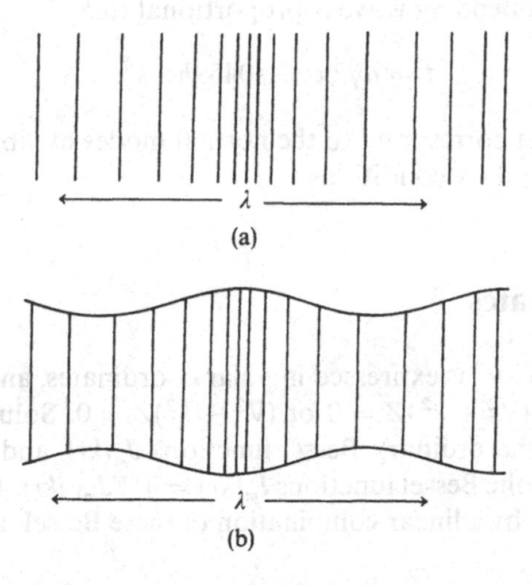

Fig. 3.7. (a) Pure longitudinal wave in an infinite solid. (b) Quasi-longitudinal wave in a bar or plate.

resist changes in volume (as do fluids), but they resist changes in shape as well. Plane transverse waves occur in bodies that are large compared to the wavelength in all three dimensions, but also in flat plates of uniform thickness (see Chapter 2 in Cremer et al., 1973). Transverse waves propagate at the same speed as torsional waves in a circular rod ($c_T = \sqrt{G/\rho}$). The shear modulus G is considerably smaller than Young's modulus E, so transverse and torsional waves propagate at roughly 60% of the speed of longitudinal waves. The radiation of sound in both cases is rather insignificant compared to the case of bending waves, which we now discuss.

The equation of motion for bending or flexural waves in a plate is

$$\frac{\partial^2 z}{\partial t^2} + \frac{Eh^2}{12\rho(1-\nu^2)}\nabla^4 z = 0, \tag{3.11}$$

where ρ is density, ν is Poisson's ratio, E is Young's modulus, and h is the plate thickness. For harmonic solutions, $z = Z(x, y)e^{j\omega t}$:

$$\nabla^4 Z - \frac{12\rho(1-\nu^2)\omega^2}{Eh^2}Z = \nabla^4 Z - k^4 Z = 0, \tag{3.12}$$

where

$$k^2 = \frac{\sqrt{12}\omega}{h}\sqrt{\frac{\rho(1-\nu^2)}{E}} = \frac{\sqrt{12}\omega}{c_L h}.$$

Bending waves in a plate are dispersive; that is, their velocity v depends upon the frequency

$$v(f) = \omega/k = \sqrt{\omega h c_L/\sqrt{12}} = \sqrt{1.8 f h c_L}. \tag{3.13}$$

The frequency of a bending wave is proportional to k^2:

$$f = \omega/2\pi = 0.0459 h c_L k^2. \tag{3.14}$$

The values of k that correspond to the normal modes of vibration depend, of course, on the boundary conditions.

3.6. Circular Plates

For a circular plate, ∇^2 is expressed in polar coordinates, and $Z(r, \phi)$ can be a solution of either $(\nabla^2 + k^2)Z = 0$ or $(\nabla^2 - k^2)Z = 0$. Solutions of the first equation contain the ordinary Bessel functions $J_m(kr)$, and solutions to the second, the hyperbolic Bessel functions $I_m(kr) = j^{-m}J_m(jkr)$. Thus, the possible solutions are given by a linear combination of these Bessel functions times an angular function:

$$Z(r, \phi) = \cos(m\phi + \alpha)[AJ_m(kr) + BI_m(kr)]. \tag{3.15}$$

If the plate is clamped at its edge $r = a$, then $Z = 0$ and $\partial Z/\partial r = 0$. The first of these conditions is satisfied if $AJ_m(ka) + BI_m(ka) = 0$, and the second if $AJ'_m(ka) + BI'_m(ka) = 0$.

The allowed values of k are labeled k_{mn}, where m gives the number of nodal diameters and n the number of nodal circles in the corresponding normal mode:

$$k_{01} = 3.189/a, \quad k_{11} = 4.612/a, \quad k_{21} = 5.904/a,$$

$$k_{02} = 6.306/a, \quad k_{12} = 7.801/a, \quad k_{22} = 9.400/a,$$

$$k_{03} = 9.425/a, \quad k_{13} = 10.965/a, \quad k_{23} = 12.566/a,$$

$$[k_{mn} \rightarrow (2n + m)\pi/2a \quad \text{as} \quad n \rightarrow \infty].$$

The corresponding mode frequencies are given in Table 3.1.

Table 3.1. Vibration frequencies of a circular plate with clamped edge.

$f_{01} = 0.4694 c_L h/a^2$	$f_{11} = 2.08 f_{01}$	$f_{21} = 3.41 f_{01}$	$f_{31} = 5.00 f_{01}$	$f_{41} = 6.82 f_{01}$
$f_{02} = 3.89 f_{01}$	$f_{12} = 5.95 f_{01}$	$f_{22} = 8.28 f_{01}$	$f_{32} = 10.87 f_{01}$	$f_{42} = 13.71 f_{01}$
$f_{03} = 8.72 f_{01}$	$f_{13} = 11.75 f_{01}$	$f_{23} = 15.06 f_{01}$	$f_{33} = 18.63 f_{01}$	$f_{43} = 22.47 f_{01}$

A plate with a free edge is more difficult to handle mathematically. The boundary conditions used by Kirchoff lead to a rather complicated expression for k_{mn}, which reduces to $(2n + m)\pi/2r$ for large ka (see Rayleigh, 1894). The (2, 0) mode is now the fundamental mode; the modal frequencies are given in Table 3.2. The mode frequencies for a plate with a simply supported (hinged) edge are given in Table 3.3.

Table 3.2. Vibration frequencies of a circular plate with free edge.

—	—	$f_{20} = 0.2413 c_L h/a^2$	$f_{30} = 2.328 f_{20}$	$f_{40} = 4.11 f_{20}$	$f_{50} = 6.30 f_{20}$
$f_{01} = 1.73 f_{20}$	$f_{11} = 3.91 f_{20}$	$f_{21} = 6.71 f_{20}$	$f_{31} = 10.07 f_{20}$	$f_{41} = 13.92 f_{20}$	$f_{51} = 18.24 f_{20}$
$f_{02} = 7.34 f_{20}$	$f_{12} = 11.40 f_{20}$	$f_{22} = 15.97 f_{20}$	$f_{32} = 21.19 f_{20}$	$f_{42} = 27.18 f_{20}$	$f_{52} = 33.31 f_{20}$

Table 3.3. Vibration frequencies of a circular plate with a simply supported edge.

$f_{01} = 0.2287 c_L h/a^2$	$f_{11} = 2.80 f_{01}$	$f_{21} = 5.15 f_{01}$
$f_{02} = 5.98 f_{01}$	$f_{12} = 9.75 f_{01}$	$f_{22} = 14.09 f_{01}$
$f_{03} = 14.91 f_{01}$	$f_{13} = 20.66 f_{01}$	$f_{23} = 26.99 f_{01}$

The frequencies in Tables 3.1–3.3 are derived mainly from calculations given by Leissa (1969). Measurements on two large brass plates by Waller (1938) are in good agreement with the data in Table 3.2. Some modes of circular plates are shown in Fig. 3.8.

Chladni (1802) observed that the addition of one nodal circle raised the frequency of a circular plate by about the same amount as adding two nodal

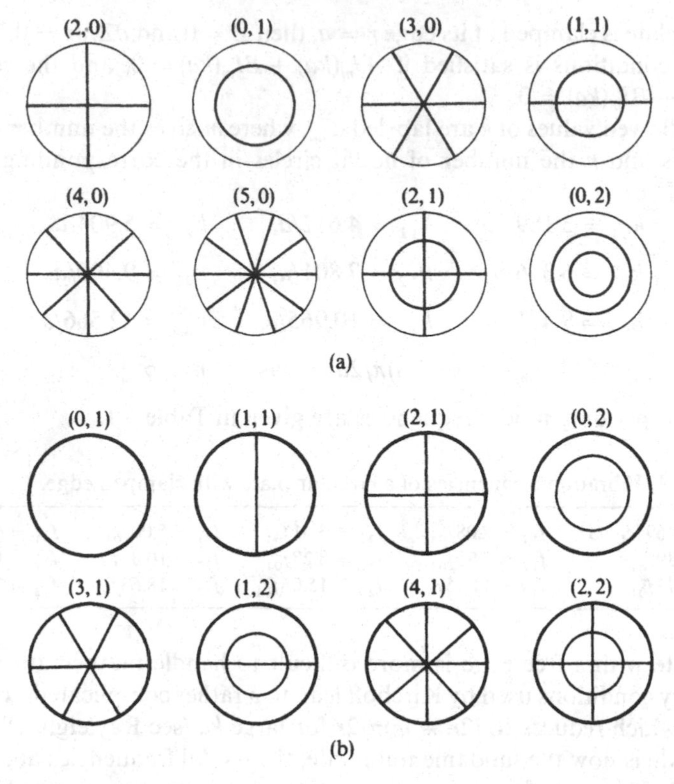

Fig. 3.8. Vibrational modes of circular plates: (a) free edge and (b) clamped or simply supported edge. The mode number (n, m) gives the number of nodal diameters and circles, respectively.

diameters, a relationship that Rayleigh (1894) calls Chladni's law. For large values of ka, $ka \simeq (m + 2n)\pi/2$, so that f is proportional to $(m + 2n)^2$. The modal frequencies in a variety of circular plates can be fitted to families of curves: $f_{mn} = c(m + 2n)^p$. In flat plates, $p = 2$, but in nonflat plates (cymbals, bells, etc.), p is generally less than 2 (Rossing, 1982c).

3.7. Elliptical Plates

The fundamental frequency of an elliptical plate of moderately small eccentricity with a clamped edge is given approximately by the formula (Leissa, 1969)

$$f \approx \frac{0.291 c_L h}{a^2} \sqrt{1 + \frac{2}{3}\left(\frac{a}{b}\right)^2 + \left(\frac{a}{b}\right)^4}, \tag{3.16}$$

where a and b are the semimajor and semiminor axes. An elliptical plate with $a/b = 2$ has frequencies 37% greater than a circular plate with the same area.

Waller (1950) shows Chladni patterns and gives relative frequencies for elliptical plates with $a/b = 2$ and $a/b = 5/4$. The nodal patterns resemble those in rectangular plates of similar shape.

3.8. Rectangular Plates

Since each edge of a rectangular plate can have any of the three boundary conditions listed in Section 3.5 (free, clamped, or simply supported), there are 27 different combinations of boundary conditions, and each leads to a different set of vibrational modes. Our discussions will be limited to three cases in which the same boundary conditions apply to all four edges.

3.8.1. Simply Supported Edges

The equation of motion is easily solved by writing the solutions as a product of three functions of single variables, as in the rectangular membrane (Section 3.1).

The displacement amplitude is given by

$$Z = A \sin \frac{(m+1)\pi x}{L_x} \sin \frac{(n+1)\pi y}{L_y}, \tag{3.17}$$

where L_x and L_y are the plate dimensions, and m and n are integers (beginning with zero). The corresponding vibration frequencies are

$$f_{mn} = 0.453 c_L h \left[\left(\frac{m+1}{L_x} \right)^2 + \left(\frac{n+1}{L_y} \right)^2 \right]. \tag{3.18}$$

The displacement is similar to that of a rectangular membrane, but the modal frequencies are not. Note that the nodal lines are parallel to the edges; this is not the case for plates with free or clamped edges, as we shall see.

It is convenient to describe a mode in a rectangular plate by (m, n), where m and n are the numbers of nodal lines in the y and x directions, respectively (not counting nodes at the edges). To do this, we use $m + 1$ and $n + 1$ in Eq. (3.17) rather than m and n, as in a rectangular membrane [Eq. (3.4)]. Thus, the fundamental mode is designated (0, 0) rather than (1, 1).

3.8.2. Free Edges

Calculating the modes of a rectangular plate with free edges was described by Rayleigh as a problem "of great difficulty." However, Rayleigh's own methods lead to approximate solutions that are close to measured values, and refinements by Ritz bring them even closer. Results of many subsequent investigations are summarized by Leissa (1969).

The limiting shapes of a rectangle are the square plate and the thin bar. The modes of a thin bar with free ends have frequencies [from Eq. (2.63)]

$$f_n = \frac{0.113h}{L^2}\sqrt{\frac{E}{\rho}}[3.0112^2, 5^2, \ldots, (2n+1)^2]. \tag{3.19}$$

The nth mode has $n + 1$ nodal lines perpendicular to the axis of the bar. As the bar takes on appreciable width, bending along one axis causes bending in a perpendicular direction. This comes about because the upper part of the bar above the neutral axis (see Fig. 2.17) becomes longer (and thus narrower), while the lower part becomes shorter (and thus wider). We have already seen how Poisson's constant ν is a measure of the lateral contraction that accompanies a longitudinal expansion in a plate (Section 3.5) and how the factor $1 - \nu^2$ appears in the expression for both longitudinal and bending wave velocities [Eqs. (3.9) and (3.13)].

Several bending modes in a rectangular plate can be derived from the bending modes of a bar. The $(m, 0)$ modes might be expected to have nodal lines parallel to one pair of sides, and the $(0, n)$ modes would have nodes parallel to the other pair of sides. Because of the coupling between bending motions in the two directions, however, the modes are not pure bar modes. The nodal lines become curved, and the plate takes on a sort of saddle shape (i.e., concave in one direction but convex in the perpendicular direction). This can be called anticlastic bending, and it is quite evident in the modes of two different rectangular plates shown in Fig. 3.9.

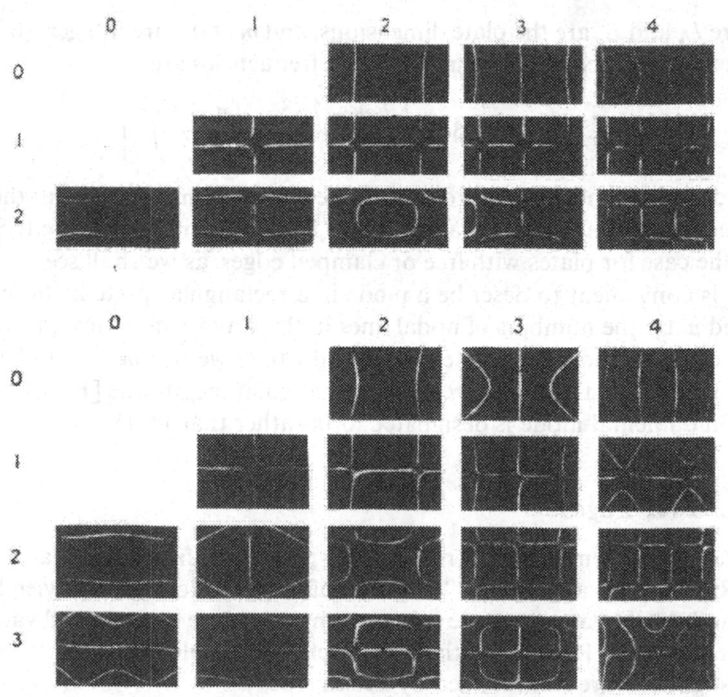

Fig. 3.9. Chladni patterns showing the vibrational modes of rectangular plates of different shapes: (a) $L_x/L_y = 2$; (b) $L_x/L_y = 3/2$ (Waller, 1949).

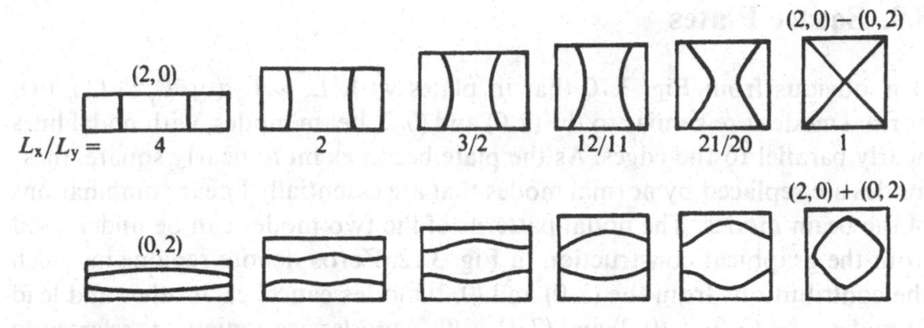

Fig. 3.10. Mixing of the (2, 0) and (0, 2) modes in rectangular plates with different L_x/L_y ratios (after Waller, 1961).

It is interesting to note how the combinations develop in a rectangle as L_x/L_y approaches unity. Fig. 3.10 shows the shapes of two modes that are descendents of the (2, 0) and (0, 2) beam modes in rectangles of varying L_x/L_y. When $L_x \gg L_y$, the (2, 0) and (0, 2) modes appear quite independent. However, as $L_y \rightarrow L_x$, the beam modes mix together to form two new modes. In the square, the mixing is complete, and two combinations are possible depending upon whether the component modes are in phase or out of phase.

Frequencies for the modes that have as their bases the (2, 0) and (0, 2) beam modes are shown in Fig. 3.11. The frequencies have been normalized to L_x, and the normalized frequency of the (2, 0) mode is seen to be relatively independent of L_y. The dashed curves are obtained from an approximate formula using the Rayleigh method, whereas the solid curves are from a more exact numerical calculation (Warburton, 1954).

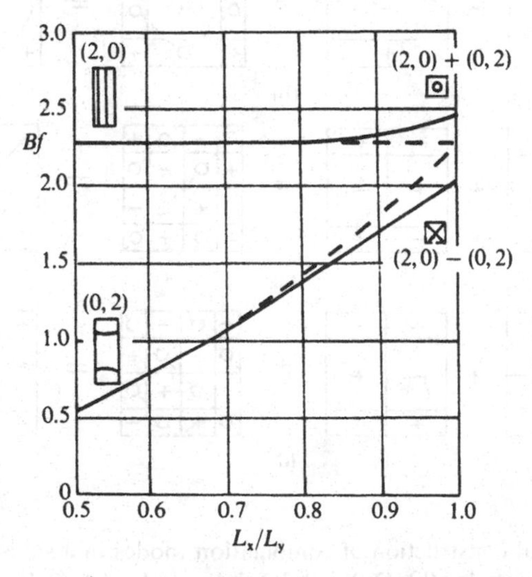

Fig. 3.11. Normalized frequencies for the (2, 0) and (0, 2) modes (and modes based on combinations of these) in rectangular plates with free edges and varying L_x/L_y ratios (from Warburton, 1954). $B = 2.21(L_x^2/h)\sqrt{\rho(1 - \nu^2)/E}$.

3.9. Square Plates

It is obvious from Fig. 3.10 that in plates with $L_x \gg L_y$ (or $L_y \gg L_x$), two normal modes are similar to the (2, 0) and (0, 2) beam modes, with nodal lines nearly parallel to the edges. As the plate becomes more nearly square, these modes are replaced by normal modes that are essentially linear combinations of the beam modes. The nodal patterns of the two modes can be understood from the graphical construction in Fig. 3.12. Zeros denote regions in which the contributions from the (2, 0) and (0, 2) modes cancel each other and lead to nodes. The (2, 0) + (0, 2) and (2, 1) − (0, 2) modes are sometimes referred to as the ring mode and X mode, respectively, on account of the shapes of their nodal patterns.

From Fig. 3.11, it is clear that the (2, 0) + (0, 2) ring mode has a higher frequency than the (2, 0) − (0, 2) X mode. In the X mode, the bending motions characteristic of the (2, 0) and (0, 2) beam modes aid each other through an elastic interaction that we call Poisson coupling, since its strength de-

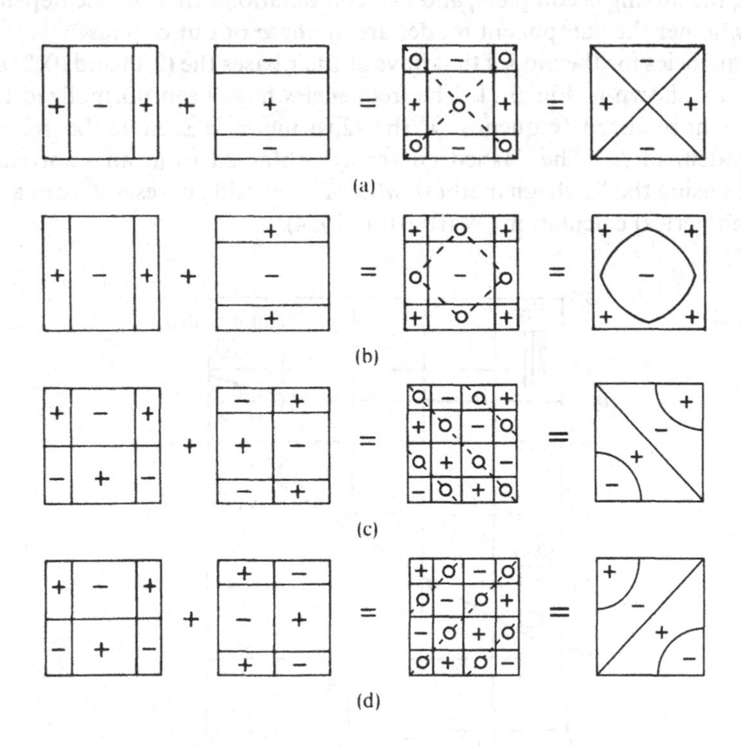

Fig. 3.12. Graphical construction of combination modes in a square isotropic plate: (a) (2, 0 − 0, 2), X mode; (b) (2, 0 + 0, 2), ring mode; (c) (2, 1 − 1, 2) mode; and (d) (2, 1 + 1, 2) mode.

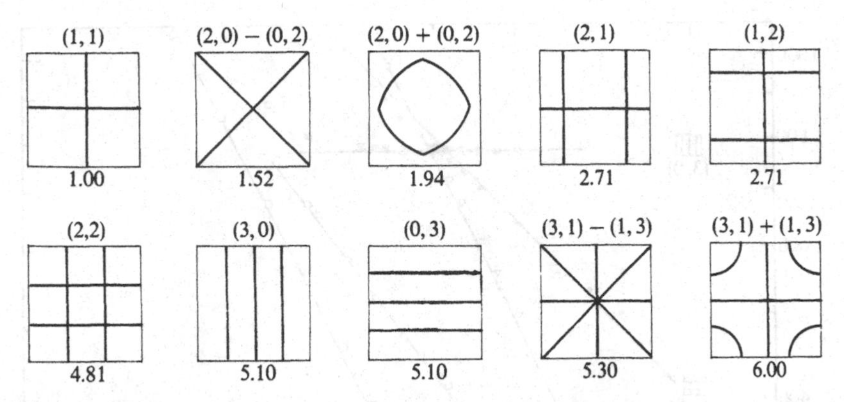

Fig. 3.13. The first 10 modes of an isotropic square plate with free edges. The modes are designated by m and n, the numbers of nodal lines in the two directions, and the relative frequencies for a plate with $\nu = 0.3$ are given below the figures.

pends upon the value of Poisson's constant. In the (2, 0) + (0, 2) ring mode, however, there is an added stiffness due to the fact that the (2, 0) and (0, 2) bending motions oppose each other. Thus, the Poisson coupling splits a modal degeneracy that otherwise would have existed in a square plate. The ratio of the (2, 0 + 0, 2) and (2, 0 − 0, 2) mode frequencies is (Warburton, 1954)

$$\frac{f+}{f-} = \sqrt{\frac{1 + 0.7205\nu}{1 - 0.7205\nu}}. \tag{3.20}$$

Also shown in Fig. 3.12 are the (2, 1) − (1, 2) and (2, 1) + (1, 2) modes, which have the same frequency as the (2, 1) and (1, 2) modes, since Poisson coupling does not aid or oppose either combination. Thus, any of these four modes can be excited depending upon where the driving force is applied. There are, in fact, a large number of degenerate modes, all linear combinations of the (2, 1) and (1, 2) modes, which can be excited.

The first 10 modes of an isotropic square plate with free edges are shown in Fig. 3.13. The mode of lowest frequency, the (1, 1) mode, is a twisting mode in which opposite corners move in phase. Its frequency is given by

$$f_{11} = \frac{c_T}{2L_y} = \frac{h}{L_x L_y}\sqrt{\frac{G}{\rho}} = \frac{h}{L^2}\sqrt{\frac{E}{2\rho(1+\nu)}} = \frac{hc_L}{L^2}\sqrt{\frac{1-\nu}{2}}, \tag{3.21}$$

where the torsional wave velocity c_T from Fig. 2.23 is used (subject to the restriction that $L_x > 6h$). In this equation, h is the thickness and G is the shear modulus.

Note that the (2, 1) and (1, 2) modes form a degenerate pair, as do the (3, 0) and (0, 3) modes. However, Poisson coupling removes the degeneracy in the case of the (3, 1)/(1, 3) pair just as it does in the (2, 0)/(0, 2) case. The general rule is that a nondegenerate pair of modes $(m, n \pm n, m)$ exists in a square plate when $m - n = \pm 2, 4, 6, \ldots$.

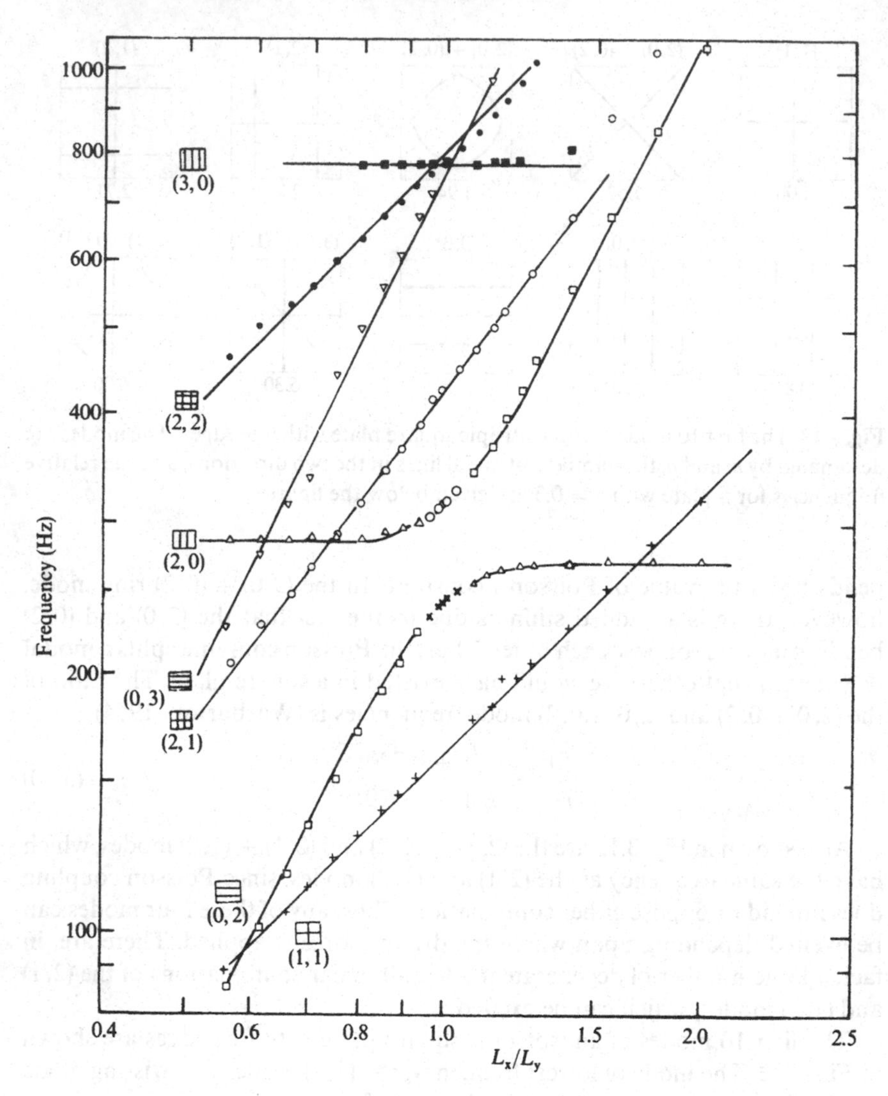

Fig. 3.14. Modal frequencies of an isotropic aluminum plate (L_x constant). Lines representing the (1, 1) and (2, 2) twisting modes have a slope of 1; lines representing the (0, 2) and (0, 3) bending modes have a slope of 2 (from Caldersmith and Rossing, 1983).

The modal frequencies in an aluminum plate with a varying length to width ratio are shown in Fig. 3.14. In this case, L_x was kept constant as L_y was varied, so the frequency of the (3, 0) mode, for example, is unchanged. The (1, 1) mode has a slope of 1, as predicted by Eq. (3.19). The (0, 3) bending mode has a slope of 2, as does the (0, 2) mode above and below the region of $L_x = L_y$. The (2, 1) mode, which combines twisting and bending motions, has a slope of about $\frac{4}{3}$.

3.10. Square and Rectangular Plates with Clamped Edges

The first eight modes of a square plate with clamped edges are shown in Fig. 3.15. There is considerable variation in the mode designation by various authors, and so we have used the same designation that was used in Section 3.8.1 for a plate with simply supported edges: *m* and *n* are the numbers of nodes in the directions of the y and x axes, respectively, not counting the nodes at the edges. The fundamental (0, 0) mode has a frequency: $f_{00} = 1.654 c_L h/L^2$, where h is the thickness, L is length, and $c_L = \sqrt{E/\rho(1 - \nu^2)}$ is the longitudinal wave velocity (Leissa, 1969). The relative frequencies of the modes are given below the patterns in Fig. 3.15.

Comparing the modes of the square plate with clamped edges to one with free edges, we note that

1. the (1, 1) mode has a frequency nearly 10 times greater than the (1, 1) mode in a free plate.
2. three other modes exist below the (1, 1) mode in the clamped plate.
3. the X mode and ring mode are only about 0.5% different in frequency, and the diameter of the ring node is smaller than it is in a free plate.
4. nondegenerate mode pairs $(m, n \pm n, m)$ exist when $m - n = \pm 2, 4, 6, \ldots$, as in the free plates, but the transition from modes characteristic of rectangular plates to those of square plates changes much more abruptly as $L_x \to L_y$ in clamped plates than in free plates (Warburton, 1954).

Relative frequencies of rectangular plates with clamped edges (from Leissa, 1969) are given in Table 3.4. The actual frequencies can be obtained by multiplying the relative frequencies by $1.654 c_L h/L_y^2$.

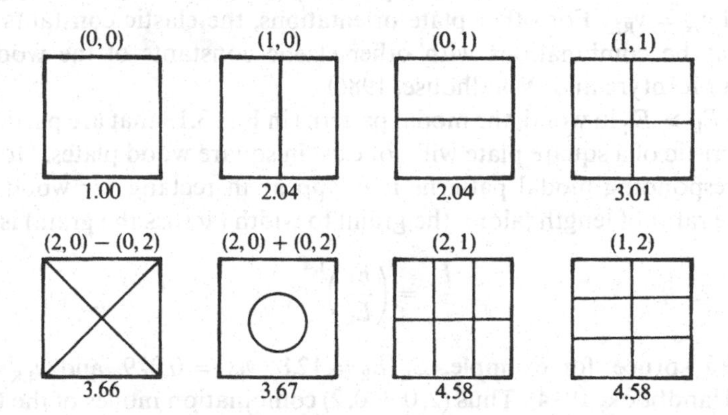

Fig. 3.15. Modal patterns for the first eight modes of a square plate with clamped edges. Relative frequencies are given below the patterns.

Table 3.4. Relative vibrational frequencies of rectangular plates with clamped edges.

Mode	Mode	$L_x/L_y = 1$	1.5	2	2.5	3	∞
(0, 0)		1.00	0.75	0.68	0.66	0.64	0.62
(0, 1)		2.04	1.88	1.82	1.79	1.78	1.72
(1, 0)	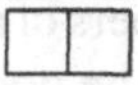	2.04	1.16	0.88			
(1, 1)	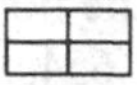	3.01	2.27	2.02	1.91	1.86	1.72

3.11. Rectangular Wood Plates

Wood can be described as an orthotropic material; it has different mechanical properties along three perpendicular axes (longitudinal, radial, and tangential, which we denote by L, R, and T). Thus, there are three elastic moduli and six Poisson's ratios, although they are related by expressions of the form (Wood Handbook, 1974)

$$\frac{\nu_{ij}}{E_i} = \frac{\nu_{ji}}{E_j}, \qquad i, j = \mathrm{L, R, T}.$$

Most plates in musical instruments are quarter-cut plates (the log is split or sawed along two radii); the growth rings lie perpendicular to the plate. For a quarter-cut plate, the axes L and R lie in the plane and the axis T in the direction of the thickness. Thus, the constants of interest are E_L, E_R, ν_{LR}, and $\nu_{RL} = \nu_{LR} E_R/E_L$.

To describe the vibrational modes of wood plates generally requires four elastic constants. These may be Young's moduli along (E_x) and across (E_y) the grain, the in-plane shear modulus G, and the larger of the two Poisson ratios ν_{xy}. For a quarter-cut plate in the xy plane, $E_x = E_L$, $E_y = E_R$, $G = G_{LR}$, $\nu_{xy} = \nu_{LR}$, and $\nu_{yx} = \nu_{RL}$. For other plate orientations, the elastic constants of the plate may be combinations with other elastic constants of the wood (see Fig. 1 in McIntyre and Woodhouse, 1986).

Since $E_L > E_R$ in wood, the modal patterns in Fig. 3.13 that are particularly characteristic of a square plate will not exist in square wood plates. However, the corresponding modal patterns may appear in rectangular wood plates when the ratio of length (along the grain) to width (across the grain) is

$$\frac{L_x}{L_y} = \left(\frac{E_L}{E_R}\right)^{1/4}.$$

For Sitka spruce, for example, $E_L/E_R = 12.8$, $\nu_{RL} = 0.029$, and $\nu_{LR} = 0.37$ (Wood Handbook, 1974). Thus $(2, 0 \pm 0, 2)$ combination modes of the type in Fig. 3.15, for example, might be expected in a rectangular plate with $L_x/L_y = 1.9$. (This is somewhat greater than the value 1.5, which has appeared several places in the literature.)

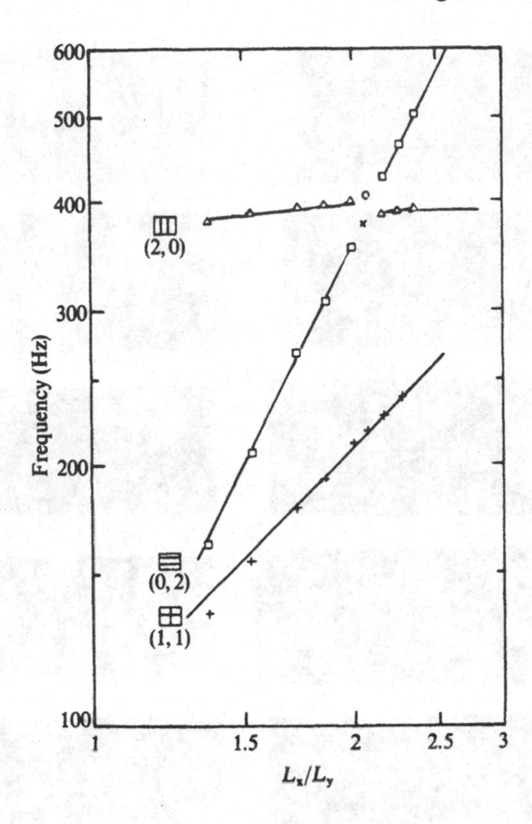

Fig. 3.16. Modal frequencies of a quarter-cut spruce plate (L_x constant). The (1, 1) twisting mode has a slope of 1; the (0, 2) bending mode has a slope of 2 (Caldersmith and Rossing, 1983).

The modal frequencies of a quarter-cut spruce plate are shown in Fig. 3.16. The (2, 0 $\pm$ 0, 2) ring mode and X mode occur at about $L_x/L_y = 2$, as expected. The curve relating the frequency of the (1, 1) mode to L_x/L_y has a slope of one, as in the aluminum plate in Fig. 3.14, and for the (2, 0) mode, the slope is two.

Some modal patterns obtained in the same spruce plate are shown in Fig. 3.17. The (2, 0) and (0, 2) beam modes appear relatively unmodified when L_x/L_y is well above or below the critical ratio of 2.08 where the X mode and ring mode appear. Note that the (2, 0) and (0, 2) modes mix oppositely in phase for L_x/L_y above and below 2.08.

Modes of vibration closely resembling the X mode and ring mode, along with the (1, 1) twisting mode, have been used by violin makers for centuries to tune the top and back plates of violins before assembling the instruments. If a plate is held with the fingers at a nodal line and tapped at an antinode, the trained ear of a skilled violin maker can ascertain whether the plate has a clear and full ring. In recent years, many violin makers, following the lead of Carleen

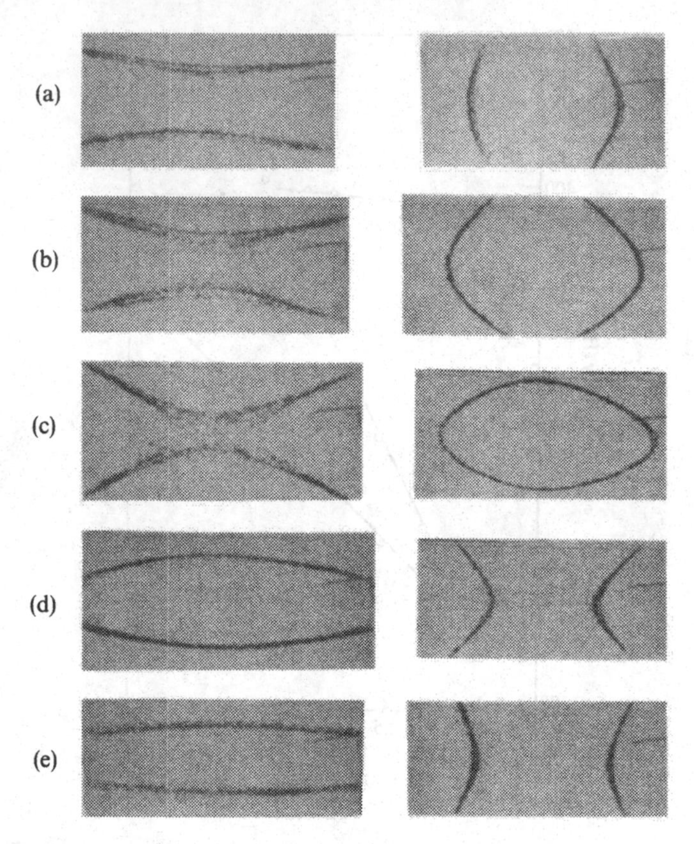

Fig. 3.17. Modal shapes of the $A(2,0) + B(0,2)$ combination modes in a quarter-cut spruce plate. In (a) and (e), $|A| \ll 1$ (opposite in sign in the two cases). In (f) and (j), $|B| \ll 1$. In (c) and (h), $A/B = \pm 1$ (Caldersmith and Rossing, 1983).

Hutchins, have used Chladni patterns to test and guide the tuning of these three important modes (see Fig. 3.18).

Many of the formulas in Chapters 2 and 3 for vibrations of bars and plates of isotropic material are easily modified for wood by substituting E_y or E_x for E and $\nu_{yx}\nu_{xy}$ for ν^2. For example, Eqs. (3.18) and (3.19) become

$$f_{mn} = 0.453h\left[c_x\left(\frac{m+1}{L_x}\right)^2 + c_y\left(\frac{n+1}{L_y}\right)^2\right], \tag{3.18$'$}$$

where

$$c_x = \sqrt{E_x/\rho(1-\nu_{xy}\nu_{yx})} \quad \text{and} \quad c_y = \sqrt{E_y/\rho(1-\nu_{xy}\nu_{yx})}$$

and

$$f_n = \frac{0.113h}{L^2}(c_x c_y)^{1/2}[(3.0112)^2, 5^2, \ldots, (2n+1)^2]. \tag{3.19$'$}$$

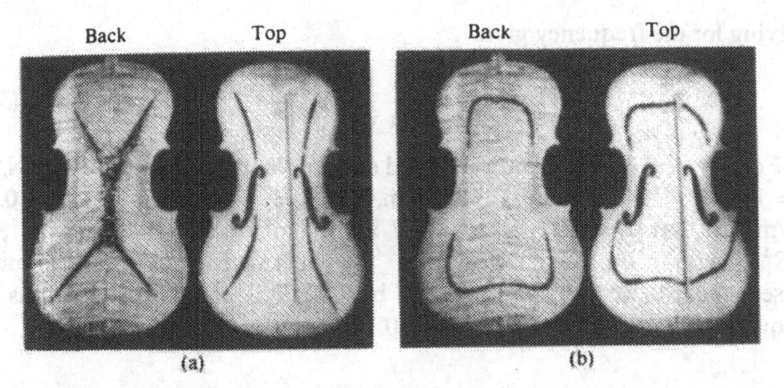

Fig. 3.18. Chladni patterns showing two modes of vibration in the top and back of a viola (Hutchins, 1977).

The torsional stiffness D_{xy} of a wooden plate depends upon E_x, E_y, and G, but it can be approximated by the geometric average $\sqrt{D_x D_y}$ of the bending stiffness in the x and y directions (Caldersmith, 1984). Thus, Eq. (3.21) can be written as

$$f_{11} = \frac{h}{L_x L_y}\sqrt{\frac{G}{\rho}} \approx \frac{h(c_y c_x)^{1/2}}{L_y L_x}\sqrt{\frac{1 - \sqrt{\nu_{yx}\nu_{xy}}}{2}}. \tag{3.21$'$}$$

The analogue to Eq. (2.73) is, approximately,

$$G = \frac{\sqrt{E_x E_y}}{2(1 + \sqrt{\nu_{xy}\nu_{yx}})}. \tag{2.73$'$}$$

3.12. Bending Stiffness in a Membrane

In Chapter 2, we described a stiff string as being slightly barlike and added a term to the equation of motion to represent the bending stiffness. We follow the same approach now by describing a stiff membrane as being slightly platelike, and we add a term to the equation of motion [Eq. (3.1) or (3.7)] to represent the bending stiffness:

$$\frac{\partial^2 z}{\partial t^2} = \frac{T}{\sigma}\nabla^2 z - \frac{h^2 E}{12\rho(1 - \nu^2)}\nabla^4 z = c^2 \nabla^2 z - S^4 \nabla^4 z, \tag{3.22}$$

where T is tension, σ is mass per unit area, h is thickness, E is Young's modulus, ρ is density, ν is Poisson's ratio, and c is the velocity of transverse waves in the membrane without stiffness.

Assuming a solution $z = AJ_m(kr)\cos m\theta \cos \omega t$ leads to the equation

$$k^2 + \frac{k^4 S^4}{c^2} - \frac{\omega^2}{c^2} = 0. \tag{3.23}$$

Solving for the frequency gives

$$f = \frac{\omega}{2\pi} = \frac{ck}{2\pi}\sqrt{1 + \frac{k^2 S^4}{c^2}}. \tag{3.24}$$

For a typical Mylar membrane used on a kettledrum, $E = 3.5 \times 10^9$ N/m², $h = 1.9 \times 10^{-4}$ m, a (radius) = 0.328 m, $\rho = 1.38 \times 10^3$ kg/m³, and $\sigma = 0.262$ kg/m², so that $c = 100$ m/s, $S^4 = 8.7 \times 10^{-3}$ m⁴/s². For $k_{11} = 10$ m⁻¹, the k^2S^4/c^2 term in Eq. (3.24) is about 10^{-4}, so the frequency of the (1, 1) mode is raised about 0.005% by the effect of bending stiffness. In other drums, the frequency change may be as large as 0.1%.

3.13. Shallow Spherical Shells

Vibrations of curved shells have been extensively studied, both theoretically and experimentally, by various investigators. In addition to the flexural modes of vibration found in a flat plate, a curved shell has many longitudinal, torsional, and thickness shear modes. Fortunately, the lowest modes in a

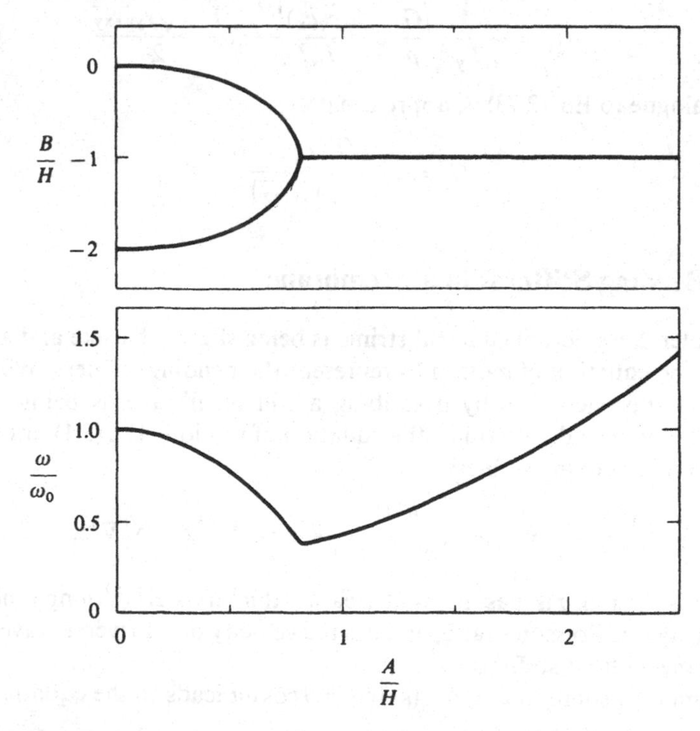

Fig. 3.19. Calculated motion center B and frequency ω for vibrations of amplitude A on a thin spherical-cap shell of dome height H. The normal mode frequency for small-amplitude vibrations is ω_0 (Fletcher, 1985).

shallow shell are mainly flexural, and thus a simple theory of transverse vibrations is quite accurate for treating the lowest modes.

Equation (3.11) for a flat plate can be suitably modified to describe a shallow spherical shell by adding a term $\nabla^2 F/R$, where R is the radius of curvature of the shell, and F is Airy's stress function:

$$\frac{\partial^2 z}{\partial t^2} + \frac{Eh^2}{12\rho(1-\nu^2)}\nabla^4 z + \frac{\nabla^2 F}{R} = 0. \tag{3.25}$$

A second equation, in which h is the shell thickness, completes the problem (Johnson and Reissner, 1958):

$$\nabla^4 F - \frac{hE\nabla^2 z}{R} = 0. \tag{3.26}$$

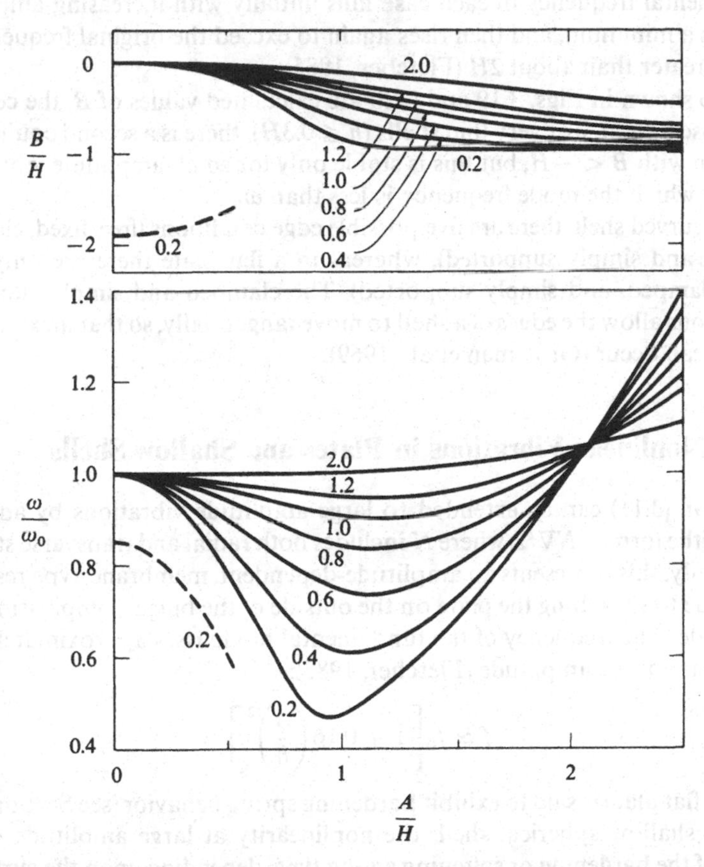

Fig. 3.20. Calculated motion center B and frequency ω for vibrations of amplitude A on a spherical-cap shell of dome height H when the normalized shell thickness h/H has the value shown as a parameter. The normal mode frequency for small-amplitude vibrations is ω_0. The broken curve shows the limited range behavior of a moderately thin everted shell (Fletcher, 1985).

The modal frequencies of a spherical shell are generally somewhat greater than those of a flat plate. In a clamped edge shell whose apex height H (above the edge plane) equals shell thickness h, for example, the fundamental mode has a frequency about 40% greater than that of a flat plate with the same radius and thickness (Kalnins, 1963). Other modal frequencies for a shell with a clamped edge are given in Kalnins (1963) and for a free edge in Reissner (1955).

In a very thin shell, in which flexural rigidity makes a minor contribution to the restoring force, the frequency of the fundamental mode initially decreases with amplitude A, as shown in Fig. 3.19. After reaching a minimum at an amplitude $A \simeq 0.8H$, it increases with amplitude (Fletcher, 1985).

When the shell is not ideally thin, the flexural rigidity cannot be ignored, the frequency can be calculated by using a variational method (Reissner, 1964). The results for values of h/H up to 2.0 are shown in Fig. 3.20. Again, the fundamental frequency in each case falls initially with increasing amplitude, reaches a minimum, and then rises again to exceed the original frequency ω_0 for A greater than about $2H$ (Fletcher, 1985).

Also shown in Figs. 3.19 and 3.20 are calculated values of B, the centroid of the oscillation. For very thin shells ($h \leq 0.3H$), there is a second equilibrium position with $B < -H$, but this is stable only for small amplitude vibrations, and for which the mode frequency is less than ω_0.

In a curved shell, there are five possible edge conditions (free, fixed, clamped, hinged, and simply supported), whereas in a flat plate there are only three (free, clamped, and simply supported). The clamped and simply supported conditions allow the edges of a shell to move tangentially, so that inextensional modes can occur (Grossman et al., 1969).

3.14. Nonlinear Vibrations in Plates and Shallow Shells

Equation (3.11) can be extended to large-amplitude vibrations by adding a term of the form $-N\nabla^2 z$, where N includes both radial and transverse stresses. Physically, this represents an amplitude-dependent, membrane-type restoring force due to stretching the plate on the outside of the bulge, compressing it on the inside. The frequency of the fundamental mode rises approximately with the square of the amplitude (Fletcher, 1985):

$$f \simeq f_0\left[1 + 0.16\left(\frac{A}{h}\right)^2\right]. \tag{3.27}$$

Thus, a flat plate is said to exhibit hardening spring behavior (see Section 1.11).

In a shallow spherical shell, the nonlinearity at large amplitude can be either of the hardening or softening spring type, depending upon the curvature. In a shell with a fixed edge, softening occurs when the apex height H exceeds the thickness h. When $H = 2h$ and $A = h$, for example, the frequency is about 10% less than its small-amplitude value (Grossman et al., 1969).

The softening effect at moderately large amplitude offsets some of the increase in vibrational frequency with curvature. Again using as an example a shell with a fixed edge having $H = 2h$, at small amplitude the fundamental frequency is approximately twice that of a flat plate, but the ratio diminishes to 1.6 when $A = h$.

Spherical, cylindrical, and conical shells all show nonlinear behavior of the softening spring type at moderately large amplitude. A shell with two independent radii of curvature R_x and R_y can show either hardening or softening behavior, however. For most ratios of R_x/R_y, shells with a rectangular edge, for example, show an initial softening followed by a hardening at very large amplitude. The hyperbolic parabola ($R_x/R_y = -1$), however, shows the same type of hardening behavior as a flat plate (Leissa and Kadi, 1971).

An interesting case of nonlinear frequency shift has been noted in a small gong used in Chinese opera. The gong, which has a slightly domed striking surface, surrounded by conical shoulders, glides upward by about two semitones after being struck (Rossing and Fletcher, 1983). That is, the frequency at the large initial amplitude is some 12% lower than at small amplitude, thus illustrating a nonlinearity of the softening spring type shown in Fig. 3.19.

Nonlinear behavior at large amplitude gives rise to harmonics of the fundamental frequency. For small amplitudes, the amplitude A_2 of the second harmonic varies with A_1^2 and that of the third harmonic A_3 with A_1^3, where A_1 is the fundamental amplitude. For $A_1 \geq H$, however, the motion becomes more nearly symmetric, and A_2 decreases with increasing A_1. Indeed, in the thin-shell approximation, A_2 is zero for $A_1 \geq 0.8H$ (Fletcher, 1985). In approximations of higher order, the influence of A_2 on A_3, etc., is taken into account, and the harmonic structure changes in a rather complex way with amplitude.

3.15. Driving Point Impedance

In Section 1.7, we discussed the mechanical impedance $\tilde{Z}$ (and its reciprocal, the mobility or mechanical admittance $\tilde{Y}$) of a simple oscillator. In Fig. 1.13(a), the real part of the admittance reaches its maximum value at resonance, while the imaginary part goes through zero. On the Nyquist plot in Fig. 1.13(c), the entire frequency span from $\omega = 0$ to $\omega \to \infty$ represents a complete circle.

In a more complex system, the impedance depends upon the location at which the force $\tilde{F}$ and the velocity $\tilde{v}$ are measured. If the force $\tilde{F}$ is applied at a single point and the velocity $\tilde{v}$ is measured at the same point, the quotient $\tilde{F}(\mathbf{n}_1)/\tilde{v}(\mathbf{n}_1) = \tilde{Z}$, is called the driving-point impedance. Measuring $\tilde{F}$ and $\tilde{v}$ at different locations gives a transfer impedance.

The driving point impedance is often measured by means of an impedance head, which incorporates both a force transducer and accelerometer, as shown in Fig. 3.21. Both transducers employ piezoelectric crystals, and the accelerometer has an inertial mass attached to the crystal, as shown. Since

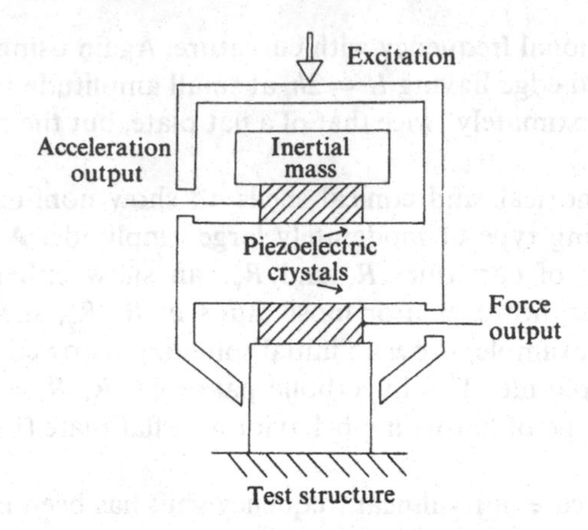

Fig. 3.21. Impedance head consisting of an accelerometer and force transducer.

attaching the impedance head adds mass m to the structure, the measured impedance Z_1' is the true driving point impedance Z_1 plus the impedance $j\omega m$ of the added mass m below the force transducer:

$$Z_1' = Z_1 + j\omega m. \tag{3.28}$$

The second term can be minimized by placing the force transducer next to the structure and making its mass as small as possible.

In practice, the impedance head is generally attached to an electromagnetic shaker, which furnishes the driving force or excitation. The output of the accelerometer is integrated to obtain a velocity signal, which is divided by the force to obtain the admittance (or vice versa for impedance). The driving force may be a swept sinusoid, random noise, or pseudorandom noise. When random noise is used, it is necessary to employ a real-time spectrum analyzer to obtain the admittance or impedance of the structure as a function of frequency (often called the frequency response function).

3.15.1. Impedance of Infinite Bars and Thin Plates

The input impedance for longitudinal waves in an infinitely long bar is simply μc_L, where μ is the mass per unit length and c_L is the longitudinal wave speed. The input impedance for flexural waves is a little more difficult to calculate, since the speed of flexural waves is frequency dependent (Section 2.15), and also because in flexure, there are exponentially decaying near fields in addition to the propagating waves (Cremer et al., 1973). The impedance is

$$\tilde{Z}(f) = \mu v(f)\frac{1+j}{2}. \tag{3.29}$$

The impedance is complex and increases with $\sqrt{f}$, as does the flexural wave speed $v(f)$, up to a certain limiting frequency.

The input impedance of a thin, isotropic, infinite plate, on the other hand, turns out to be real and independent of the frequency (Cremer et al., 1973):

$$Z = 8\sqrt{B\sigma} = 8\sqrt{\frac{E\sigma}{1-\nu^2}\frac{h^3}{12}}, \tag{3.30}$$

where B is the flexural stiffness, σ is the mass per unit area, h is the thickness, and ν is Poisson's ratio.

3.15.2. Impedance of Finite Bars and Plates

When the driving point impedance or admittance of a finite structure is plotted as a function of frequency, a series of maxima and minima are added to the curves for the corresponding infinite structure. The normalized driving-point impedance at one end of a bar with free ends is shown in Fig. 3.22. The heavily damped curve ($\delta = 1$) approximates Eq. (3.29) for an infinite bar, while the lightly damped curve ($\delta = 0.01$) has sharp maxima and minima.

The minima in Fig. 3.22 correspond to normal modes of the bar, whereas

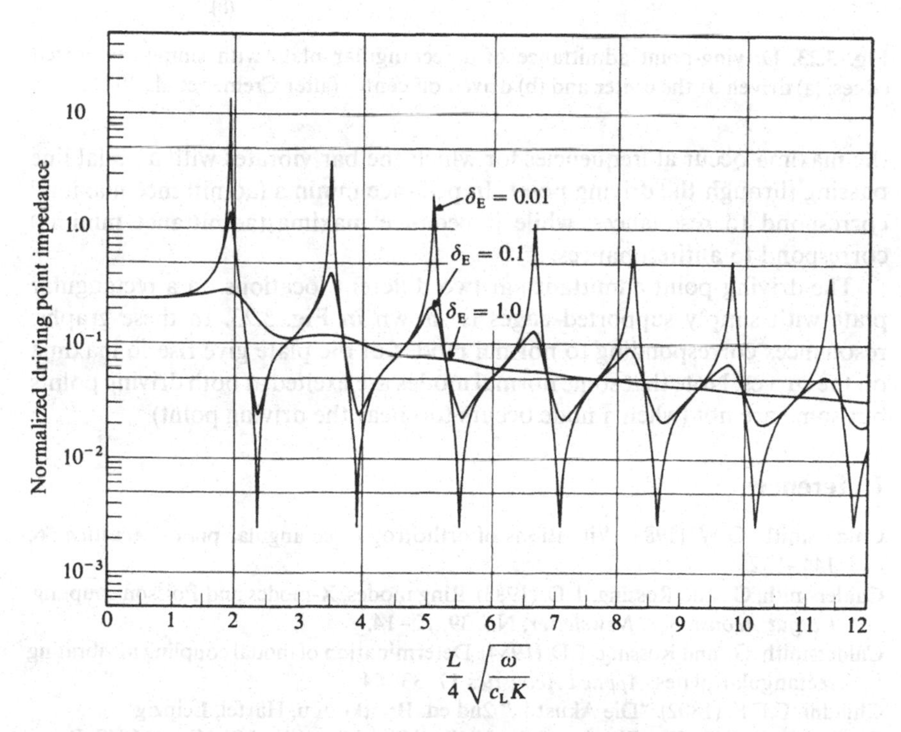

Fig. 3.22. Normalized driving-point impedance at one end of a bar with free ends. Note that the horizontal axis is proportional to $\sqrt{f}$; the normalized impedance, which includes a factor $1/f$, decreases with frequency even though the actual driving-point impedance increases as $\sqrt{f}$. Three different values of damping are shown (Snowdon, 1965).

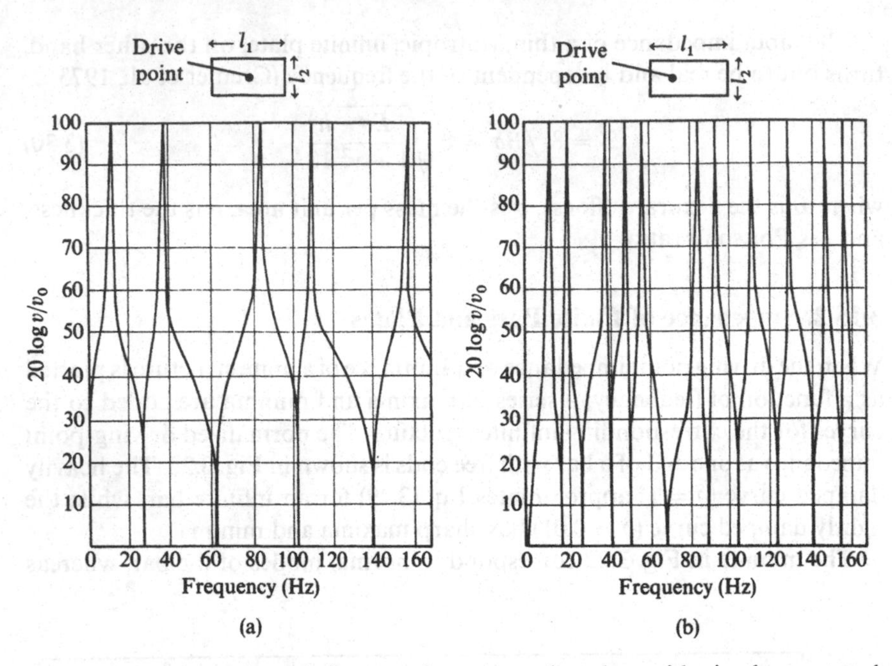

Fig. 3.23. Driving-point admittance of a rectangular plate with simply-supported edges: (a) driven at the center and (b) driven off center (after Cremer et al., 1973).

the maxima occur at frequencies for which the bar vibrates with a nodal line passing through the driving point. Impedance minima (admittance maxima) correspond to resonances, while impedance maxima (admittance minima) correspond to antiresonances.

The driving-point admittance at two different locations on a rectangular plate with simply supported edges is shown in Fig. 3.23. In these graphs, resonances corresponding to normal modes of the plate give rise to maxima on the curves. Note that some normal modes are excited at both driving points but some are not (when a node occurs too near the driving point).

References

Caldersmith, G.W. (1984). Vibrations of orthotropic rectangular plates. *Acustica* **56**, 144–152.

Caldersmith, G., and Rossing, T.D. (1983). Ring modes, X-modes and Poisson coupling. *Catgut Acoust. Soc. Newsletter*, No. 39, 12–14.

Caldersmith, G., and Rossing, T.D. (1984). Determination of modal coupling in vibrating rectangular plates. *Applied Acoustics* **17**, 33–44.

Chladni, E.F.F. (1802). "Die Akustik," 2nd ed. Breitkopf u. Härtel, Leipzig.

Cremer, L. (1984). "The Physics of the Violin." Translated by J.S. Allen, M.I.T. Press, Cambridge, Massachusetts.

Cremer, L., Heckl, M., and Ungar, E.E. (1973). "Structure-Borne Sound," Chapter 4. Springer-Verlag, Berlin and New York.

Fletcher, N.H. (1985). Nonlinear frequency shifts in quasispherical-cap shells: Pitch glide in Chinese gongs. *J. Acoust. Soc. Am.* **78**, 2069–2073.

French, A.P. (1971). "Vibrations and Waves," p. 181 ff. Norton, New York.

Grossman, P.L., Koplik, B., and Yu, Y-Y. (1969). Nonlinear vibrations of shallow spherical shells. *J. Appl. Mech.* **36**, 451–458.

Hearmon, R.F.S. (1961). "Introduction to Applied Anisotropic Elasticity." Oxford Univ. Press, London.

Hutchins, C.M. (1977). Another piece of the free plate tap tone puzzle. *Catgut Acoust. Soc. Newsletter*, No. 28, 22.

Hutchins, C.M. (1981). The acoustics of violin plates. *Scientific American* **245** (4), 170.

Johnson, M.W., and Reissner, E. (1958). On transverse vibrations of shallow spherical shells. *Q. Appl. Math.* **15**, 367–380.

Kalnins. A. (1963). Free nonsymmetric vibrations of shallow spherical shells. *Proc. 4th U.S. Cong. Appl. Mech.*, 225–233. Reprinted in "Vibrations: Beams, Plates, and Shells" (A. Kalnins and C.L. Dym, eds.), Dowden, Hutchinson and Ross, Stroudsburg, Pennsylvania 1976.

Kinsler, L.E., Frey, A.R., Coppens, A.B., and Sanders, J.V. (1982). "Fundamentals of Acoustics," 3rd ed., Chapter 4. Wiley, New York.

Leissa, A.W. (1969). "Vibration of Plates," NASA SP-160, NASA, Washington, D.C.

Leissa, A.W., and Kadi, A.S. (1971). Curvature effects on shallow spherical shell vibrations, *J. Sound Vibr.* **16**, 173–187.

McIntyre, M.E., and Woodhouse, J. (1984/1985/1986). On measuring wood properties, Parts 1, 2, and 3, *J. Catgut Acoust. Soc.* No. 42, 11–25; No. 43, 18–24; and No. 45, 14–23.

Morse, P.M. (1948). "Vibration and Sound," Chapter 5. McGraw-Hill, New York. Reprinted 1976, Acoustical Soc. Am., Woodbury, New York.

Morse, P.M., and Ingard, K.U. (1968). Chapter 5. "Theoretical Acoustics," McGraw-Hill, New York. Reprinted 1986, Princeton Univ. Press, Princeton, New Jersey.

Rayleigh, Lord (1894). "The Theory of Sound," Vol. 1, 2nd ed., Chapters 9 and 10. Macmillan, New York. Reprinted by Dover, New York, 1945.

Reissner, E. (1946). On vibrations of shallow spherical shells. *J. Appl. Phys.* **17**, 1038–1042.

Reissner, E. (1955). On axi-symmetrical vibrations of shallow spherical shells. *Q. Appl. Math.* **13**, 279–290.

Rossing, T.D. (1982a). "The Science of Sound," Chapters 2 and 13. Addison-Wesley, Reading, Massachusetts.

Rossing, T.D. (1982b). The physics of kettledrums. *Scientific American* **247** (5), 172–178.

Rossing, T.D. (1982c). Chladni's law for vibrating plates. *American J. Physics* **50**, 271–274.

Rossing, T.D., and Fletcher, N.H. (1983). Nonlinear vibrations in plates and gongs. *J. Acoust. Soc. Am.* **73**, 345–351.

Snowdon, J.C. (1965). Mechanical impedance of free–free beams. *J. Acoust. Soc. Am.* **37**, 240–249.

Ver, I.L., and Holmer, C.I. (1971). Interaction of sound waves with solid structures. In "Noise and Vibration Control" (L.L. Beranek, ed.). McGraw-Hill, New York.

Waller, M.D. (1938). "Vibrations of free circular plates. Part I: Normal modes. *Proc. Phys. Soc.* **50**, 70–76.

Waller, M.D. (1949). Vibrations of free rectangular plates. *Proc. Phys. Soc. London* **B62**, 277–285.

Waller, M.D. (1950). Vibrations of free elliptical plates. *Proc. Phys. Soc. London* **B63**, 451–455.
Waller, M.D. (1961). "Chladni Figures: A study in Symmetry." Bell, London.
Warburton, G.B. (1954). The vibration of rectangular plates. *Proc. Inst. Mech. Eng.* **A168**, 371–384.
Wood Handbook (1974). Mechanical properties of wood. In "Wood Handbook: Wood as an Engineering Material." U.S. Forest Products Laboratory, Madison, Wisconsin.

CHAPTER 4

Coupled Vibrating Systems

4.1. Coupling Between Two Identical Vibrators

In Chapter 1, we considered the free vibrations of a two-mass system with three springs of equal stiffness; we found that there were two normal modes of vibration. Such a system could have been viewed as consisting of two separate mass/spring vibrators coupled together by the center spring (see Fig. 1.20). If the coupling spring were made successively weaker, the two modes would become closer and closer in frequency.

A similar behavior is exhibited by two simple pendulums connected by a spring, as in Fig. 4.1. Each pendulum, vibrating independently at small amplitude, has a frequency given by

$$f_0 = \frac{1}{2\pi}\sqrt{\frac{g}{l}},$$

where l is the length of the pendulum and g is the acceleration of gravity. The coupled system has two normal modes of vibration given by

$$f_1 = \frac{1}{2\pi}\sqrt{\frac{g}{l}} \quad \text{and} \quad f_2 = \frac{1}{2\pi}\sqrt{\left(\frac{g}{l} + \frac{2K_c}{m}\right)}, \tag{4.1}$$

where K_c is the spring constant and m is mass. The pendulums move in phase in the mode of lower frequency and in opposite phase in the mode of higher frequency.

The coupling can be expressed in terms of a coupling frequency $\omega_c = \sqrt{K_c/m}$, in which case $\omega_1^2 = \omega_0^2$ and $\omega_2^2 = \omega_0^2 + 2\omega_c^2$. If either pendulum is clamped in place, the other pendulum oscillates at a frequency $\omega^2 = \omega_0^2 + \omega_c^2$. This is not a normal mode frequency, however. If the system is started with pendulum A at its rest position and pendulum B in a displaced position, for example, the resulting motion changes with time. During each swing, pendulum B gives up some of its motion to pendulum A until pendulum B finds itself at rest and pendulum A has all the motion. Then, the process reverses. The exchange of energy between the two pendulums takes place at a rate that depends on the coupling frequency ω_c.

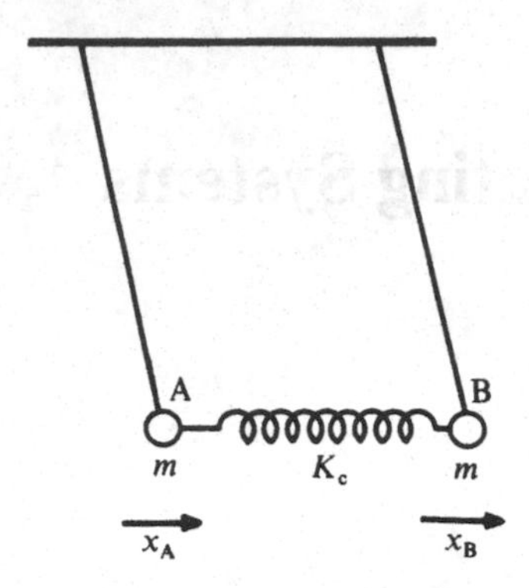

Fig. 4.1. Two simple pendulums coupled by a spring.

4.2. Normal Modes

Solution of the equations of motion to obtain the normal modes is relatively easy in this case. The equations of motion for small vibrations are

$$m\ddot{x}_A + \frac{mg}{l}x_A + K(x_A - x_B) = 0, \qquad (4.2a)$$

and

$$m\ddot{x}_B + \frac{mg}{l}x_B + K(x_B - x_A) = 0. \qquad (4.2b)$$

These can be rewritten in terms of ω_0 and ω_c, previously defined:

$$\ddot{x}_A + (\omega_0^2 + \omega_c^2)x_A - \omega_c^2 x_B = 0, \qquad (4.3a)$$

and

$$\ddot{x}_B + (\omega_0^2 + \omega_c^2)x_B - \omega_c^2 x_A = 0. \qquad (4.3b)$$

If the two equations are added together, we obtain

$$\frac{d^2}{dt^2}(x_A + x_B) + \omega_0^2(x_A + x_B) = 0. \qquad (4.4a)$$

If they are subtracted, we obtain

$$\frac{d^2}{dt^2}(x_A - x_B) + (\omega_0^2 + 2\omega_c^2)(x_A - x_B) = 0. \qquad (4.4b)$$

These are equations for simple harmonic motion. In the first, the variable is $(x_A + x_B)$, and the frequency is $\omega_0/2\pi$. In the second, the variable is $(x_A - x_B)$, and the frequency is $\sqrt{\omega_0^2 + 2\omega_c^2}/2\pi$. These represent the two normal modes described previously.

It is sometimes desirable to define normal coordinates q_1 and q_2 along which displacements can take place independently. In this case, $\mathbf{q}_1 = \mathbf{x}_A + \mathbf{x}_B$

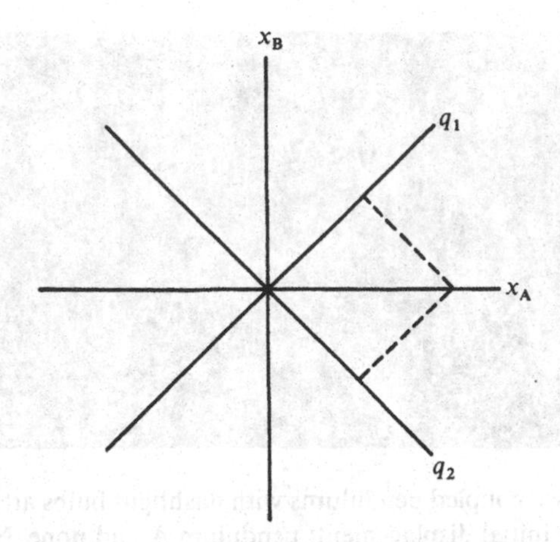

Fig. 4.2. Relationship of the normal coordinates (q_1, q_2) of two coupled oscillators to the individual coordinates (x_A, x_B).

and $\mathbf{q}_2 = \mathbf{x}_A - \mathbf{x}_B$; the normal coordinates are rotated 45° from the old, as shown in Fig. 4.2. Thus, the transformation from (x_A, x_B), the coordinates of the individual pendulum, to (q_1, q_2), the normal coordinates, can be found geometrically. In mode 1, the system oscillates along the coordinate q_1 with amplitude Q_1 and angular frequency ω_1, and in mode 2 it oscillates along q_2 with amplitude Q_2 and angular frequency ω_2. [The actual frequencies are given by Eq. (4.1).] Thus, x_A and x_B can be written as

$$x_A = \frac{Q_1}{\sqrt{2}} \cos \omega_1 t + \frac{Q_2}{\sqrt{2}} \cos \omega_2 t, \tag{4.5a}$$

and

$$x_B = \frac{Q_1}{\sqrt{2}} \cos \omega_1 t - \frac{Q_2}{\sqrt{2}} \cos \omega_2 t. \tag{4.5b}$$

Q_1 and Q_2 are determined by the initial displacements.

If the system is given an initial displacement $x_A(0) = A_0$, $x_B(0) = 0$, then $Q_1 = A_0/\sqrt{2}$, $Q_2 = A_0/\sqrt{2}$, and Eqs. (4.5) become

$$x_A = \frac{A_0}{2}(\cos \omega_1 t + \cos \omega_2 t), \tag{4.6a}$$

and

$$x_B = \frac{A_0}{2}(\cos \omega_1 t - \cos \omega_2 t). \tag{4.6b}$$

These equations can be rewritten:

$$x_A = A_0 \cos\left(\frac{\omega_2 - \omega_1}{2} t\right) \cos\left(\frac{\omega_2 + \omega_1}{2} t\right), \tag{4.7a}$$

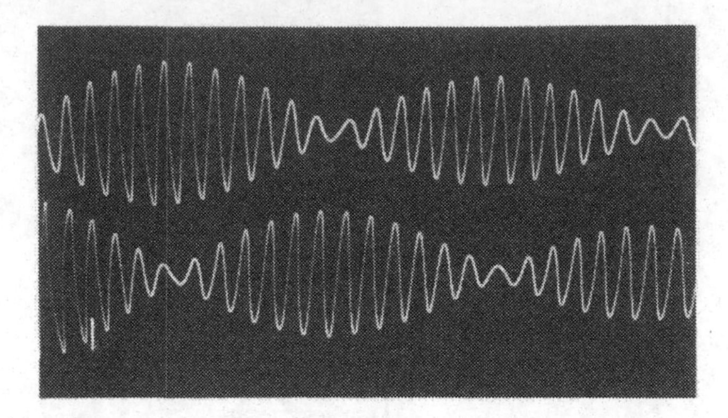

Fig. 4.3. Motion of two coupled pendulums with flashlight bulbs attached to the bobs. Pendulum B had an initial displacement; pendulum A had none. Note the exchange of kinetic energy at a rate ω_m/π, and also note the effect of damping (from French, 1971).

and

$$x_B = A_0 \sin\left(\frac{\omega_2 - \omega_1}{2}t\right) \sin\left(\frac{\omega_2 + \omega_1}{2}t\right). \tag{4.7b}$$

These can be interpreted as oscillations at a frequency $\bar{\omega} = (\omega_2 + \omega_1)/2$ with the amplitude modulated at a frequency $\omega_m = (\omega_2 - \omega_1)/2$. Note that the oscillations of the two pendulums at frequency $\bar{\omega}$ are 90° different in phase, and so are the oscillations in the amplitude at frequency ω_m. The photograph of two coupled pendulums in Fig. 4.3 illustrates this.

4.3. Weak and Strong Coupling

In the case we have been discussing, the average frequency and modulation frequency are

$$\bar{f} = \frac{\omega_2 + \omega_1}{4\pi} = \frac{1}{4\pi}\left(\sqrt{\omega_0^2 + 2\omega_c^2} + \omega_0\right),$$

and

$$f_m = \frac{\omega_2 - \omega_1}{4\pi} = \frac{1}{4\pi}\left(\sqrt{\omega_0^2 + 2\omega_c^2} - \omega_0\right). \tag{4.8}$$

In the weak coupling case, $\omega_c \ll \omega_0$, so we can write

$$\bar{f} = \frac{1}{4\pi}\left[\omega_0\left(1 + \frac{2\omega_c^2}{\omega_0^2}\right)^{1/2} + \omega_0\right] \simeq \frac{\omega_0}{2\pi}\left(1 + \frac{\omega_c^2}{2\omega_0^2}\right), \tag{4.9a}$$

and

$$f_m = \frac{1}{4\pi}\left[\omega_0\left(1 + \frac{2\omega_c^2}{\omega_0^2}\right)^{1/2} - \omega_0\right] \simeq \frac{\omega_c^2}{4\pi\omega_0}. \tag{4.9b}$$

When the coupling is weak, we can neglect the energy stored in the coupling

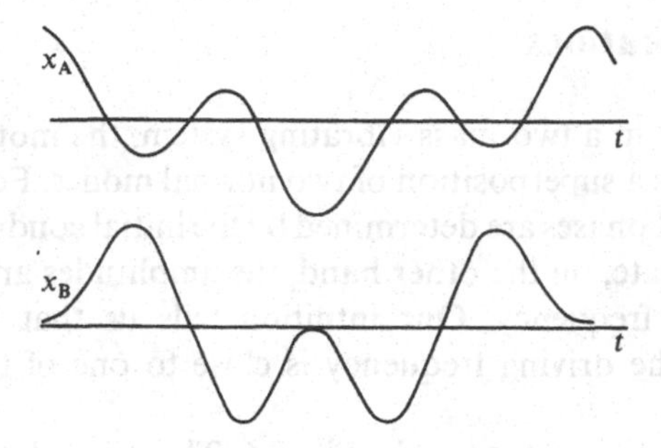

Fig. 4.4. Motion of two strongly coupled pendulums with $\omega_c = 2\omega_0$.

spring and characterize the motion as an interchange of energy between pendulum A and pendulum B at a rate given by f_m. The pendulum whose amplitude is increasing is the one that is lagging in phase by 90°, as expected for a vibrator absorbing power from a driving force at resonance. The two pendulums alternate as the driver and the driven. When it comes to rest, the driver suddenly changes phase by 180° so that it can become the driven vibrator.

In the case where $\omega_c = 2\omega_0$ (a strong coupling), $\bar{f} = 2f_0$ and $f_m = f_0$. The modulation frequency is half the average frequency, so the excursions are alternately large and small. The kinetic energy is exchanged rapidly between the two bobs, as shown in Fig. 4.4. The normal mode frequencies in this case are in a 3 : 1 ratio: $f_1 = f_0$ and $f_2 = 3f_0$. The initial conditions that give the motion in Fig. 4.4 are $x_A(0) = A_0$ and $x_B(0) = 0$; that is, bob B is held at its rest point while bob A is displaced and both are released together.

Note that the motions shown in Fig. 4.4 are described by Eqs. (4.6a) and (4.6b), but they could also be obtained graphically from Fig. 4.2 by marking off appropriate time intervals between $A_0/2$ and $-A_0/2$ along both the q_1 and q_2 axes. The displacements x_A and x_B at each time are found by projecting $q_1(t)$ and $q_2(t)$ on the appropriate axis and adding them.

In the case of two coupled pendulums, the lower frequency ω_1 is independent of the coupling strength; this is not true of all coupled oscillations. What is generally true, however, is that the separation between ω_1 and ω_2 increases with the coupling strength, as shown in Fig. 4.5.

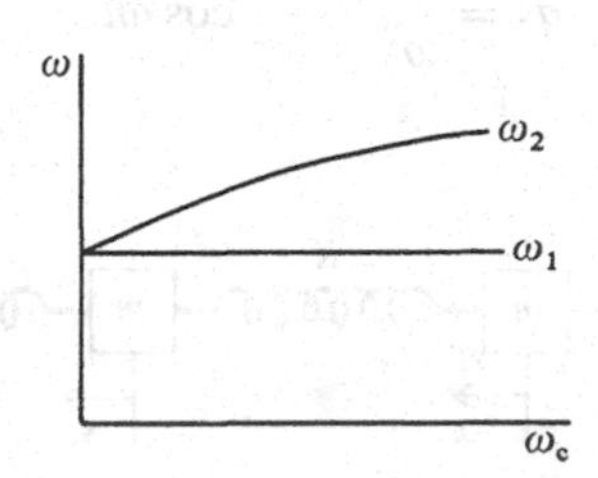

Fig. 4.5. Dependence of the mode frequencies on the coupling strength for two coupled pendulums.

4.4. Forced Vibrations

We have seen how in a two-mass vibrating system, the motion of each mass can be described as a superposition of two normal modes. For free vibrations, the amplitudes and phases are determined by the initial conditions. In a system driven in steady state, on the other hand, the amplitudes and phases depend upon the driving frequency. Our intuition tells us that large amplitudes will occur when the driving frequency is close to one of the normal mode frequencies.

Consider the two-mass system in Fig. 4.6. The normal mode angular frequencies are $\omega_1 = \sqrt{K/m} = \omega_0$ and $\omega_2 = \sqrt{(K/m) + (2K_c/m)} = \sqrt{\omega_0^2 + 2\omega_c^2}$. Suppose that a driving force $F_0 \cos \omega t$ is applied to mass A. The equations of motion are

$$\ddot{x}_A + (\omega_0^2 + \omega_c^2)x_A - \omega_c^2 x_B = \frac{F_0}{m}\cos \omega t, \tag{4.10a}$$

and

$$\ddot{x}_B + (\omega_0^2 + \omega_c^2)x_B - \omega_c^2 x_A = 0. \tag{4.10b}$$

Again, we introduce normal coordinates $q_1 = x_A + x_B$ and $q_2 = x_A - x_B$; we add and subtract Eqs. (4.10a) and (4.10b) to obtain

$$\ddot{q}_1 + \omega_1^2 q_1 = \frac{F_0}{m}\cos \omega t, \tag{4.11a}$$

and

$$\ddot{q}_2 + \omega_2^2 q_2 = \frac{F_0}{m}\cos \omega t, \tag{4.11b}$$

where $\omega_1^2 = \omega_0^2$ and $\omega_2^2 = \omega_0^2 + 2\omega_c^2$. Note that the same driving force appears in both normal mode equations.

These are the equations of two harmonic oscillators with natural frequencies ω_1 and ω_2 and with no damping. The steady-state solutions are [see Eq. (1.39)]

$$q_1 = \frac{F_0/m}{\omega_1^2 - \omega^2}\cos \omega t, \tag{4.12a}$$

and

$$q_2 = \frac{F_0/m}{\omega_2^2 - \omega^2}\cos \omega t. \tag{4.12b}$$

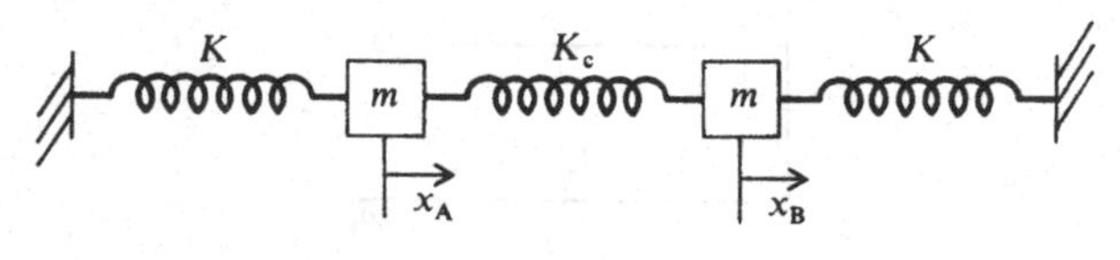

Fig. 4.6. Two-mass oscillator (compare Fig. 4.1.).

The displacements of the masses A and B are

$$x_A = \frac{q_1 + q_2}{2} = \frac{F_0}{2m} \frac{\omega_1^2 - \omega^2 + \omega_2^2 - \omega^2}{(\omega_1^2 - \omega^2)(\omega_2^2 - \omega^2)} \cos \omega t$$

$$= \frac{F_0}{m} \frac{(\omega_1^2 + \omega_c^2) - \omega^2}{(\omega_1^2 - \omega^2)(\omega_2^2 - \omega^2)} \cos \omega t, \tag{4.13a}$$

and

$$x_B = \frac{q_1 - q_2}{2} = \frac{F_0}{2m} \frac{\omega_2^2 - \omega^2 - \omega_1^2 + \omega^2}{(\omega_1^2 - \omega^2)(\omega_2^2 - \omega^2)} \cos \omega t$$

$$= \frac{F_0}{m} \frac{\omega_c^2}{(\omega_1^2 - \omega^2)(\omega_2^2 - \omega^2)} \cos \omega t. \tag{4.13b}$$

The steady-state displacement amplitudes as functions of ω, from Eq. (4.13), are shown in Fig. 4.7(a). Below the first resonance ω_1, the displacements x_A and x_B are in the same direction, whereas above the second resonance ω_2, they are in opposite directions. With no damping, the amplitudes approach infinity at the resonance and the phase jumps abruptly. If damping were included, the phase would vary smoothly as one goes through the resonances.

Note that at a frequency ω_A lying between ω_1 and ω_2, x_A goes to zero. This is called an antiresonance, and it occurs at the natural frequency at which mass B would oscillate if mass A were fixed. We can easily calculate this frequency using Eq. (4.13a). x_A goes to zero when

$$\omega^2 = \omega_1^2 + \omega_c^2 = \omega_A^2 = \omega_2^2 - \omega_c^2. \tag{4.14}$$

At this frequency, $\omega_1^2 - \omega^2 = -\omega_c^2$ and $\omega_2^2 - \omega^2 = \omega_c^2$, so Eq. (4.13b) gives the result

$$\left.\frac{x_B}{\cos \omega t}\right|_{\omega_A} = \frac{F_0}{m} \frac{\omega_c^2}{(-\omega_c^2)(\omega_c^2)} = \frac{-F_0}{m} \frac{1}{\omega_c^2} = \frac{-F_0}{K_c}.$$

This result appears somewhat paradoxical, for it says that at $\omega = \omega_A$ the driven mass does not move but the other mass does. There are several ways to deal with this paradox (for example, any real driving force cannot have a single frequency ω, because this would imply that it has existed since $t = -\infty$), but it is perhaps best to note that the paradox does not exist in a real system with damping, however small. In such a system, x_A and x_B both have minimum values at ω_A, but neither one is zero.

The response of a similar system with damping is shown in Fig. 4.8. If ω_1 and ω_2 are separated by an amount that is several times greater than their line widths, then the two normal modes are excited independently at these frequencies. Below ω_1, both masses move in phase with the driving force. At ω_1, they lag the driving force by 90° (see Fig. 1.8), and above ω_1, the phase difference is 180°. Between ω_A and ω_2, mass A again moves in phase with the driving force, while above ω_2 mass B does so.

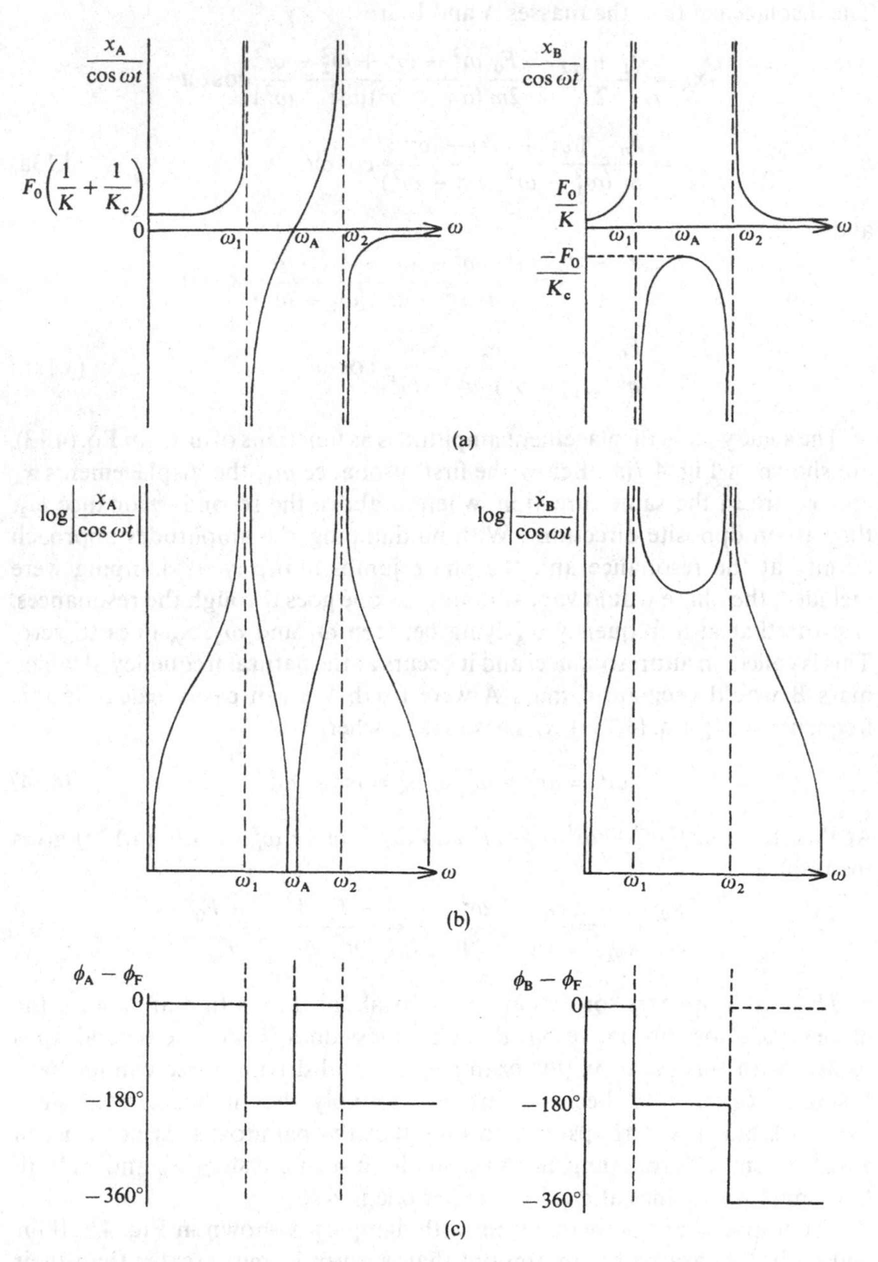

Fig. 4.7. Frequency response of the two-mass coupled system in Fig. 4.6 to a driving force applied to mass A. (a) Amplitudes of x_A and x_B. (b) Absolute values of amplitude on a logarithmic scale. (c) Phases of masses A and B with respect to the driving force.

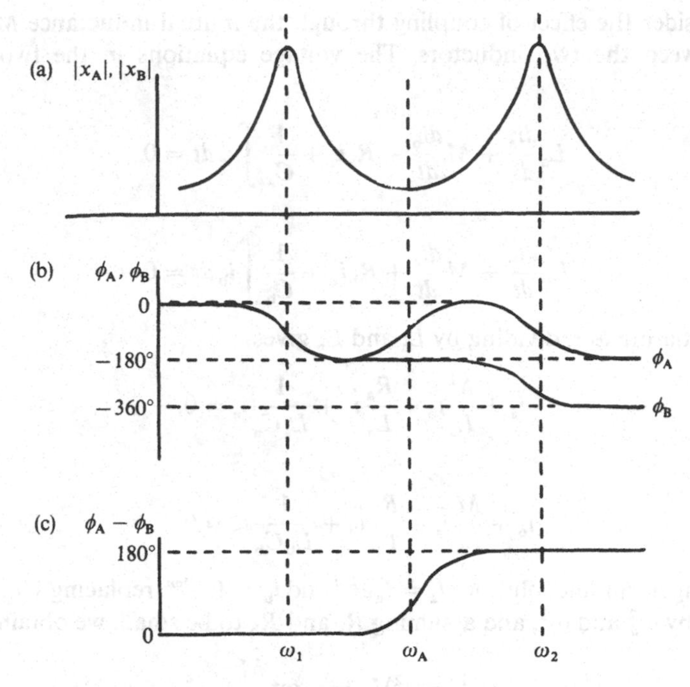

Fig. 4.8. (a) Amplitude of mass A or mass B as functions of driving frequency when a force is applied to mass A. (b) Phases of masses A and B with respect to the driving force. (c) Phase difference between x_A and x_B.

4.5. Coupled Electrical Circuits

We have already seen in Chapter 1 the usefulness of equivalent electrical circuits in understanding mechanical oscillatory systems. We will now consider some examples of coupled electrical circuits that are the bases of useful equivalent circuits.

The electrical circuit shown in Fig. 4.9 consists of two *RLC* circuits having natural frequencies ω_a and ω_b when there is no coupling:

$$\omega_a = \frac{1}{\sqrt{L_a C_a}} \quad \text{and} \quad \omega_b = \frac{1}{\sqrt{L_b C_b}}.$$

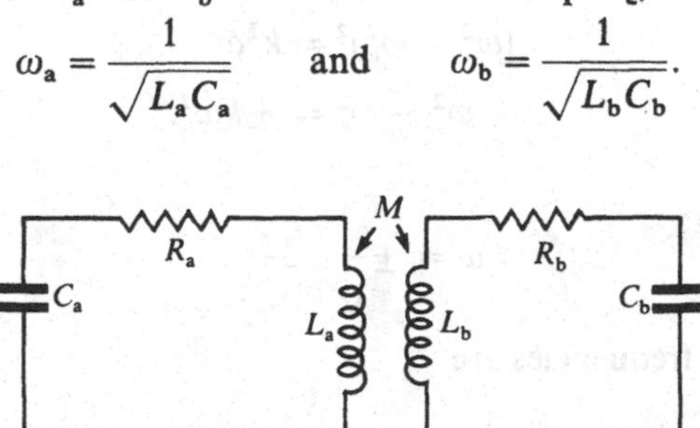

Fig. 4.9. Two *RLC* circuits coupled by mutual inductance *M*.

We consider the effect of coupling through the mutual inductance M existing between the two inductors. The voltage equations in the two loops are

$$L_a \frac{di_a}{dt} + M\frac{di_b}{dt} + R_a i_a + \frac{1}{C_a}\int i_a\, dt = 0,$$

and (4.15)

$$L_b \frac{di_b}{dt} + M\frac{di_a}{dt} + R_b i_b + \frac{1}{C_b}\int i_b\, dt = 0.$$

Differentiating and dividing by L_a and L_b gives

$$\ddot{i}_a + \frac{M}{L_a}\ddot{i}_b + \frac{R_a}{L_a}\dot{i}_a + \frac{1}{L_a C_a} i_a = 0,$$

and

$$\ddot{i}_b + \frac{M}{L_b}\ddot{i}_a + \frac{R_b}{L_b}\dot{i}_b + \frac{1}{L_b C_b} i_b = 0.$$

Assuming harmonic solutions $i_a = I_a e^{j\omega t}$ and $i_b = I_b e^{j\omega t}$, replacing $1/L_a C_a$ and $1/L_b C_b$ by ω_a^2 and ω_b^2, and assuming R_a and R_b to be small, we obtain

$$(\omega^2 - \omega_a^2)I_a = -\omega^2 \frac{M}{L_a} I_b,$$

and (4.16)

$$(\omega^2 - \omega_b^2)I_b = -\omega^2 \frac{M}{L_b} I_a.$$

Multiplying these two equations together gives

$$(\omega^2 - \omega_a^2)(\omega^2 - \omega_b^2) = \omega^4 \frac{M^2}{L_a L_b} = k^2\omega^4, \tag{4.17}$$

where $k^2 = M^2/L_a L_b$ is the coupling coefficient. Solving for ω gives the resonance frequencies.

A particularly simple case occurs when $\omega_a^2 = \omega_b^2$. Then,

$$(\omega^2 - \omega_a^2)^2 = k^2\omega^4,$$

$$\omega^2 - \omega_a^2 = \pm k\omega^2,$$

and

$$\omega = \pm\frac{\omega_a}{\sqrt{1 \pm k}}.$$

The two positive frequencies are

$$\omega_1 = \frac{\omega_a}{\sqrt{1+k}} \quad \text{and} \quad \omega_2 = \frac{\omega_a}{\sqrt{1-k}} \tag{4.18}$$

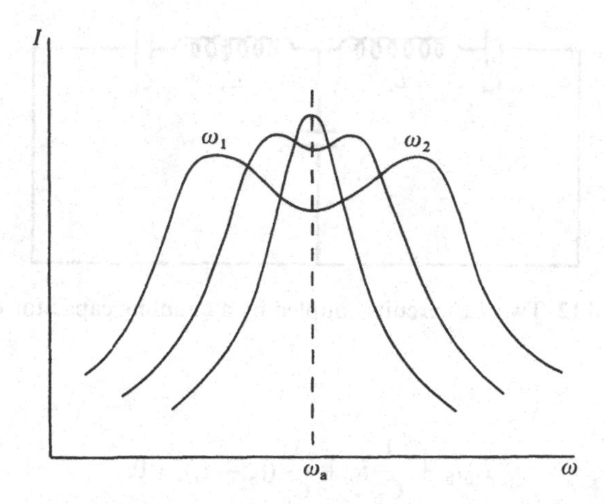

Fig. 4.10. Behavior of the circuit of Fig. 4.9 for three values of the coupling constant k.

The current amplitudes for three different values of coupling are shown in Fig. 4.10. With tight coupling, a pronounced dip occurs between the peaks at ω_1 and ω_2.

If the two circuits are identical ($L_a = L_b = L$ and $C_a = C_b = C$),

$$\omega_1 = \frac{1}{\sqrt{C(L+M)}} \quad \text{and} \quad \omega_2 = \frac{1}{\sqrt{C(L-M)}}. \tag{4.19}$$

These values are shown in Fig. 4.11.

Next, we consider the circuit shown in Fig. 4.12, which consists of two LC circuits plus a coupling capacitor C_c. The differential equations are

$$L_a \ddot{i}_a + \frac{1}{C_a} i_a + \frac{1}{C_c}(i_a - i_b) = 0,$$

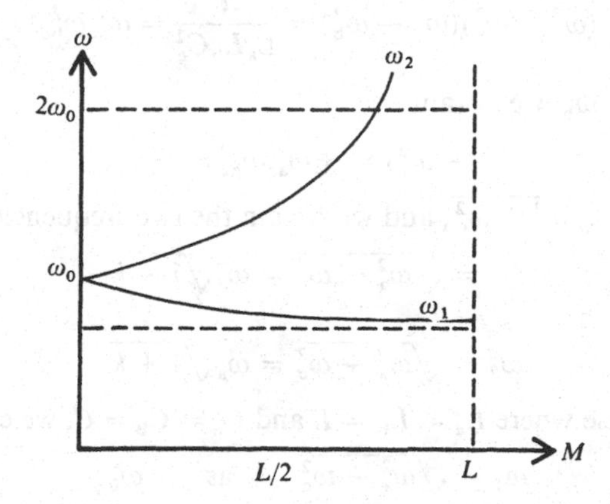

Fig. 4.11. Variation of modal frequencies with mutual inductance M of identical coupled circits.

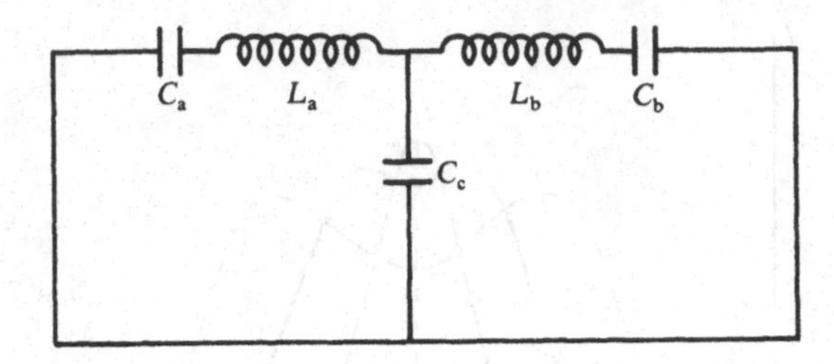

Fig. 4.12. Two LC circuits coupled by a coupling capacitor C_c.

and

$$L_b \ddot{i}_b + \frac{1}{C_b} i_b + \frac{1}{C_c}(i_b - i_a) = 0. \tag{4.20}$$

Again, assuming harmonic solutions, $i_a = I_1 e^{j\omega t}$ and $i_b = I_2 e^{j\omega t}$, leads to

$$-\omega^2 I_a + \frac{1}{L_a}\left(\frac{1}{C_a} + \frac{1}{C_c}\right) I_a - \frac{1}{L_a C_c} I_b = 0,$$

and

$$-\omega^2 I_b + \frac{1}{L_b}\left(\frac{1}{C_b} + \frac{1}{C_c}\right) I_b - \frac{1}{L_b C_c} I_a = 0,$$

from which

$$(\omega^2 - \omega_a^2) I_a = -\omega_{ac}^2 I_b,$$

and (4.21)

$$(\omega^2 - \omega_b^2) I_b = -\omega_{bc}^2 I_a,$$

where $\omega_a^2 = (1/L_a C_a) + (1/L_a C_c)$, $\omega_b^2 = (1/L_b C_b) + (1/L_b C_c)$, $\omega_{ac}^2 = 1/L_a C_c$, and $\omega_{bc}^2 = 1/L_b C_c$. Multiplying the above equations together gives

$$(\omega^2 - \omega_a^2)(\omega^2 - \omega_b^2) = \frac{1}{L_a L_b C_c^2} = \omega_{ac}^2 \omega_{bc}^2. \tag{4.22}$$

In the case $\omega_a = \omega_b$, we obtain

$$(\omega^2 - \omega_a^2) = \pm \omega_{ac}\omega_{bc} = \pm \omega_c^2,$$

from which $\omega = \sqrt{\omega_a^2 \pm \omega_c^2}$, and we obtain the two frequencies

$$\omega_1 = \sqrt{\omega_a^2 - \omega_c^2} = \omega_a \sqrt{1 - k},$$

and (4.23)

$$\omega_2 = \sqrt{\omega_a^2 + \omega_c^2} = \omega_a \sqrt{1 + k}.$$

In the special case where $L_a = L_b = L$ and $C_a = C_b = C$, we can designate

$$\omega_1 = \sqrt{\omega_a^2 - \omega_c^2} \quad \text{as} \quad \omega_0,$$

so (4.24)

$$\omega_2 = \sqrt{\omega_0^2 + 2\omega_c^2},$$

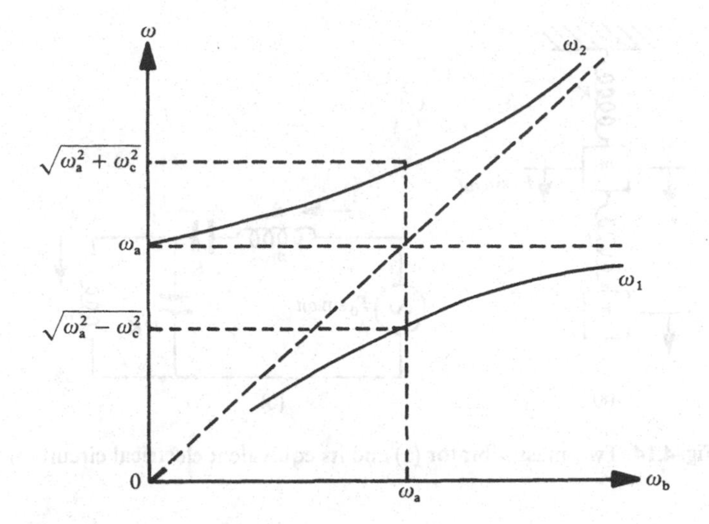

Fig. 4.13. Normal mode frequencies of the circuit in Fig. 4.12. The dashed lines indicate the uncoupled case ($k \to 0$ or $C_c \to \infty$).

as in the mechanical systems shown in Figs. 4.1 and 4.6. Thus, the circuit in Fig. 4.12 is the electrical analogue for both these systems, where the coupling capacitor C_c takes the place of the spring K_c.

It is instructive to plot ω as a function of ω_b from Eq. (4.22). When $\omega_c = 0$ (no coupling), we obtain two straight lines (the dashed lines in Fig. 4.13). For $\omega_c > 0$, we obtain two curves that approach the dashed lines asymptotically. At $\omega_b = \omega_a$, $\omega = \sqrt{\omega_a^2 \pm \omega_c^2}$ as in Eq. (4.23). Note that in both limits $\omega_b \ll \omega_a$ and $\omega_b \gg \omega_a$ the normal mode frequencies ω_1 and ω_2 approach those of the uncoupled modes ω_a and ω_b, and the normal modes resemble those of the uncoupled LC circuits.

4.6. Forced Vibration of a Two-Mass System

A coupled system with wide application is the two-mass system in which a sinusoidal driving force is applied to mass m_1, as shown in Fig. 4.14. This system is the prototype, for example, of a bass reflex loudspeaker system, a guitar with the ribs fixed, and the dynamic absorber used to damp machine vibrations. The equivalent electrical circuit is also shown in Fig. 4.14.

The equations of motion for the two-mass system in Fig. 4.14 are

$$m_1\ddot{x}_1 + K_1x_1 + K_2(x_1 - x_2) = F_0 \sin \omega t,$$

and

$$m_2\ddot{x}_2 + K_2(x_2 - x_1) = 0.$$

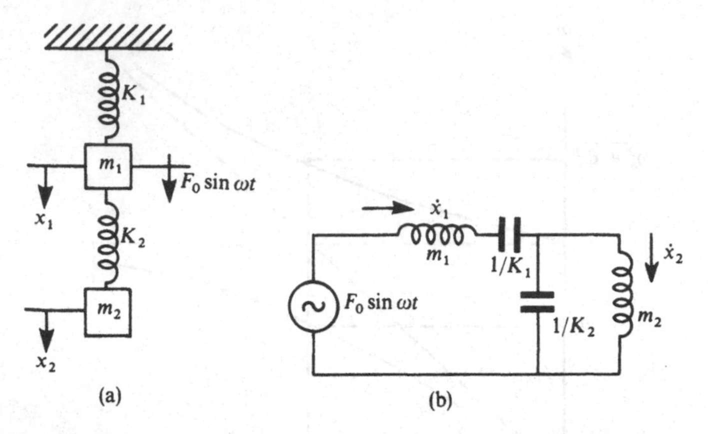

Fig. 4.14. Two-mass vibrator (a) and its equivalent electrical circuit (b).

Solutions of these equations lead to expressions for the displacement amplitudes:

$$X_1 = \frac{(K_2 - \omega^2 m_2)F_0}{(K_1 + K_2 - \omega^2 m_1)(K_2 - \omega^2 m_2) - K_2^2}$$

$$= \frac{F_0(\omega_2^2 - \omega^2)}{m_1[\omega_1^2(1 + K_2/K_1) - \omega^2](\omega_2^2 - \omega^2) - K_2\omega_2^2}, \tag{4.25a}$$

and

$$X_2 = \frac{k_2 F_0}{(K_1 + K_2 - \omega^2 m_1)(K_2 - \omega^2 m_2) - K_2^2}$$

$$= \frac{F_0\omega_2^2}{m_1[\omega_1^2(1 + K_2/K_1) - \omega^2](\omega_2^2 - \omega^2) - K_2\omega_2^2}. \tag{4.25b}$$

Like the two-mass system discussed in Section 4.4, this system has two resonances and one antiresonance. X_1 goes to zero at the antiresonance frequency $\omega = \omega_2$, and both X_1 and X_2 approach infinity when the denominator goes to zero.

A dynamic absorber consists of a small mass m_2 and spring K_2 attached to the primary vibrating system and selected so that $\omega_1 = \omega_2$. Then, the amplitude of mass m_1 goes to zero at the original resonance frequency ω_1. Similarly, a bass reflex speaker is often designed so that the resonance frequency ω_2 of the enclosure is the same as that of the loudspeaker cone ω_1. Then, the cone is restrained from moving at its resonance frequency.

In a guitar or violin, m_1 and K_1 represent the mass and spring constant of the top plate, m_2 is the effective mass of the air in the sound hole, and K_2 is the spring constant of the enclosed air. In most instruments, ω_2 is substantially less than ω_1, however.

4.7. Systems with Many Masses

Consider the system in Fig. 4.15 with four masses and five springs. External forces f_i may act on any or all of the masses, so we have four equations of motion:

$$m_1\ddot{x}_1 + K_1x_1 - K_2(x_2 - x_1) = F_1,$$

$$m_2\ddot{x}_2 + K_2(x_2 - x_1) - K_3(x_3 - x_2) = F_2,$$

$$m_3\ddot{x}_3 + K_3(x_3 - x_2) - K_4(x_4 - x_3) = F_3,$$

and

$$m_4\ddot{x}_4 + K_4(x_4 - x_3) + K_5x_4 = F_4. \tag{4.26}$$

Adding the four equations gives

$$\sum_{i=1}^{4} m_i\ddot{x}_i = \sum_{i=1}^{4} F_i - K_1x_1 - K_5x_4. \tag{4.27}$$

Assuming harmonic solutions, $x_i = X_i \sin \omega t$ leads to four algebraic equations, in which the coefficients of X_1 to X_4 form the determinant:

$$\Delta = \begin{vmatrix} d_1 & -K_2 & 0 & 0 \\ -K_2 & d_2 & -K_3 & 0 \\ 0 & 0 & d_3 & -K_4 \\ 0 & 0 & -K_4 & d_4 \end{vmatrix},$$

where $d_1 = K_1 + K_2 - m_1\omega^2$, $d_2 = K_2 + K_3 - m_2\omega^2$, $d_3 = K_3 + K_4 - m_3\omega^2$, and $d_4 = K_4 + K_5 - m_4\omega^2$. The amplitude X_1 is given by

$$X_1 = \frac{1}{\Delta}\begin{vmatrix} F_1 & -K_2 & 0 & 0 \\ F_2 & d_2 & -K_3 & 0 \\ F_3 & -K_3 & d_3 & -K_4 \\ F_4 & 0 & -K_4 & d_4 \end{vmatrix},$$

and the other three amplitudes are given by analogous expressions (Jacobson and Ayre, 1958).

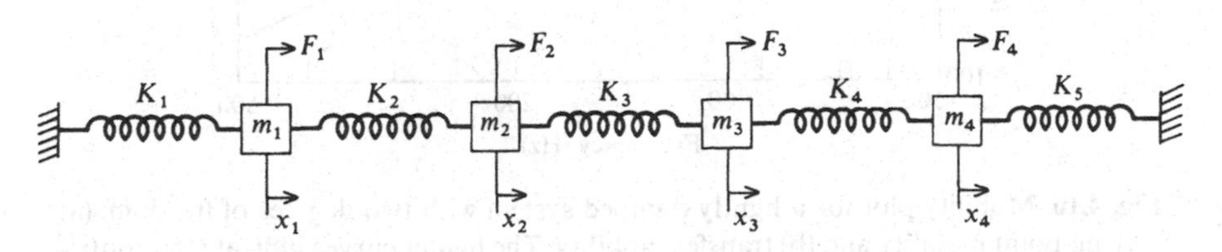

Fig. 4.15. Four-mass vibrating system.

4.8. Graphical Representation of Frequency Response Functions

There are several ways to represent the frequency response function of a vibrating system (see the appendix to this chapter). One useful representation is a mobility plot illustrated in Fig. 4.16 for a lightly damped system with two degrees of freedom [a linear two-mass system as in Fig. 1.20(a), for example]. If a force F is applied to mass m_1, the driving-point mobility is $\dot{x}_1/F$ and the transfer mobility is $\dot{x}_2/F$.

Note that individual mode curves in Fig. 4.16 have slopes of ± 6 dB/octave above and below resonance, as discussed in Appendix A.1. Their phases relative to the driving force are not indicated, however. In Fig. 4.16(a), both curves have the same phase (and thus are additive) below the first resonance

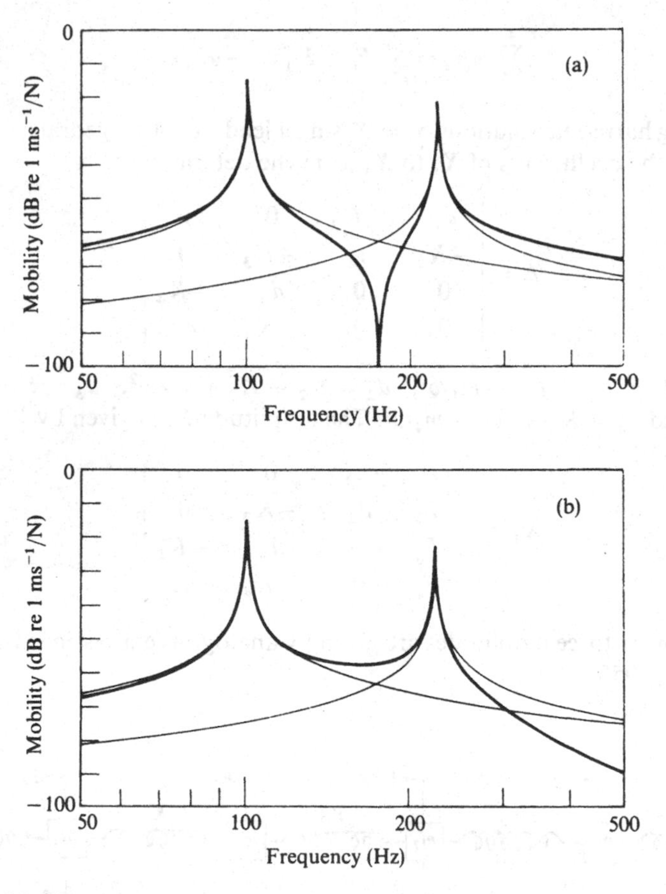

Fig. 4.16. Mobility plot for a lightly damped system with two degrees of freedom: (a) driving-point mobility and (b) transfer mobility. The lighter curves indicate the contributions from the two normal modes of the system (after Ewins, 1984).

and above the second one. Between the two resonances, they are subtractive, however, which leads to an antiresonance where they cross. The opposite situation occurs in the transfer mobility curves in Fig. 4.16(b), where the curves are additive between the resonances but subtractive elsewhere. [This same behavior is noted in Fig. 4.7(b).]

There is a general rule that if two consecutive modes have modal constants with the same sign, then there will be an antiresonance at some frequency between the natural frequencies of the two modes. If they have opposite signs, there will not be an antiresonance, but just a minimum, as in Fig. 4.16(b). The modal constant is the product of two eigenvector elements, one at the drive point and one at the response point. In the case of driving-point mobility, the two points are the same, so the modal constant must be positive. This leads to alternating resonances and antiresonances, as appear in the driving-point admittance (mobility) curves in Fig. 3.22 and also in the driving-point impedance (reciprocal of mobility) curve in Fig. 3.21.

Figure 4.17 shows an example of a transfer mobility curve of a system with four degrees of freedom (four normal modes). The compliance of this structure is positive over part of the frequency range and negative over part of it. The modal constant in this example changes sign from the first to the second mode (resulting in an antiresonance) but not from the second to the third mode for the driving and the observation points selected.

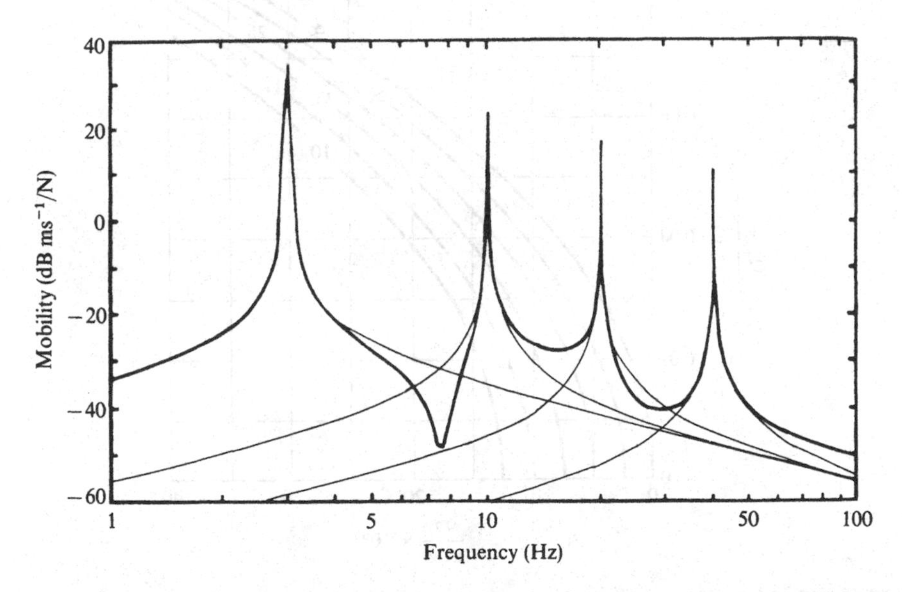

Fig. 4.17. Example of a transfer mobility curve for a lightly damped structure with four degrees of freedom. The lighter curves indicate the contributions from the four normal modes, which contribute to the frequency response function (after Ewins, 1984).

4.9. Vibrating String Coupled to a Soundboard

In Section 2.12, we considered a string terminated by a nonrigid end support (such as the bridge of a string instrument). If the support is masslike, the resonance frequencies will be raised slightly (as if the string were shortened). On the other hand, if the support is springlike, the resonance frequencies will be lowered (as if the string were lengthened).

When the string is terminated at a structure that has resonances of its own (such as the soundboard of a piano or the top plate of a guitar), the situation becomes a little more complicated. Below each resonance of the structure, the termination appears springlike, while above each resonance it appears masslike. Thus, the structural resonances tend to push the string resonances away from the structural resonance frequency, or to split each string resonance into two resonances, one above and one below the soundboard resonance. More correctly described, two new modes of vibration have been created, both of which are combinations of a string mode and a structural mode. This is similar to the behavior of the coupled mechanical and electrical vibrating systems we have considered (see Fig. 4.10 and 4.11).

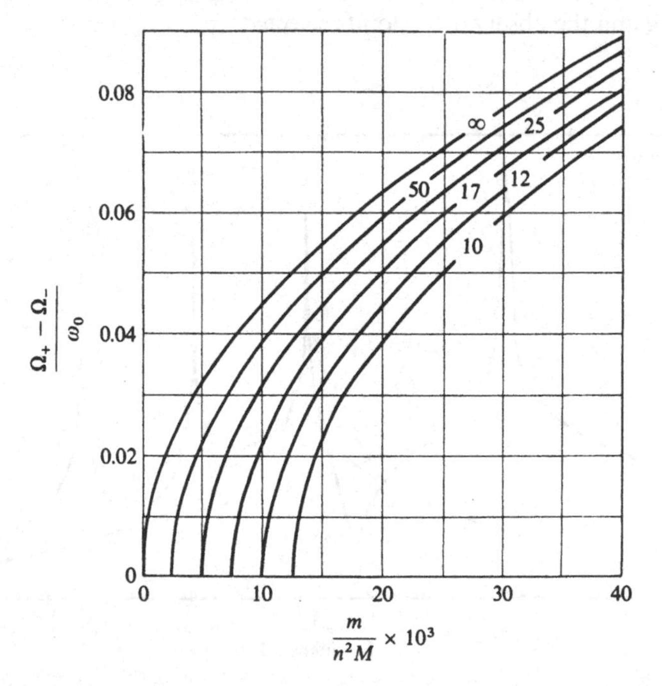

Fig. 4.18. Normal mode splitting as a function of the ratio of string mass m to soundboard mass M times the mode number n on the string when the resonance frequencies of the string and soundboard mode are the same. Ω_+ and Ω_- are the mode angular frequencies in the coupled system, and Q-values for the soundboard appear on each curve (Gough, 1981).

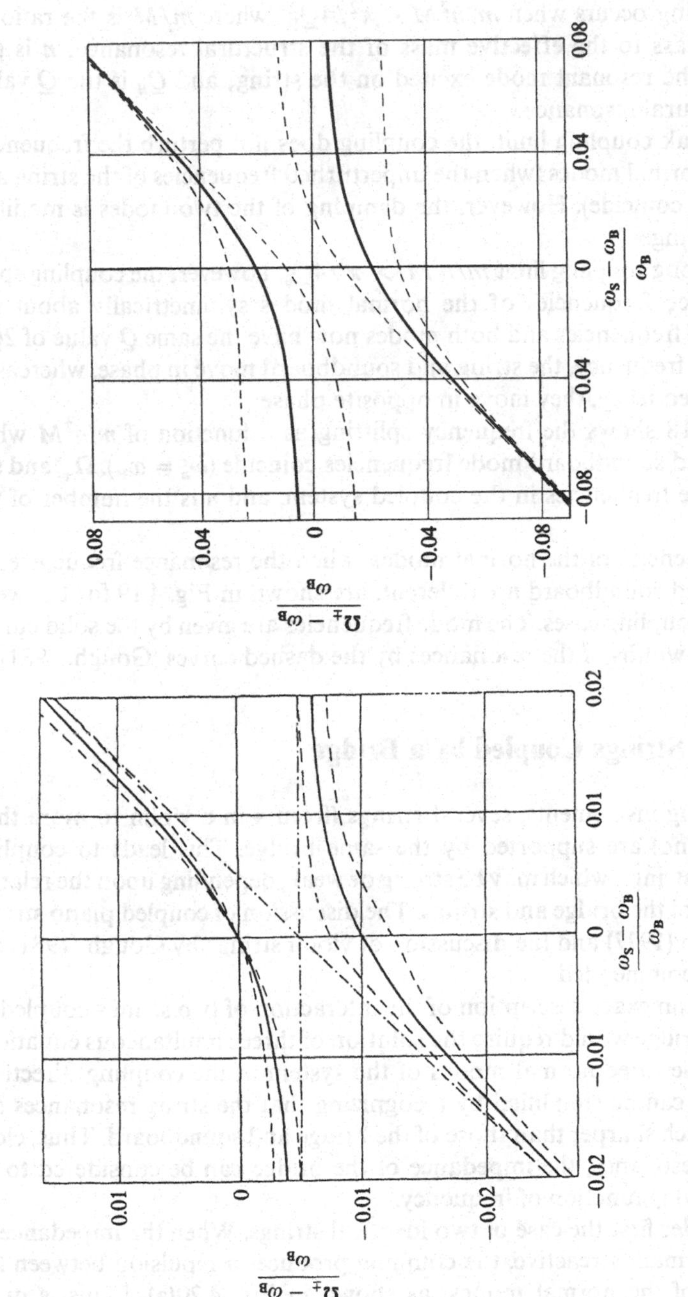

Fig. 4.19. Normal mode frequencies of a string coupled to a soundboard as a function of their uncoupled frequencies ω_s and ω_B: (a) weak coupling and (b) strong coupling (Gough, 1981).

Gough (1981) discusses systems both with weak coupling and strong coupling when $\omega_n = \omega_B$ (string and structural resonances at the same frequency). Weak coupling occurs when $m/n^2 M < \pi^2/4Q_B^2$, where m/M is the ratio of the string mass to the effective mass of the structural resonance, n is the number of the resonant mode excited on the string, and Q_B is the Q value of the structural resonance.

In the weak coupling limit, the coupling does not perturb the frequencies of the two normal modes (when the unperturbed frequencies of the string and soundboard coincide). However, the damping of the two modes is modified by the coupling.

In the strong coupling limit $m/n^2M > \pi^2/4Q_B^2$, however, the coupling splits the resonance frequencies of the normal modes symmetrically about the unperturbed frequencies and both modes now have the same Q value of $2Q_B$. At the lower frequency, the string and soundboard move in phase, whereas at the higher frequency, they move in opposite phase.

Figure 4.18 shows the frequency splitting as a function of m/n^2M when the string and soundboard mode frequencies coincide ($\omega_n = \omega_B$). Ω_+ and Ω_- are the mode frequencies in the coupled system, and n is the number of the string mode.

The frequencies of the normal modes, when the resonance frequencies of the string and soundboard are different, are shown in Fig. 4.19 for the weak and strong coupling cases. The mode frequencies are given by the solid curves and the half-widths of the resonances by the dashed curves (Gough, 1981).

4.10. Two Strings Coupled by a Bridge

In most string instruments, several strings (from 4 in a violin to more than 200 in a piano) are supported by the same bridge. This leads to coupling between the strings, which may be strong or weak, depending upon the relative impedances of the bridge and strings. The discussion of coupled piano strings by Weinreich (1977) and the discussion of violin strings by Gough (1981) are especially recommended.

Although an exact description of the interaction of two strings coupled to a common bridge would require the solution of three simultaneous equations describing the three normal modes of the system in the coupling direction, the problem can be simplified by recognizing that the string resonances are generally much sharper than those of the bridge and soundboard. Thus, close to a string resonance, the impedance of the bridge can be considered to be a slowly varying function of frequency.

We consider first the case of two identical strings. When the impedance of the bridge is mainly reactive, the coupling produces a repulsion between the frequencies of the normal modes, as shown in Fig. 4.20(a). [This is quite similar to the behavior of the string–soundboard coupling in Fig. 4.19(b).] When the impedance of the bridge is mainly resistive, however (as it will be at

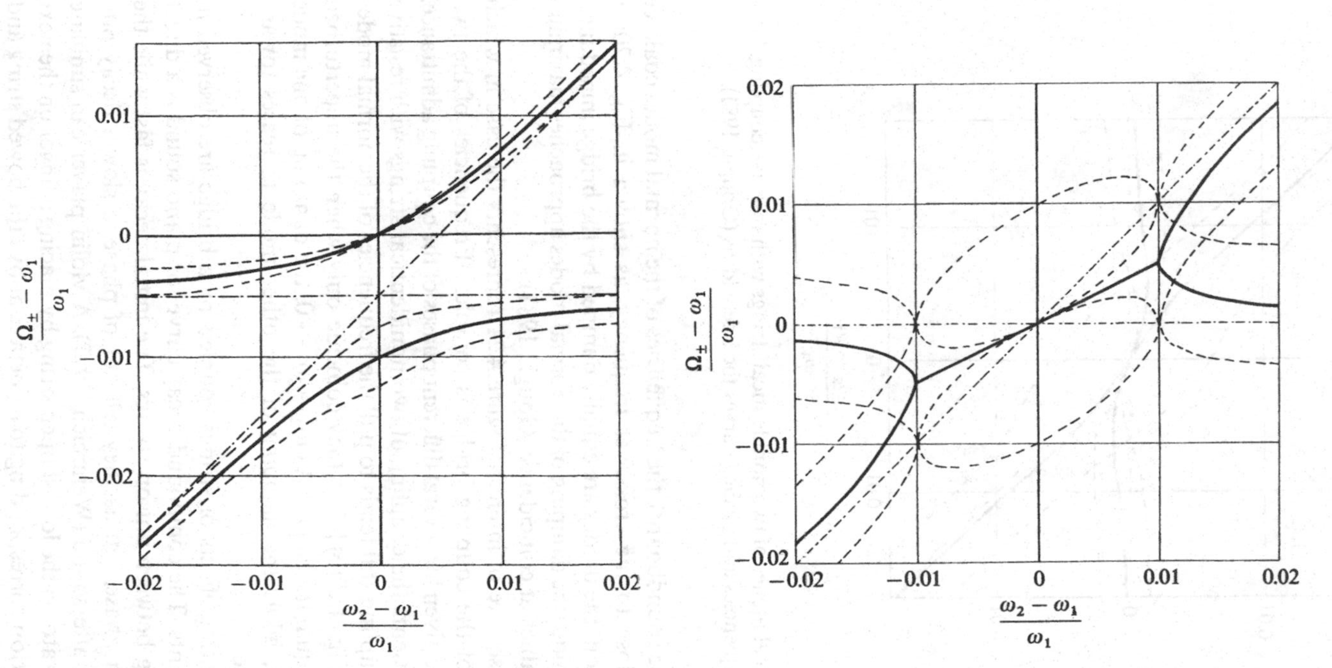

Fig. 4.20. The normal modes of two identical strings coupled together by a bridge: (a) reactive coupling and (b) resistive coupling. ω_2 and ω_1 are the unperturbed string frequencies; Ω_+ and Ω_- are the frequencies of the coupled system. The solid curves are the frequencies and the dashed curves give the half-widths. The dash–dot lines give the frequencies of the strings without coupling (Gough, 1981).

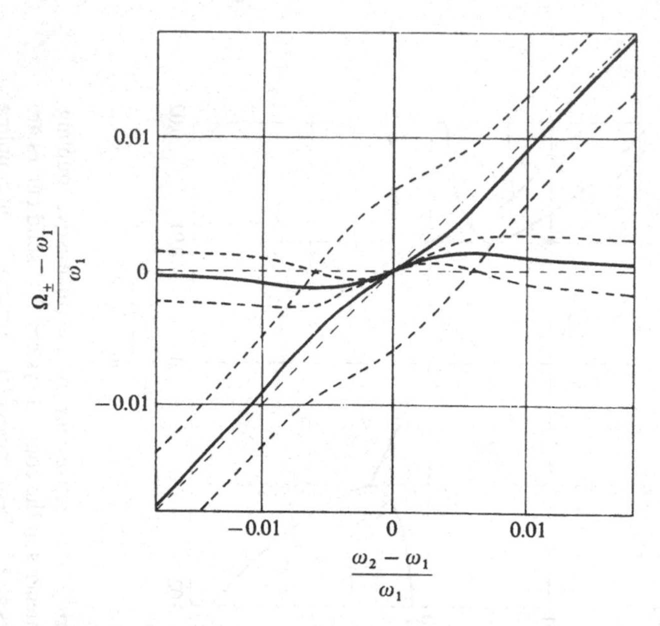

Fig. 4.21. The normal modes of two nonidentical strings with resistive coupling. Solid curves give the frequencies and dashed curves the line widths (Gough, 1981).

a resonance of the soundboard), the frequencies of the normal modes coalesce over a region close to the crossover frequency, as shown in Fig. 4.20(b). Outside this region, the modes are equally damped by the bridge impedance, but inside this region, the damping of the normal modes approaches maximum and minimum values at coincidence (Gough, 1981).

Additional cases are of interest. Figure 4.21 represents the case in which the admittance of the coupling bridge is complex. Frequencies of the two normal modes are given for several different phases of the coupling admittance. Figure 4.22 represents the coupling of two nonidentical strings with resistive coupling. The coupling still tends to pull the frequencies of the normal modes together [as in Fig. 4.20(b)], but they coincide only when the unperturbed frequencies are equal ($\omega_1 = \omega_2$). As $\omega_2 - \omega_1 \to 0$, the damping of one mode approaches zero, while the damping of the other mode increases toward a maximum value.

Several interesting effects of strings coupled by a bridge are observed in musical instruments. The compound decay curve for piano sound is a direct result of coupling between unison strings. The initial decay is fast while the strings vibrate in phase, but as they fall out of phase, a slower decay rate characterizes the aftersound (Weinreich, 1977). A violin player can simulate the effect of a vibrato on the lowest open string by placing a finger on the next string at the position corresponding to an octave above the bowed string and rocking the finger back and forth to vary the intensity of the octave harmonic (Gough, 1981).

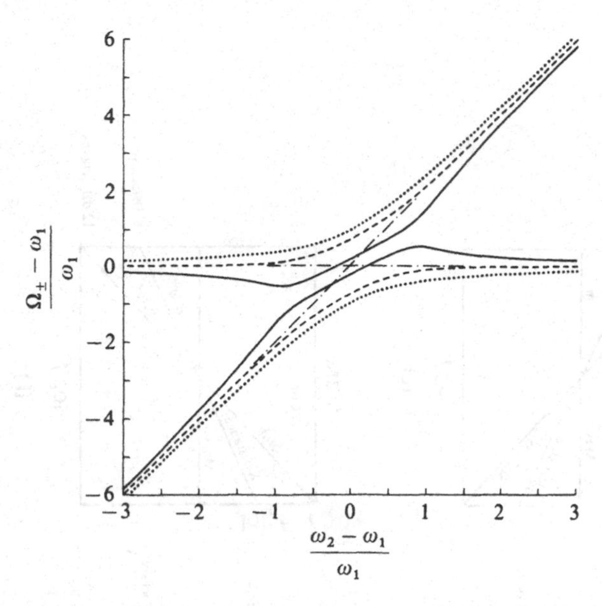

Fig. 4.22. The normal modes of a two-string system coupled by a bridge with a complex impedance. The dotted curve represents the greatest reactance, the solid curve the least (after Weinreich, 1977).

APPENDIX

A.1. Structural Dynamics and Frequency Response Functions

To determine the dynamical behavior of a structure in the laboratory, we frequently apply a force **F** at some point (x, y, z) and determine the response of the structure at the same point or some other point (x', y', z'). To describe the response, we may measure displacement **r**, velocity **v**, or acceleration **a**. In the simplest case, **F**, **r**, **v**, and **a** are in the same direction, so we speak of F, x, v, and a. Some examples of methods used to measure these variables are as follows:

- F can be measured with a load cell or force transducer;
- a can be measured with an accelerometer;
- v can be measured with a phonograph cartridge and stylus or it can be determined by integrating a; probing the near-field sound with a microphone provides a pretty good estimate of v at a point on a surface nearby; and
- x can be determined by holographic interferometry or by integrating v or a.

From these measured variables, we can construct one or more frequency response functions of interest. These include mobility (v/F), accelerance (a/F),

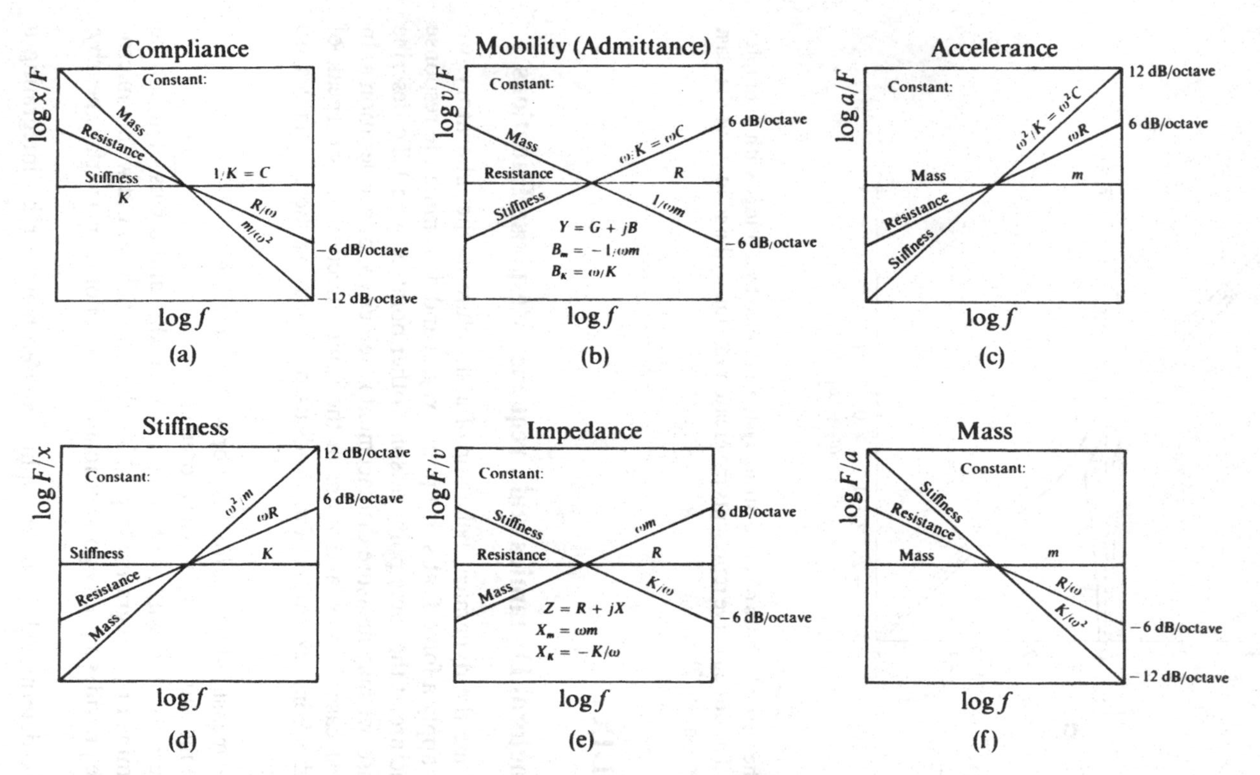

Fig. A4.1. Six examples of frequency response functions showing (a) compliance, (b) mobility, (c) accelerance, (d) stiffness, (e) impedance, and (f) dynamic mass as functions of frequency. Individual curves are drawn for single elements having constants mass, resistance, and stiffness, respectively.

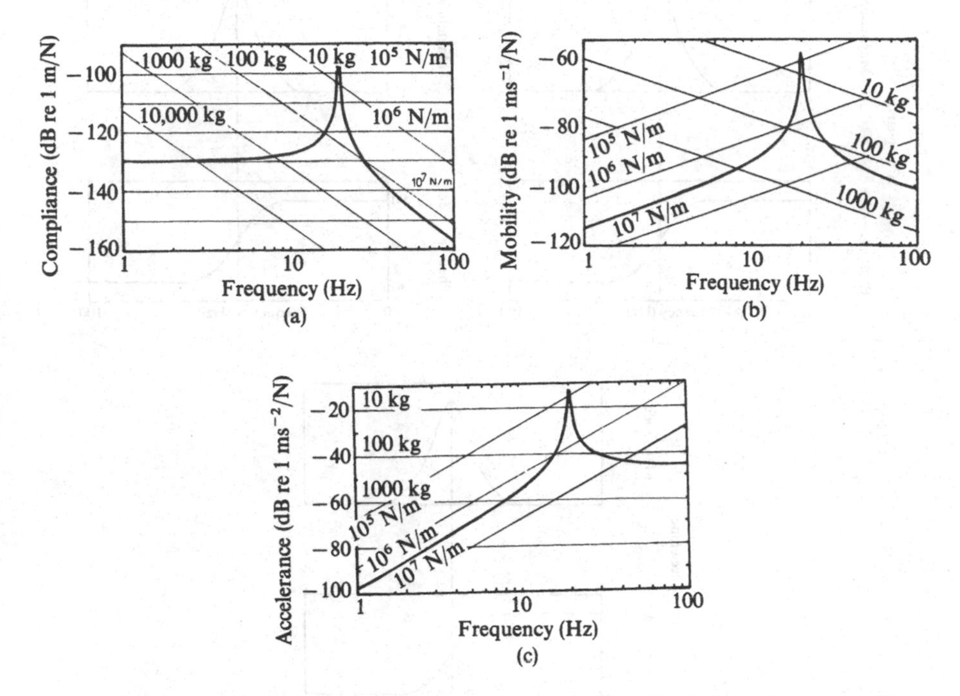

Fig. A4.2. Three examples of frequency response functions of a lightly damped oscillator with a single degree of freedom (after Ewins, 1984).

compliance (x/F), impedance (F/v), dynamic mass (F/a), and stiffness (F/x). The logarithmic minigraphs in Fig. A.4.1 illustrate how these relate to the static parameters stiffness K, mass m, and resistance R. If $(x, y, z) = (x', y', z')$, we use the prefix driving point (e.g., driving-point mobility); otherwise, we use the prefix transfer.

Combining all three elements (mass, stiffness, and resistance) into an oscillator having a single degree of freedom with light damping leads to the frequency response functions in Fig. A.4.2 (only three of the six functions are shown). Note the ± 6 dB/octave slopes at high and low frequency in the mobility plot, the -12 dB/octave slope in the compliance plot, and the $+12$ dB/octave slope in the accelerance plot, consistent with Fig. A.4.1.

Figures A.4.1 and A.4.2 show logarithms of absolute values of the parameters of interest and therefore are devoid of information about phase. To represent the phase as well as magnitude of the frequency response function of interest, a second graph is often added, as in Fig. A.4.3. At a resonance, the phase changes by 180° (0 to 180° in the compliance plot, 90° to 270° in the mobility plot, or 180° to 360° in the accelerance plot), as indicated in Figs. 1.12 or 4.8.

Two other ways of indicating phase are shown in Figs. A.4.4 and A.4.5. In Fig. A.4.4, the real and imaginary parts of the frequency response functions are separately plotted as functions of frequency. In Fig. A.4.5, the imaginary

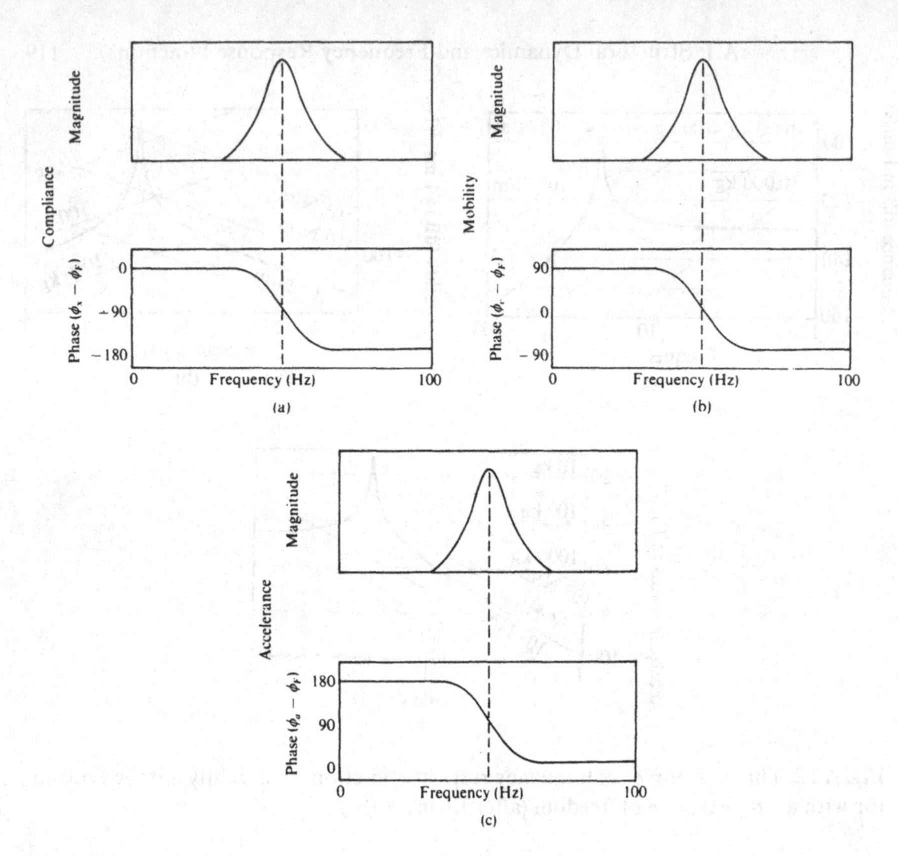

Fig. A4.3. Plots of frequency response functions showing both magnitude and phase.

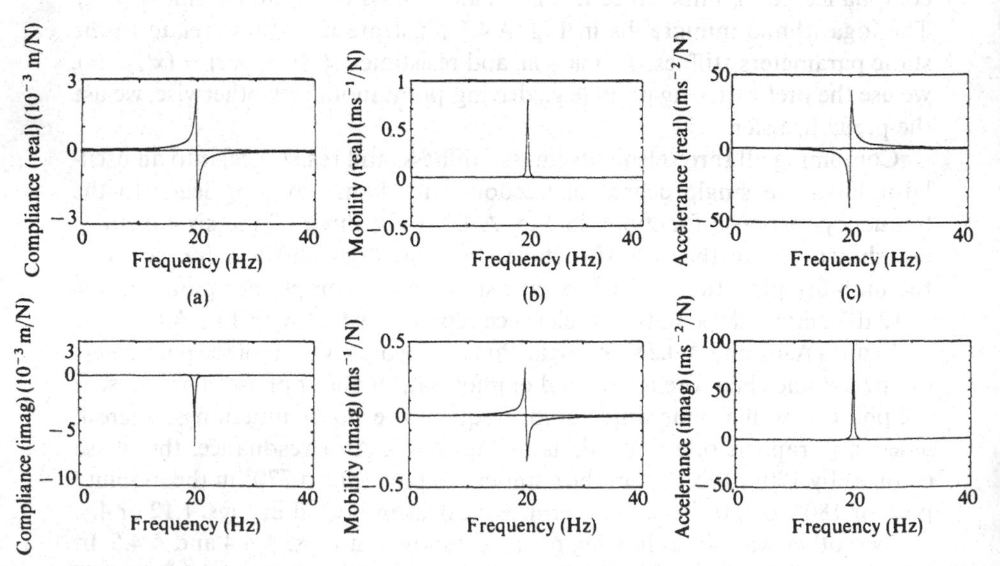

Fig. A4.4. Real and imaginary parts of several frequency response functions: (a) compliance, (b) mobility, and (c) accelerance.

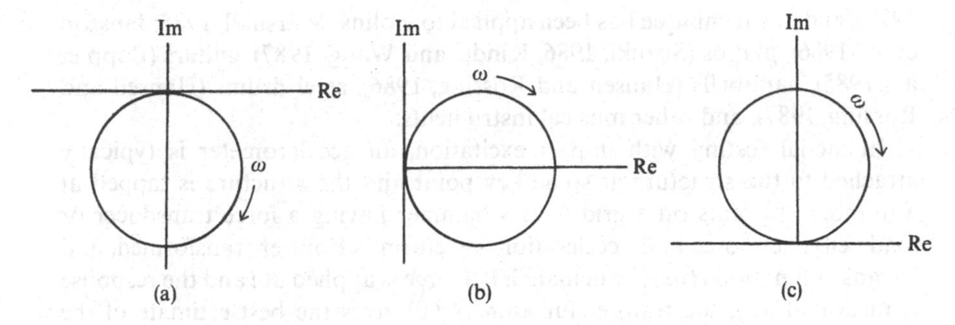

Fig. A4.5. Nyquist plots show real and imaginary parts of (a) compliance, (b) mobility, and (c) accelerance.

part is plotted as a function of the real part, with frequency shown as a parameter on the curve (Nyquist plots).

A.2. Modal Analysis

Modal analysis may be defined as the process of describing the dynamic properties of an elastic structure in terms of its normal modes of vibration. Many papers have appeared recently describing the application of modal analysis to a wide range of structures from fresh apples to large aircraft. Papers on modal analysis generally deal with either mathematical modal analysis or modal testing.

In mathematical modal analysis, one attempts to uncouple the structural equations of motion by means of some suitable transformation, so that the uncoupled equations can be solved. The frequency response of the structure can then be found by summing the respective modal responses in accordance with their degree of participation in the structural motion.

In experimental modal testing, one excites the structure at one or more points and determines the response at one or more points. From these sets of data, the natural frequencies (eigenfrequencies), mode shapes (eigenfunctions), and damping parameters are determined, often by the use of multidimensional curve-fitting routines on a digital computer. In fact it is the availability of digital computers and sophisticated software that accounts for the growing popularity of modal analysis.

Modal testing may be done with sinusoidal, random, pseudorandom, or impulsive excitation. In the case of sinusoidal excitation, the force may be applied at a single point or at several locations. The response may be measured mechanically (with accelerometers or velocity sensors), optically, or indirectly by observing the radiated sound field. Several good reviews of modal testing with impact excitation have appeared in the literature (Ewins, 1984; Allemang and Brown, 1987; Ramsey, 1975/1976; Marshall, 1986; Halvorsen and Brown,

1977), and this technique has been applied to violins (Marshall, 1985; Jansson et al., 1986), pianos (Suzuki, 1986; Kindel and Wang, 1987), guitars (Popp et al., 1985), handbells (Hansen and Rossing, 1986), steel drums (Hansen and Rossing, 1987), and other musical instruments.

In modal testing with impact excitation, an accelerometer is typically attached to the structure at some key point and the structure is tapped at a number of points on a grid with a hammer having a force transducer or load cell. Each force and acceleration waveform is Fourier transformed, and a transfer function $H(\omega)$ is calculated. If a force is applied at i and the response is measured at j, the transfer function $H_{ij}(\omega)$ gives the best estimate of the frequency response function

$$H_{ij}(\omega) = \frac{S_i^*(\omega)S_j(\omega)}{S_i^*(\omega)S_i(\omega)} = \frac{G_{ij}}{G_{ii}},$$

where $S_i^*(\omega)$ is the complex conjugate of the force spectrum $S_i(\omega)$, G_{ii} is the power spectrum of the exciting force, and G_{ij} is the cross spectrum of the force and response.

Several different algorithms may be used to extract the modal parameters from the measured frequency response functions. Single-degree-of-freedom methods include a "peak picking" method, which uses the imaginary (quadrature) component of the response function as the modal coordinate, and a "circle fit" method, which fits the best circle to the data in the Argand plane [whose coordinates are the real and imaginary parts of the response function, as in the Nyquist plot in Fig. A.4.5(a)]. Multidegree-of-freedom methods generally use a least-squares method to select the modal parameters that minimize the differences between the measured frequency response function and the function found by summing the contribution from the individual modes. Still other methods, such as the complex exponential method and the Ibrahim method, do the curve fitting in the time domain (Ewins, 1984).

A.3. Finite Element Analysis

The finite element method is a powerful numerical analysis method that can be used to calculate the vibrational modes of elastic structures. The method assumes that a structure or system can be modeled by an assemblage of building blocks (called elements) connected only at discrete points (called nodes). A complex structure is often divided into a number of familiar substructures, such as plates, beams, shells, and lumped masses.

This concept of modeling a system as a collection of discrete points was introduced by Courant (1943). His suggestion, that the Rayleigh–Ritz approach using an assumed response function for the system could be applied to triangular elements, became the mathematical basis for finite element analysis. The actual finite element terminology was later introduced by Clough (1960) and others. Clough's method became widely adopted because digital computers were available to perform the complex numerical calculations.

Finite element analysis is essentially an extension of matrix structural analysis methods that have been applied to beams and trusses for some time. This analysis is based on a set of equations of the form

$$(\mathbf{u})(\mathbf{K}) = (\mathbf{F}),$$

where $(\mathbf{u})$ is a displacement vector, $(\mathbf{F})$ is a force vector, and $(\mathbf{K})$ is the stiffness matrix in which a typical element K_{ij} gives the force F_i at the ith node due to a unit displacement u_j at the jth node.

General purpose finite element codes, such as NASTRAN, ANSYS, SAP, and ADINA, include routines for solving the equations of motion in matrix form.

$$(\mathsf{M})(\ddot{u}) + (\mathsf{R})(\dot{u}) + (\mathsf{K})(u) = (F)\cos(\omega t + \gamma),$$

where (M), (R), and (K) are the mass, damping and stiffness matrices. Smaller programs adapted from these large, general purpose systems have made it possible to apply finite element methods to structures of modest size using microcomputers.

To improve the efficiency of finite element calculations, so-called "eigenvalue economizer" routines are often used. These routines reduce the size of the dynamical matrix by condensing it around master nodes (Rieger, 1986). Guidance in the selection of such nodes can often be obtained from the results of modal testing. In this and others ways, modal analysis and finite element analysis have become complementary methods for studying the dynamical behavior of large and small structures, including musical instruments.

References

Allemang, R.J., and Brown, D.L. (1987). Experimental modal analysis. In "*Handbook of Experimental Mechanics*" (A.S. Kabayashi ed.). Prentice-Hall, Englewood Cliffs, New Jersey.

Clough, R.W. (1960). The finite method in plane stress analysis. *Proc. 2nd ASCE Conf. on Electronic Computation*, Pittsburgh, Pennsylvania.

Courant, R. (1943). Variational methods for the solution of problems of equilibrium and vibrations. *Bull. Am. Math. Soc.* **49**, 1.

Ewins, D.J. (1984). "Modal Testing: Theory and Practice." Research Studies Press, Letchworth, England.

French, A.P. (1971) "Vibrations and Waves," W.W. Norton, New York.

Gough, C.E. (1981). The theory of string resonances on musical instruments. *Acustica* **49**, 124–141.

Halvorsen, W.G., and Brown, D.L. (1977). Impulse techniques for structural frequency response testing. *Sound and Vibration* **11** (11), 8–21.

Hansen, U.J., and Rossing, T.D. (1986). Modal analysis of a handbell (abstract). *J. Acoust. Soc. Am.* **79**, S92.

Hansen, U.J., and Rossing, T.D. (1987). Modal analysis of a Caribbean steel drum (abstract). *J. Acoust. Soc. Am.* **82**, S68.

Jacobson, L.S., and Ayre, R.S. (1958). "Engineering Vibrations." McGraw-Hill, New York.

Jansson, E., Bork, I., and Meyer, J. (1986). Investigation into the acoustical properties of the violin. *Acustica* **62**, 1–15.

Kindel, J. and Wang, I. (1987), Modal analysis and finite element analysis of a piano soundboard, *Proc. 5th Int'l Conf. on Modal Analysis* (IMAC), 1545–1549.

Marshall, K.D. (1986). Modal analysis of a violin. *J. Acoust. Soc. Am.* **77**, 695–709.

Marshall, K.D. (1987). Modal analysis: A primer on theory and practice. *J. Catgut Acoust. Soc.* **46**, 7–17.

Popp, J., Hansen, U., Rossing. T.D., and Strong, W.Y. (1985). Modal analysis of classical and folk guitars (abstract). *J. Acoust. Soc. Am.* **77**, S45.

Ramsey, K.A. (1975/1976). Effective measurements for structural dynamics testing. *Sound and Vibration* **9** (11), 24–34; **10** (4), 18–31.

Rieger, N.F. (1986). The relationship between finite element analysis and modal analysis. *Sound and Vibration* **20** (1), 20–31.

Suzuki, H. (1986). Vibration and sound radiation of a piano soundboard. *J. Acoust. Soc. Am.* **80**, 1573–1582.

Weinreich, G. (1977). Coupled piano strings. *J. Acoust. Soc. Am.* **62**, 1474–1484.

CHAPTER 5

Nonlinear Systems

Many of the mechanical elements comprising a musical instrument behave approximately as linear systems. By this we mean that the acoustic output is a linear function of the mechanical input, so that the output obtained from two inputs applied simultaneously is just the sum of the outputs that would be obtained if they were applied separately. For this statement to be true for the instrument as a whole, it must also be true for all its parts, so that deflections must be proportional to applied forces, flows to applied pressures, and so on. Mathematically, this property is reflected in the requirement that the differential equations describing the behavior of the system are also linear, in the sense that the dependent variable occurs only to the first power. An example is the equation for the displacement y of a simple harmonic oscillator under the action of an applied force $F(t)$:

$$m\frac{d^2y}{dt^2} + R\frac{dy}{dt} + Ky = F(t), \tag{5.1}$$

where m, R, and K are respectively the mass, damping coefficient, and spring coefficient, all of which are taken to be constants. Then, if $y_1(t)$ is the solution for $F(t) = F_1(t)$, and $y_2(t)$ that for $F(t) = F_2(t)$, the solution for $F = F_1 + F_2$ will be $y = y_1 + y_2$.

A little consideration shows, of course, that this description must be an oversimplification (Beyer, 1974). Mass is indeed conserved (apart from relativistic effects), but spring coefficients cannot remain constant when displacements approach the original dimensions of the system, nor can we expect damping behavior to remain unchanged when turbulence or other complicated effects intervene. It is therefore important to know how to treat such nonlinearities mathematically so that we can make use of these techniques when we come to examine the behavior of real musical instruments. Mathematically, the problem is one of solving equations like Eq. (5.1) when at least one of the coefficients m, R, or K depends on the dependent variable y. In interesting practical cases, we will nearly always be concerned with small deviations from linearity, so that these coefficients can be expanded as rapidly convergent power series in y or dy/dt, provided y is small compared with the dimensions of the system.

One sort of problem exemplified by Eq. (5.1) is a percussion instrument or a plucked string instrument in which the forcing function $F(t)$ is external, of limited duration, and given quite explicitly. The behavior of such impulsively excited oscillators or resonators is relatively simple, as we shall see presently.

More complex, more interesting, and inherently more nonlinear is the situation encountered in steady-tone instruments, such as bowed strings or windblown pipes. Here, the forcing function $F(t)$ consists of a steady external part (the bow velocity or the blowing pressure) whose effect on the system is somehow determined by the existing amplitude of the oscillation. Thus, for example, the force between a bow and a string depends upon their relative velocities, while the flow through a reed valve depends upon the pressure difference across it. In this case, Eq. (5.1) is generalized to

$$m\frac{d^2y}{dt^2} + R\frac{dy}{dt} + Ky = F\left(y, \frac{dy}{dt}, t\right), \tag{5.2}$$

where again m, R, and K may be weak functions of y. For such a system, as we shall see presently, the whole behavior depends quite crucially upon the various nonlinearities present in the coefficients and in the function F.

5.1. A General Method of Solution

The systems we shall meet in musical instruments will ultimately prove to be much more complex than described by an equation like Eq. (5.2) since musical oscillators such as strings, plates or air columns generally have infinitely many possible vibrational modes rather than just a single one, but we can learn a great deal as a necessary preliminary by studying the nonlinear oscillator described by Eq. (5.2). In fact, musical instruments generally consist of a nearly linear resonator, the string, plate, or air column, described by the left-hand side of Eq. (5.2), excited by a generator F, which has quite nonlinear behavior. The nonlinearity in F is crucial to the description of the system.

For convenience of notation, we rewrite Eq. (5.2) in the form

$$\ddot{y} + \omega_0^2 y = g(y, \dot{y}, t), \tag{5.3}$$

where a dot signifies d/dt, ω_0 is the resonant frequency,

$$\omega_0 = \left(\frac{K}{m}\right)^{1/2}, \tag{5.4}$$

and

$$g = \frac{F - R\dot{y}}{m} \equiv f - 2\alpha\dot{y}, \tag{5.5}$$

where we have written $f \equiv F/m$ and $2\alpha \equiv R/m$.

If the damping, nonlinearity, and driving force on the system are all small, so that $g \to 0$ in Eq. (5.3), then the solution has the sinusoidal form

$$y(t) = a \sin(\omega_0 t + \phi), \tag{5.6}$$

where a is the amplitude and ϕ is the phase of the oscillation. If g is not identically zero but is small compared with the terms on the left side of Eq. (5.3), then it is reasonable to suppose that the true solution may have a form like

$$y(t) = a(t) \sin[\omega_0 t + \phi(t)], \tag{5.7}$$

where a and ϕ are both slowly varying functions of time. Now from Eq. (5.7), it is clear that

$$\dot{y} = \dot{a} \sin(\omega_0 t + \phi) + a(\omega_0 + \dot{\phi}) \cos(\omega_0 t + \phi). \tag{5.8}$$

This is a rather complicated expression, and it is clear that a given behavior of y and $\dot{y}$ as functions of time could be described in several ways depending on how this functional dependence was partitioned between $\dot{a}$ and $\dot{\phi}$. There is a great simplification if we can arrange matters so that

$$\dot{y} = a\omega_0 \cos(\omega_0 t + \phi), \tag{5.9}$$

which requires that

$$\dot{a} \sin(\omega_0 t + \phi) + a\dot{\phi} \cos(\omega_0 t + \phi) = 0. \tag{5.10}$$

If we assume Eq. (5.10) to have been satisfied and substitute Eqs. (5.7) and (5.9) into Eq. (5.3), then we find

$$\dot{a}\omega_0 \cos(\omega_0 t + \phi) - a\omega_0 \dot{\phi} \sin(\omega_0 t + \phi) = g. \tag{5.11}$$

Since Eqs. (5.10) and (5.11) must be satisfied simultaneously, we can solve for $\dot{a}$ and $\dot{\phi}$ to obtain

$$\dot{a} = \frac{g}{\omega_0} \cos(\omega_0 t + \phi), \tag{5.12}$$

and

$$\dot{\phi} = -\frac{g}{a\omega_0} \sin(\omega_0 t + \phi), \tag{5.13}$$

where g is written in terms of y and $\dot{y}$ given by Eqs. (5.7) and (5.9).

The essence of the approximation is now to neglect all terms in Eqs. (5.12) and (5.13) except those that vary slowly in comparison with ω_0. The resulting trends are then denoted by $\langle \dot{a} \rangle$ and $\langle \dot{\phi} \rangle$, respectively. When Eqs. (5.12) and (5.13) are substituted back into Eq. (5.7), we find that the effective frequency at a given time t is $\omega_0 + \langle \dot{\phi} \rangle$, and the amplitude is

$$a(t) = a(t_0) + \int_{t_0}^{t} \langle \dot{a} \rangle \, dt. \tag{5.14}$$

Within this approximation, it is thus possible to calculate the entire behavior of the system once its initial state is known. We shall use this formalism quite extensively in our later discussion.

5.2. Illustrative Examples

If this approach is to be convincing, then it is necessary, of course, that it reproduce, with adequate accuracy, the standard results for simple cases. To see that this does happen without undue labor, let us examine a few typical examples.

First, consider the simple damped system with no external forcing. For this case, in Eq. (5.5), $f = 0$ and $g = -2\alpha\dot{y}$. Thus, from Eq. (5.12), $\langle\dot{a}\rangle = -\alpha a$, while from Eq. (5.13), $\langle\dot{\phi}\rangle = 0$. The solution is therefore

$$a(t) \approx a(t_0)e^{-\alpha(t-t_0)}\sin(\omega_0 t + \phi_0), \tag{5.15}$$

which is in adequate agreement with the exact solution, provided $\alpha \ll \omega_0$.

Next, take the case of a damped harmonic oscillator driven at some frequency ω that is close to its resonance. Here,

$$g = f\sin\omega t - 2\alpha\dot{y}, \tag{5.16}$$

so that, from Eqs. (5.12) and (5.13),

$$\langle\dot{a}\rangle = \frac{f}{2\omega_0}\sin[(\omega - \omega_0)t - \phi] - \alpha a, \tag{5.17}$$

and

$$\langle\dot{\phi}\rangle = -\frac{f}{2a\omega_0}\cos[(\omega - \omega_0)t - \phi]. \tag{5.18}$$

The actual motion of the system clearly depends upon the amplitude and phase of its initial state, but the important thing to check is the final steady state as $t \to \infty$. If $\langle\dot{\phi}\rangle$ is to be constant, then from Eq. (5.18), we must have

$$\phi = (\omega - \omega_0)t + \phi_0, \tag{5.19}$$

so that

$$\langle\dot{\phi}\rangle = -\frac{f}{2a\omega_0}\cos\phi_0. \tag{5.20}$$

The requirement that Eqs. (5.19) and (5.20) be consistent gives

$$\cos\phi_0 = -\frac{2a\omega_0(\omega - \omega_0)}{f}, \tag{5.21}$$

so that the oscillator vibrates with frequency $\omega_0 + \langle\dot{\phi}\rangle = \omega$, in synchronism with the external force. The steady amplitude for which $\langle\dot{a}\rangle = 0$ is given by Eq. (5.17) as

$$a = -\frac{f}{2\alpha\omega_0}\sin\phi_0, \tag{5.22}$$

or, using Eq. (5.21),

$$|a| = \frac{f}{2\alpha\omega_0}\left[1 - \frac{4a^2\omega_0^2(\omega - \omega_0)^2}{f^2}\right]^{1/2}$$

$$\approx \frac{f}{2\alpha\omega_0}\left[1 - \frac{(\omega - \omega_0)^2}{2\alpha^2}\right], \tag{5.23}$$

which is the response near the top of a normal resonance curve. The phase difference ϕ_0 between the force and the resulting displacement is also correctly given, being $-90°$ at resonance (as shown in Fig. 1.12).

The method set out in Section 5.1 is called, for obvious reasons, the method of slowly varying parameters (Van der Pol, 1934; Bogoliubov and Mitropolsky, 1961; and Morse and Ingard, 1968). In nearly all cases, it allows us to calculate the entire time evolution of the system from a given initial state simply by integrating Eqs. (5.17) and (5.18), a procedure that must generally be carried out numerically. The results again agree quite closely with the exact solutions, where these can be obtained, subject only to the restrictions that ω is close to ω_0 and that small terms of higher frequency are neglected in the solution.

The method can also be used in an obvious way to follow the behavior of nonlinear oscillators. A simple example is the case in which the spring parameter K of the oscillator gets progressively stiffer or weaker as the displacement y increases, for example, as $K \to K + \beta m y^2$. This example was treated by a different method in Chapter 1. Referring now to Eqs. (5.3)–(5.5), we find that

$$g = f(t) - 2\alpha y - \beta y^2. \tag{5.24}$$

The resulting form of Eq. (5.3) is known as Duffing's equation. Some of its more interesting properties have been discussed by Prosperetti (1976) and by Ueda (1979). This expression [Eq. (5.24)] can be simply inserted in the $\langle \cdots \rangle$ forms of Eqs. (5.12) and (5.13), and these equations integrated to give the behavior.

If $f(t)$ is a sinusoidal excitation of frequency ω, then we can readily calculate the steady response curve of the oscillator. Two cases were shown in Fig. 1.22, that for $\beta = b/m > 0$, which corresponds to a spring that hardens with increasing amplitude, and that for $\beta = b/m < 0$, which corresponds to a softening spring. The overhanging part of the curve represents an unstable situation and the oscillator exhibits amplitude transitions with hysteresis, as shown by the broken lines in Fig. 5.1. These transitions are given by integration of Eqs. (5.17) and (5.18).

It is useful at this stage to recall the treatment of Duffing's equation given in Chapter 1. The solution given by Eqs. (1.79) and (1.80) similarly describes the steady-state resonance curve of Fig. 5.1, or the earlier Fig. 1.22, and indeed gives information about the third harmonic as well. Our present method, however, allows us, in addition, to calculate the approach to the steady state or to follow other time-varying phenomena. If $f(t)$ is zero after an initial impulsive supply of energy, for example, then the equations show that, as the amplitude decays toward zero in a more or less exponential manner, the

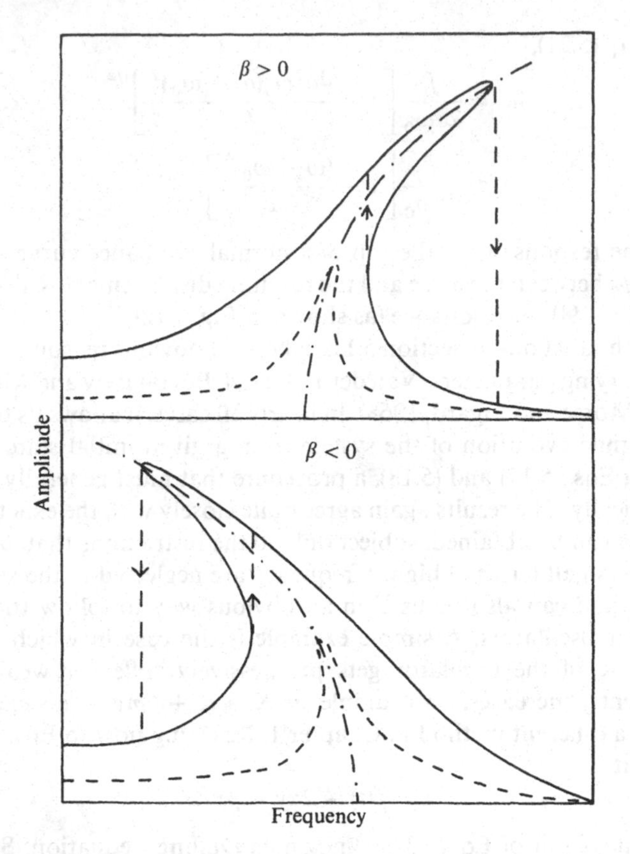

Fig. 5.1. The steady frequency response of a nonlinear oscillator of the Duffing type excited by an external sinusoidal force. The dotted curves show the response for a smaller exciting force, while the vertical broken lines show transitions as the frequency is swept with a constant exciting force.

frequency glides from its large-amplitude value toward its limiting small-amplitude value along the spine of the curve.

5.3. The Self-Excited Oscillator

Of particular interest in musical instruments, and indeed in many fields of electronic technology as well, is an oscillator that is arranged so as to modulate some external steady flow of air, electricity, or some other quantity, with a part of the resulting modulated flow then being fed back in an appropriate phase to excite the oscillator. The system equation has the general form

$$\ddot{y} + \omega_0^2 y = g(y, \dot{y}), \tag{5.25}$$

where the form of g depends on the arrangement of the system, and the magnitude of g is related to the magnitude of the external force, which provides energy to drive the oscillator.

The best known case is the Van der Pol (1934) oscillator, for which

$$g = \dot{y}a(1 - y^2). \tag{5.26}$$

Inserting this into Eq. (5.3), we see that the damping of the oscillator is negative for $y^2 < 1$ and positive for $y^2 > 1$. Oscillations of small amplitude thus tend to grow, while oscillations of large amplitude are damped. There is a stable oscillation regime or "limit cycle" of amplitude $a = 2$ for which the energy losses when $y > 1$ just balance the energy gains when $y < 1$. The oscillations are fairly closely sinusoidal, provided a is not very much greater than unity but, of course, there is some admixture of higher odd harmonics.

The behavior of a Van der Pol oscillator and indeed of other similar self-excited systems is well encompassed by our formalism, and simple insertion of the appropriate form of $g(y, \dot{y})$ into the $\langle \cdots \rangle$ forms of Eqs. (5.12) and (5.13) provides a prescription from which the development of the system can be calculated. Discussion of specific cases will be left until the underlying physical systems have been introduced, but a few general comments are in order. The most important are the observations that the quiescent state $y = 0$ is always a possible solution, but that this state is unstable to small fluctuations $\delta y, \delta\dot{y}$, if the part of $g(\delta y, \delta\dot{y})$ in phase with $\delta\dot{y}$ is positive. If this is true, then such fluctuations will lead to growth of the displacement y. The second important set of generalizations inquires whether or not the oscillation settles into a stable limit cycle. This depends on the detail of the nonlinearity and thus upon the physics of the system. Clearly, all musically useful systems do settle to a stable cycle of nonzero amplitude.

5.4. Multimode Systems

All real systems of finite extent have an infinite number of possible vibration modes and, in musically useful resonators, these modes are generally nearly linear in behavior and have well-separated characteristic frequencies, often in nearly harmonic relationship. This does not mean that musical oscillating systems are nearly linear, but rather that such systems usually consist of a nearly linear multimode resonator excited by some nonlinear feedback mechanism.

Suppose that the equation describing wavelike propagation in the oscillator has the form

$$\mathscr{L}\psi - \frac{\partial^2\psi}{\partial t^2} = 0, \tag{5.27}$$

where $\mathscr{L}$ is some linear differential operator typically involving ∇^2 or ∇^4. If we separate variables by writing solutions to this equation as products of spatial functions and a time variation like $\sin\omega t$, then the eigenvalue equation is

$$\mathscr{L}\psi_n + \omega_n^2\psi_n = 0 \tag{5.28}$$

where the eigenfrequencies ω_n are determined from the requirement that the eigenfunctions ψ_n should satisfy appropriate boundary conditions, usually $\psi_n = 0$ or $\nabla\psi_n = 0$, at the surfaces of the resonator.

If we extend Eq. (5.27) to include on the right side a force of unit magnitude (in appropriate units) and frequency ω applied at the point $\mathbf{r}_0$ then this equation becomes

$$\mathscr{L}G_\omega + \omega^2 G_\omega = -\delta(\mathbf{r} - \mathbf{r}_0), \tag{5.29}$$

where $\delta(\mathbf{r} - \mathbf{r}_0)$ is the Dirac delta function, and we have written ψ as $G_\omega(\mathbf{r}, \mathbf{r}_0)$, which is the Green function for the system at frequency ω, taken to obey the same boundary conditions as do the ψ_n. If we assume that G_ω can be expanded as

$$G_\omega(\mathbf{r}, \mathbf{r}_0) = \sum_n a_n \psi_n(\mathbf{r}), \tag{5.30}$$

and substitute this into Eq. (5.29), then, using Eq. (5.28),

$$\sum_n a_n(\omega_n^2 - \omega^2)\psi_n(\mathbf{r}) = \delta(\mathbf{r} - \mathbf{r}_0). \tag{5.31}$$

If we multiply both sides by $\psi_n(\mathbf{r})$ and integrate over the whole volume of the resonator using the usual orthonormality condition

$$\int \psi_n(\mathbf{r})\psi_m(\mathbf{r})\,d\mathbf{r} = \delta_{nm}, \tag{5.32}$$

then we find an expression for a_n that, substituted back into Eq. (5.30), gives

$$G_\omega(\mathbf{r}, \mathbf{r}_0) = \sum_n \frac{\psi_n(\mathbf{r}_0)\psi_n(\mathbf{r})}{\omega_n^2 - \omega^2}. \tag{5.33}$$

Clearly, this Green function has simple poles at $\omega = \pm\omega_n$, where the ω_n are the resonance frequencies of the system.

Since the resonator system is assumed to be linear, and since $G(\mathbf{r}, \mathbf{r}_0)$ is the response caused by a force of unit magnitude and frequency ω applied at $\mathbf{r}$, we can write the general response Ψ to a set of forces of different frequencies ω and phases θ and distributed over the resonator like the functions $F_\omega(\mathbf{r})$ as the simple sum

$$\Psi(\mathbf{r}, t) = \sum_\omega \left[\int G_\omega(\mathbf{r}, \mathbf{r}_0)F_\omega(\mathbf{r}_0)\,d\mathbf{r}_0\right]\sin(\omega t + \theta). \tag{5.34}$$

This formulation does not include damping. It could be included as an extension to the linear theory and has the effect of moving the poles of G_ω slightly off the real axis, but it is more conveniently incorporated along with nonlinear effects at a later stage.

The essential feature of Eq. (5.34) for our present purpose is expressed by Eq. (5.33), which shows that the mode ψ_n with frequency ω_n is excited principally by force components with frequency ω lying close to ω_n. Indeed, if we write the excitation of this nth mode in the form

$$y_n = a_n \sin(\omega t + \phi_n)\psi_n(\mathbf{r}), \tag{5.35}$$

when it is being driven at frequency ω by a force $F_\omega(\mathbf{r}_0)$ applied at the point $\mathbf{r}_0$, then Eqs. (5.33) and (5.34) are equivalent to the sum of the results from the set of differential equations:

$$\ddot{y}_n + \omega_n^2 y_n = \sum_\omega F_\omega \sin(\omega t + \theta_\omega). \tag{5.36}$$

If we retain in Eq. (5.36) only those force components F_ω with frequencies near ω_n, then we can apply the method of slowly varying parameters as developed previously for a simple oscillator. The forcing term on the right side can be written to include the damping term $-2\alpha_n \dot{y}_n$ and, in a general situation, it will contain driving forces with frequencies near ω_n derived from nonlinear terms involving many of the other modes of the system. In general, therefore, Eq. (5.36) has the form

$$\ddot{y}_n + \omega_n^2 y_n = g_n(y_n, y_m, \ldots; \dot{y}_n, \dot{y}_m, \ldots), \tag{5.37}$$

and we omit from g_n, as explicitly evaluated, all terms except those for which the frequency is close to ω_n.

Equation (5.37) is a shorthand way of writing an infinite set of coupled differential equations, one for each of the modes ω_n. In practice, however, the very high frequency modes will have small excitation amplitudes, so that the system can be reduced to a finite and indeed relatively small set of N coupled equations describing those modes that are appreciably excited. Each of these equations can be manipulated to give $\langle \dot{a}_n \rangle$ and $\langle \dot{\phi}_n \rangle$, and the resulting $2N$ equations can be easily integrated numerically to define the behavior of the system.

Clearly, the linear physics of a musical system resides in the study of the modes ψ_n and their characteristic frequencies ω_n, while the nonlinear physics involves elucidation of the coupling functions g_n. Both these matters are quite specific to individual systems, so we will not consider particular examples at this stage of our discussion.

As with simple oscillators, so multimode systems can be divided into those that are purely dissipative and can be excited only by an impulsive or time-varying external force, and those that are self-exciting with only a steady external supply of energy. Gongs are typical examples of the former class and can exhibit many interesting phenomena as energy is passed back and forth between the modes by the agency of the nonlinear coupling terms g_n. Musical instruments producing steady tones, such as winds or bowed strings, belong to the second class.

5.5. Mode Locking in Self-Excited Systems

For many musical resonators, such as stretched strings or air columns in pipes, the normal mode frequencies are very nearly, but not quite, in integral ratio. Precisely harmonic systems do not exist. Now, if the coupling function g_n in

Eq. (5.37) is linear or has no nonlinear terms involving combinations of different modes, then the system becomes essentially uncoupled and each mode takes on an excitation frequency close to its natural frequency ω_n. These frequencies are never precisely in integer ratios, so the resulting total excitation has a nonrepeating waveform.

Sustained tones from real musical instruments do, however, have precisely repeating waveforms, apart from deliberate vibrato effects, and so their individual modes must be somehow locked into precise frequency and phase relationships despite the inharmonicities of the natural resonances. It is important to see how this is accomplished (Fletcher, 1978).

Consider a system for which just two modes are appreciably excited by the feedback mechanism and suppose that their natural frequencies ω_n and ω_m are related approximately, but not exactly, as the ratio of the two small integers n and m:

$$m\omega_n \approx n\omega_m. \tag{5.38}$$

Now, from Eq. (5.38), the leading nonlinear term by which mode m can provide a driving force at nearly the frequency of mode n involves the amplitudes a_n and a_m in the form $a_m^n a_n^{m-1}$. Similarly, mode n can influence mode m in proportion to $a_n^m a_m^{n-1}$. The coefficients of these terms depend upon the Taylor expansion of the nonlinear driving functions g_n and g_m, respectively, and generally decrease sharply as n and m increase. These two functions will be of the same general form but may differ in detail. The directions in which the frequencies of modes m and n are pushed by their $\langle\dot{\phi}\rangle$ terms in Eq. (5.13) will depend upon the combinations of phase angles ϕ_n and ϕ_m involved, and these will vary rapidly if the two modes are not locked together. Once locking occurs, however, this represents a stable situation. This argument can be generalized to the case of more than two interacting modes.

The conditions favoring mode locking are thus that the inharmonicity of the modes not be too great, that the integers n and m linking the modes be as small as possible, that the mode amplitudes be large, and that the nonlinearity of the coupling function be as large as possible. When these conditions are fulfilled, then all modes contributing to an instrumental sound will rapidly settle down to give a phase and frequency locked repetitive waveform. This situation is specially favored when one of the dominant modes excited by the feedback mechanism is the fundamental, for then $m = 1$ and the integer combination is as simple as possible.

Conversely, the conditions favoring nonlocking of modes and thus the production of complex multiphonic effects are great mode inharmonicity (often produced by peculiar venting arrangements in wind instruments), a low excitation level, and preferential excitation of modes other than the fundamental. The complexity of the possible variations on this theme makes it desirable to consider it in the context of particular instruments.

References

Beyer, R.T. (1974). "Nonlinear Acoustics," pp. 60–90. U.S. Naval Sea Systems Command.

Bogoliubov, N.N., and Mitropolsky, Y.A. (1961). "Asymptotic Methods in the Theory of Non-linear Oscillations." Hindustan, New Delhi, and Gordon & Breach, New York.

Fletcher, N.H. (1978). Mode locking in nonlinearly excited inharmonic musical oscillators. *J. Acoust. Soc. Am.* **64**, 1566–1569.

Morse, P.M., and Ingard, K.U. (1968). "Theoretical Acoustics," pp. 828–882. McGraw-Hill, New York. Reprinted 1986, Princeton Univ. Press, Princeton, New Jersey.

Prosperetti, A. (1976). Subharmonics and ultraharmonics in the forced oscillations of weakly nonlinear systems. *Am. J. Phys.* **44**, 548–554.

Ueda, Y. (1979). Randomly transitional phenomena in the system governed by Duffing's equation. *J. Statistical Phys.* **20**, 181–196.

Van der Pol, B. (1934). The nonlinear theory of electric oscillations. *Proc. I.R.E.* **22**, 1051–1086.

Part II
Sound Waves

CHAPTER 6

Sound Waves in Air

The sensation we call sound is produced primarily by variations in air pressure that are detected by their mechanical effect on the tympana (ear drums) of our auditory system. Motion of each tympanum is communicated through a linked triplet of small bones to the fluid inside a spiral cavity, the cochlea, where it induces nerve impulses from sensory hair cells in contact with a thin membrane (the basilar membrane). Any discussion of details of the physiology and psychophysics of the hearing process (Stevens and Davis, 1938; Gulick, 1971) would take us too far afield here. The important point is the dominance of air pressure variation in the mechanism of the hearing process. Direct communication of vibration through the bones of the head to the cochlea is possible, if the vibrating object is in direct contact with the head, and intense vibrations at low frequencies can be felt by nerve transducers in other parts of the body, for example in the case of low organ notes, but this is not part of the primary sense of hearing.

The human sense of hearing extends from about 20 Hz to about 20 kHz, though the sensitivity drops substantially for frequencies below about 100 Hz or above 10 kHz. This frequency response is understandably well matched to human speech, most of the energy of which lies between 100 Hz and 10 kHz, with the information content of vowel sounds concentrated in the range of 300 Hz–3 kHz and the information content of consonants mostly lying above about 1 kHz. Musical sounds have been evolved to stimulate the sense of hearing over its entire range, but again most of the interesting information lies in the range of 100 Hz–3 kHz.

Since the ears respond to pressure only in their immediate vicinity, we devote this and the following chapter to a discussion of the way in which pressure variations—sound waves—propagate through the air and to the way in which vibrating objects couple to the air and excite sound waves.

6.1. Plane Waves

Waves will propagate in any medium that has mass and elasticity, or their equivalents in nonmechanical systems. Solid materials, which have both shear and compressive elasticity, allow the propagation of both shear (transverse) and compressive (longitudinal) waves so that their behavior can be very complicated (Morse and Feshbach, 1953, pp. 142–151). Fluids, and in particular gases such as air, have no elastic resistance to shear, though they do have a viscous resistance, and the only waves that can propagate in them are therefore longitudinal, with the local motion of the air being in the same direction as the propagation direction of the wave itself.

When sound waves are generated by a small source, they spread out in all directions in a nearly spherical fashion. We shall look at spherical waves in detail a little later. It is simplest in the first place to look at a small section of wave at a very large distance from the source where the wave fronts can be treated as planes normal to the direction of propagation. In the obvious mathematical idealization, we take these planes to extend to infinity so that the whole problem has only one space coordinate x measuring distance in the direction of propagation.

Referring to Fig. 6.1, suppose that ξ measures the displacement of the air during passage of a sound wave, so that the element ABCD of thickness dx moves to A′B′C′D′. Taking S to be the area normal to x, the volume of this element then becomes

$$V + dV = S\,dx\left(1 + \frac{\partial \xi}{\partial x}\right). \tag{6.1}$$

Now suppose that p_a is the total pressure of the air. Then the bulk modulus K is defined quite generally by the relation

$$dp_a = -K\frac{dV}{V}. \tag{6.2}$$

We can call the small, varying part dp_a of p_a the sound pressure or acoustic pressure and write it simply as p. Comparison of Eq. (6.2) with Eq. (6.1), noting that V is just $S\,dx$, then gives

$$p = -K\frac{\partial \xi}{\partial x}. \tag{6.3}$$

Finally, we note that the motion of the element ABCD must be described by

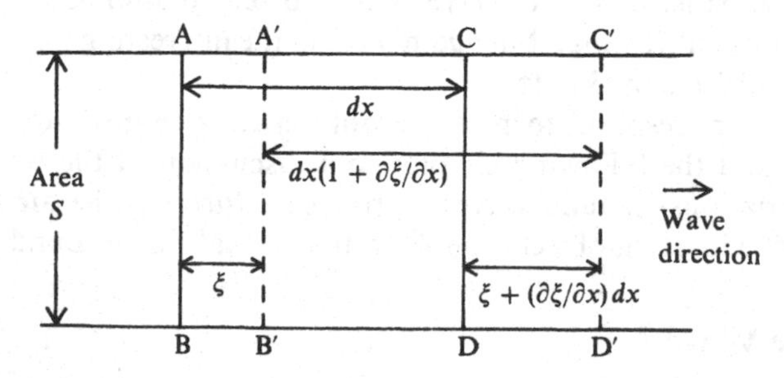

Fig. 6.1. In passage of a plane wave of displacement ξ, the fluid on plane AB is displaced to A′B′ and that on CD to C′D′.

Newton's equations so that, setting the pressure gradient force in the x direction equal to mass times acceleration,

$$-S\left(\frac{\partial p}{\partial x}dx\right) = \rho S\, dx \frac{\partial^2 \xi}{\partial t^2},$$

or

$$-\frac{\partial p}{\partial x} = \rho \frac{\partial^2 \xi}{\partial t^2}. \tag{6.4}$$

Then, from Eqs. (6.3) and (6.4),

$$\frac{\partial^2 \xi}{\partial t^2} = \frac{K}{\rho}\frac{\partial^2 \xi}{\partial x^2}, \tag{6.5}$$

or, differentiating Eq. (6.5) again with respect to x and Eq. (6.3) twice with respect to t,

$$\frac{\partial^2 p}{\partial t^2} = \frac{K}{\rho}\frac{\partial^2 p}{\partial x^2}. \tag{6.6}$$

Equations (6.5) and (6.6) are two different versions of the one-dimensional wave equation, one referring to the acoustic displacement ξ and the other to the acoustic pressure p. They apply equally well to any fluid if appropriate values are used for the bulk modulus K and density ρ. For the case of wave propagation in air, we need to decide whether the elastic behavior is isothermal, and thus described by the equation

$$p_a V = \text{constant} = nkT, \tag{6.7}$$

where T is the absolute temperature, or whether it is adiabatic, and so described by

$$p_a V^\gamma = \text{constant}, \tag{6.8}$$

where $\gamma = C_p/C_v = 1.4$ is the ratio of the specific heats of air at constant pressure and at constant volume, respectively, and p_a, as before, is the average atmospheric pressure.

Clearly, the temperature tends to rise in those parts of the wave where the air is compressed and to fall where it is expanded. The question is, therefore, whether appreciable thermal conduction can take place between these two sets of regions in the short time available as the peaks and troughs of the wave sweep by. It turns out (Fletcher, 1974) that at ordinary acoustic wavelengths the pressure maxima and minima are so far apart that no appreciable conduction takes place, and the behavior is therefore adiabatic. Only at immensely high frequencies does the free-air propagation tend to become isothermal. For sound waves in pipes or close to solid objects, on the other hand, the behavior also becomes isothermal at very low frequencies—below about 0.1 Hz for a 20 mm tube. Neither of these cases need concern us here.

Taking logarithms of Eq. (6.8) and differentiating, we find, using Eq. (6.2),

$$K = \gamma p_a, \tag{6.9}$$

so that Eq. (6.6) becomes

$$\frac{\partial^2 p}{\partial t^2} = c^2 \frac{\partial^2 p}{\partial x^2}, \tag{6.10}$$

where

$$c^2 = \frac{K}{\rho} = \frac{\gamma p_a}{\rho}, \tag{6.11}$$

and similarly for ξ from Eq. (6.5). As we shall see in a moment, the quantity c is the propagation speed of the sound wave.

It is easy to verify, by differentiation, that possible solutions of the wave equation [Eq. (6.10)] have the form

$$p(x, t) = f_1(x - ct) + f_2(x + ct), \tag{6.12}$$

where f_1 and f_2 are completely general continuous functions of their arguments. We can also see that $f_1(x - ct)$ represents a wave of arbitrary spatial shape $f_1(x - x_0)$ or of arbitrary time behavior $f_1(ct_0 - ct)$ propagating in the $+x$ direction with speed c. Similarly, $f_2(x + ct)$ represents a different wave propagating in the $-x$ direction, also with speed c. In the case of air, or any other nearly ideal gas, Eqs. (6.7) and (6.11) show that

$$c(T) = \left(\frac{T}{T_0}\right)^{1/2} c(T_0), \tag{6.13}$$

where $c(T)$ is the speed of sound at absolute temperature T. There is, however, no variation of c with atmospheric pressure. For air at temperature ΔT degrees Celsius and 50% relative humidity,

$$c \approx 332(1 + 0.00166\,\Delta T)\ \mathrm{m\ s^{-1}}, \tag{6.14}$$

giving $c \approx 343\ \mathrm{m\ s^{-1}}$ at room temperature.

The wave equation [Eq. (6.10)] was discussed in detail in Chapter 2 in relation to waves on a string, and its two-dimensional counterpart in Chapter 3. There is no need to repeat this discussion here except to remind ourselves that it is usual to treat Eq. (6.10) in the frequency domain where the solutions have the form

$$p = Ae^{-jkx}e^{j\omega t} + Be^{jkx}e^{j\omega t}, \tag{6.15}$$

where $k = \omega/c$ and the A and B terms represent waves traveling to the right and the left, respectively. [If we adopt the conventions of quantum mechanics and write time dependence as $\exp(-i\omega t)$, as for example in Morse (1948), then j should be replaced by $-i$.]

If we consider a wave of angular frequency ω traveling in the $+x$ direction, then we can set $B = 0$ and $A = 1$ in Eq. (6.15) and write

$$p = e^{-jkx}e^{j\omega t} \rightarrow \cos(-kx + \omega t), \tag{6.16}$$

where the second form of writing is just the real part of the first. From Eq. (6.5), ξ has a similar form, though with a different amplitude and perhaps a phase factor. We can connect p and ξ through Eq. (6.4), from which

$$jkp = j\rho\omega\frac{\partial\xi}{\partial t}, \tag{6.17}$$

or, if we write u for the acoustic fluid velocity $\partial\xi/\partial t$ and remember that $k = \omega/c$, then

$$p = \rho c u. \tag{6.18}$$

The acoustic pressure and acoustic fluid velocity (or particle velocity) in the propagation direction are therefore in phase in a plane wave.

This circumstance makes it useful to define a quantity z called the wave impedance (or sometimes the specific acoustic impedance):

$$z = \frac{p}{u} = \rho c. \tag{6.19}$$

It is clearly a property of the medium and its units are Pa m^{-1} s or kg m^{-2} s^{-1}, sometimes given the name rayls (after Lord Rayleigh). For air at temperature ΔT°C and standard pressure,

$$\rho c \approx 428(1 - 0.0017\,\Delta\mathrm{T})\ \mathrm{kg\,m^{-2}\,s^{-1}}. \tag{6.20}$$

In much of our discussion, we will need to treat waves in 3 space dimensions. The generalization of Eq. (6.10) to this case is

$$\frac{\partial^2 p}{\partial t^2} = c^2\nabla^2 p. \tag{6.21}$$

This differential equation can be separated in several coordinate systems to give simple treatments of wave behavior (Morse and Feshbach, 1953, pp. 499–518, 655–666). Among these are rectangular coordinates, leading simply to three equations for plane waves of the form of Eq. (6.10), and spherical polar coordinates, which we consider later in this chapter.

Before leaving this section, however, we should emphasize that we have consistently neglected second-order terms by assuming $p \ll p_a$, so that the resulting wave equation [Eq. (6.10) or Eq. (6.21)] is linear. This is of great assistance in development of the theory and turns out to be an adequate approximation even in very intense sound fields such as exist, for example, inside a trumpet. At even higher intensities, however, and well below the shock-wave limit, nonlinear terms begin to have a detectable effect (Beyer, 1974). It will not be necessary for us to use such extensions of the theory in this book.

6.2. Spherical Waves

When we assume a time dependence $\exp j\omega t$, the wave equation, Eq. (6.21), takes the form

$$\nabla^2 p + k^2 p = 0, \tag{6.22}$$

where $k = \omega/c$. This is known as the Helmholtz equation and is separable, and therefore relatively easily treated, in rectangular, spherical polar, and cylindrical polar coordinates. In spherical coordinates,

$$\nabla^2 p = \frac{1}{r^2}\frac{\partial}{\partial r}\left(r^2\frac{\partial p}{\partial r}\right) + \frac{1}{r^2 \sin\theta}\frac{\partial}{\partial\theta}\left(\sin\theta\frac{\partial p}{\partial\theta}\right) + \frac{1}{r^2\sin^2\theta}\frac{\partial^2 p}{\partial\phi^2}, \tag{6.23}$$

and the solution to Eq. (6.22) is the sum of a series of products of radial functions multiplied by spherical harmonics. The intensity pattern in the wave can therefore be very complicated. Of particular interest, however, is the simplest case in which p has no dependence on θ or ϕ but spreads uniformly from a single point at the origin.

For such a simple spherical wave, Eq. (6.23) becomes

$$\nabla^2 p = \frac{1}{r^2}\frac{\partial}{\partial r}\left(r^2\frac{\partial p}{\partial r}\right), \tag{6.24}$$

and we can simplify matters even further by writing $p = \psi/r$, giving for Eq. (6.22)

$$\frac{\partial^2\psi}{\partial r^2} + k^2\psi = 0, \tag{6.25}$$

which is just the one-dimensional wave equation. The general solution for p is therefore a superposition of an outgoing and an incoming wave given by

$$p = \left(\frac{A}{r}e^{-jkr} + \frac{B}{r}e^{jkr}\right)e^{j\omega t}. \tag{6.26}$$

To find the acoustic particle velocity u we use the equivalent of Eq. (6.4) in the form

$$\rho\frac{\partial u}{\partial t} = -\nabla p = -\frac{\partial p}{\partial r}, \tag{6.27}$$

where the second form of writing is possible since p depends only on r and t. Explicitly then, from Eq. (6.27) for the case of an outgoing wave ($B = 0$),

$$u = \frac{A}{r\rho c}\left(1 + \frac{1}{jkr}\right)e^{-jkr}e^{j\omega t}. \tag{6.28}$$

In the far field, when r is much greater than one wavelength ($kr \gg 1$), u is simply $p/\rho c$ as in the plane wave case, as we clearly expect. Within about one sixth of a wavelength of the origin, however, kr become less than unity. The velocity u then becomes large and shifted in phase relative to p.

The wave impedance for a spherical wave depends on distance from the origin, measured in wavelengths (i.e., on the parameter kr), and has the value

$$z = \frac{p}{u} = \rho c\left(\frac{jkr}{1 + jkr}\right). \tag{6.29}$$

Near the origin, $|z| \ll \rho c$, while $|z| \to \rho c$ for $kr \gg 1$.

The behavior of spherical waves with angular dependence involves more complex mathematics for its solution (Morse, 1948, pp. 314–321), and we will not take it up in detail here, though we will meet the topic again in the next chapter.

6.3. Sound Pressure Level and Intensity

Because there is a factor of about 10^6 between the acoustic pressure at the threshold of audibility and the limit of intolerable overload for the human ear, and because within that range subjective response is more nearly logarithmic than linear (actually, it is more complicated than this—see Gulick, 1971, pp. 108–134; Rossing, 1982, Chapter 6), it is convenient to do something similar for acoustic pressures. We therefore define the sound pressure level (SPL or L_p) for an acoustic pressure p to be

$$L_p = 20 \log_{10}\left(\frac{p}{p_0}\right), \tag{6.30}$$

measured in decibels (dB) above the reference pressure p_0. By convention both p and p_0 are rms values and p_0 is taken to be 20 μPa, which is approximately the threshold of human hearing in its most sensitive range from 1 to 3 kHz. On this SPL scale, 1 Pa is approximately 94 dB and the threshold of pain is about 120 dB. The normal range for music listening is about 40 to 100 dB. It is usual, however, to specify sound pressure levels in the environment after applying a filter (Type-A weighting) to allow for the decreased sensitivity of human hearing at low and high frequencies. The SPL is then given in dB(A).

The sensitivity of normal human hearing is shown in Fig. 6.2, as originally

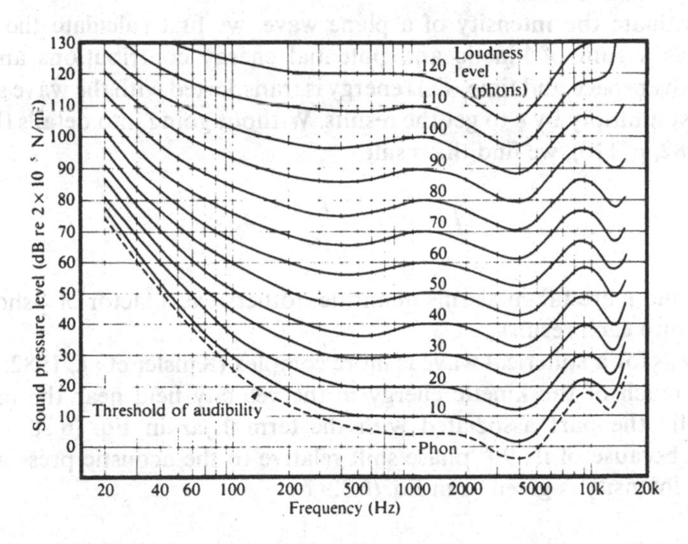

Fig. 6.2. Equal-loudness curves for human hearing. The broken curve is the approximate threshold of hearing for pure tones, and the 120-phon curve is the threshold of discomfort.

determined by Fletcher and Munson (1933). Each contour passes through points of subjectively equal loudness and is labeled with a loudness level in phons, taken by definition to be the SPL associated with that contour at a frequency of 1 kHz. The normal threshold of hearing for pure tones is a few decibels above the 0 phon contour, and the threshold of pain, or at least discomfort, is at about 120 phon. The A-weighting curve used in specifying environmental (or musical) noise (or sound) levels is approximately the inverse of the 40 phon contour.

Because human hearing responds to acoustic pressure at a point, or rather at two points corresponding to two ears that have little acoustic coupling, the sound pressure level is usually the relevant quantity to specify. The SPL, however, depends on the environment and in particular on the reverberant quality of the space in which the sound source and listener are situated. In many cases, it is more useful physically to know the acoustic energy carried through a surface by sound waves. This quantity is called the acoustic intensity I and is measured in watts per square meter. Again, a logarithmic scale is convenient, and we define the intensity level (IL or L_I) to be

$$L_I = 10 \log_{10}\left(\frac{I}{I_0}\right) \tag{6.31}$$

in decibels. [The factor is 10 rather than 20 as in Eq. (6.30) since I is proportional to p^2.] The reference intensity I_0 is taken as 10^{-12} W m^{-2}, which makes the SPL and the IL very nearly equal for a plane wave. For a standing wave, of course, the sound pressure level may be large, but the intensity will be small since the intensities in the two waves tend to cancel because of their opposite propagation directions.

To evaluate the intensity of a plane wave, we first calculate the energy density as a sum of kinetic and potential energy contributions and then average over space and time. This energy is transported with the wave speed c, so we just multiply by c to get the results. Without going into details (Kinsler et al., 1982, p. 110), we find the result

$$I = \rho c u^2 = \frac{p^2}{\rho c} = pu, \tag{6.32}$$

where p and u are taken as rms quantities (otherwise, a factor of $\frac{1}{2}$ should be inserted into each result).

Analysis for a spherical wave is more complex (Kinsler et al., 1982, p. 112) because much of the kinetic energy in the velocity field near the origin—specifically the part associated with the term $1/jkr$ in Eq. (6.28)—is not radiated because of its 90° phase shift relative to the acoustic pressure. The radiated intensity is given as in Eq. (6.32) by

$$I = \frac{p^2}{\rho c}, \tag{6.33}$$

where an rms value of p is implied, but the other forms of the result in Eq. (6.32) do not apply.

The total power P radiated in a spherical wave can be calculated by integrating $I(r)$ over a spherical surface of radius r, giving

$$P = \frac{4\pi r^2 p(r)^2}{\rho c}. \tag{6.34}$$

From Eq. (6.26), P is independent of r, as is obviously required. To get some feeling for magnitudes, a source radiating a power of 1 mW as a spherical wave produces an intensity level, or equivalently a sound pressure level, of approximately 79 dB at a distance of 1 m. At a distance of 10 m, assuming no reflections from surrounding walls or other objects, the SPL is 59 dB. These figures correspond to radiation from a typical musical instrument, though clearly a great range is possible. The disparity between this figure and the powers of order 100 W associated with amplifiers and loudspeakers is explained by the facts that the amplifier requires adequate power to avoid overload during transients and the normal operating level is only a few watts, while the loudspeaker itself has an efficiency of only about 1% in converting electrical power to radiated sound.

6.4. Reflection and Transmission

When a wave encounters any variation in the properties of the medium in which it is propagating, its behavior is disturbed. Gradual changes in the medium extending over many wavelengths lead mostly to a change in the wave speed and propagation direction—the phenomenon of refraction. When the change is more abrupt, as when a sound wave in air strikes a solid object, such as a person or a wall, then the incident wave is generally mostly reflected or scattered and only a small part is transmitted into or through the object. That part of the wave energy transmitted into the object will generally be dissipated by internal losses and multiple reflections unless the object is very thin, like a lightweight wall partition, when it may be reradiated from the opposite surface.

It is worthwhile to examine the behavior of a plane pressure wave $A\exp(-jkx)$ moving from a medium of wave impedance z_1 to one of impedance z_2. In general, we expect there to be a reflected wave $B\exp(jkx)$ and a transmitted wave $C\exp(-jkx)$. The acoustic pressures on either side of the interface must be equal, so that, taking the interface to be at $x = 0$,

$$A + B = C. \tag{6.35}$$

Similarly, the displacement velocities must be the same on either side of the interface, so that, using Eq. (6.19) and noting the sign of k for the various waves,

$$\frac{A - B}{z_1} = \frac{C}{z_2}. \tag{6.36}$$

We can now solve Eqs. (6.35) and (6.36) to find the reflection coefficient:

$$\frac{B}{A} = \frac{z_2 - z_1}{z_2 + z_1}, \tag{6.37}$$

and the transmission coefficient:

$$\frac{C}{A} = \frac{2z_2}{z_2 + z_1}. \tag{6.38}$$

These coefficients refer to pressure amplitudes. If $z_2 = z_1$, then $B = 0$ and $C = A$ as we should expect. If $z_2 > z_1$, then, from Eq. (6.37), the reflected wave is in phase with the incident wave and a pressure maximum is reflected as a maximum. If $z_2 < z_1$, then there is a phase change of 180° between the reflected wave and the incident wave and a pressure maximum is reflected as a minimum. If $z_2 \gg z_1$ or $z_2 \ll z_1$, then reflection is nearly total. The fact that, from Eq. (6.38), the transmitted wave will have a pressure amplitude nearly twice that of the incident wave if $z_2 \gg z_1$ is not a paradox, as we see below, since this wave carries a very small energy.

Perhaps even more illuminating than Eqs. (6.37) and (6.38) are the corresponding coefficients expressed in terms of intensities, using Eq. (6.32). If the incident intensity is $I_0 = A^2/z_1$, then the reflected intensity I_r is given by

$$\frac{I_r}{I_0} = \left(\frac{z_2 - z_1}{z_2 + z_1}\right)^2, \tag{6.39}$$

and the transmitted intensity I_t by

$$\frac{I_t}{I_0} = \frac{4z_2 z_1}{(z_2 + z_1)^2}. \tag{6.40}$$

Clearly, the transmitted intensity is nearly zero if there is a large acoustic mismatch between the two media and either $z_2 \gg z_1$ or $z_2 \ll z_1$.

We can use the same approach to calculate the reflection and transmission coefficients when a sound wave is incident obliquely on the plane interface between two fluid media. We then have an incident wave $A\exp(-\mathbf{k}_i.\mathbf{r})$, a specularly reflected wave $B\exp(-\mathbf{k}_r.\mathbf{r})$, and a transmitted wave $C\exp(-\mathbf{k}_t.\mathbf{r})$ and we must use the matching conditions on pressure and vector displacement at the interface to determine not only the ratio of the coefficients A, B, and C, but also the directions of the propagation vectors $\mathbf{k}_r$ and $\mathbf{k}_t$ relative to the incident vector $\mathbf{k}_i$. The algebra is complicated, without being difficult, so we simply quote the results.

Suppose that θ_i, θ_r and θ_t are the angles between the normal to the interface and the directions of $\mathbf{k}_i$, $\mathbf{k}_r$ and $\mathbf{k}_t$ respectively, just as in geometrical optics. Then we find that $\theta_r = \theta_i$ as for specular reflection, while

$$\frac{\sin \theta_t}{\sin \theta_i} = \frac{k_1}{k_2} = \frac{c_2}{c_1} \tag{6.41}$$

where k_1, c_1 refer to the first medium and k_2, c_2 to the second. Equation (6.41) is familiar as Snell's Law in geometrical optics. The reflection coefficient is given by

$$\frac{I_r}{I_i} = \left(\frac{z_2 \cos \theta_i - z_1 \cos \theta_t}{z_2 \cos \theta_i + z_1 \cos \theta_t}\right)^2 \tag{6.42}$$

and the transmission coefficient by

$$\frac{I_t}{I_0} = \frac{4 z_1 z_2 \cos^2 \theta_i}{(z_2 \cos \theta_i + z_1 \cos \theta_t)^2} \tag{6.43}$$

where Eq. (6.41) must be used to express $\cos \theta_t$ in terms of θ_i. Clearly Eqs. (6.42) and (6.43) reduce to the normal-incidence formulae (6.39) and (6.40) when $\theta_i = 0$. The results (6.41)–(6.43) contain all the familiar phenomena of geometrical optics, such as refraction and total reflection. We should note in passing that if the incident beam is localized, then the width of the transmitted beam differs from that of the incident beam by a factor $\cos \theta_t / \cos \theta_i$. We must therefore multiply the intensity transmission coefficient, given by Eq. (6.43), by this factor to obtain the total power transmission coefficient for such a localized beam.

We should also note that the expressions are also formally correct if the wave impedances z_i are complex quantities $r_i + jx_i$, allowing the possibility of wave absorption in the two fluids.

The results (6.41)–(6.43) cannot be applied to solids, because the existence of a shear modulus means that longitudinal sound waves are partially converted to transverse waves at the boundary, so that the whole analysis becomes much more complex. If the solid is isotropic and incidence on the interface is normal, however, the analysis leading to Eqs. (6.39) and (6.40), as well as the results themselves, remain valid, since propagation remains longitudinal in the solid. This allows us to work out a number of useful practical results for transmission of sound through solid panels, by neglecting the shear stiffness of the solid. Suppose we have a solid panel of the solid medium, characterized by a wave impedance z_2, separating two semi-infinite regions of fluid with wave impedances z_1 and z_3. The algebra is a little complicated because we have an incident and a reflected wave in region 1, waves travelling in both directions in region 2, and a single transmitted wave in region 3, making five unknown complex amplitudes A_i in expressions such as $A_i \exp(\pm jk_i x)$ for the pressure waves, even for normal incidence. We require that the pressures and particle velocities match over the two surfaces of the panel, which gives four complex equations and enables us to determine all the wave amplitudes in terms of the incident amplitude and the properties of the three media. From these amplitudes we can calculate the intensities in each wave from the relation $I_i = p_i u_i = p_i^2 / z_i$, where z_i is the wave impedance in the medium concerned and we assume the amplitudes to be rms values. The calculated transmission coefficient is then

$$\frac{I_t}{I_0} = \frac{4z_1 z_2^2 z_3}{z_2^2(z_1 + z_3)^2 \cos^2 k_2 l + (z_2^2 + z_1 z_2)^2 \sin^2 k_2 l} \tag{6.44}$$

where l is the thickness of the panel.

One of the most important practical applications of this result is to the case in which we have a solid panel with air on both sides. Here $z_3 = z_1 \ll z_2$ and Eq. (6.44) simplifies to

$$\frac{I_t}{I_0} \approx \left(\frac{2z_1}{z_2 \sin k_2 l}\right)^2. \tag{6.45}$$

If the panel is thin compared with the sound wavelength, then $\sin k_2 l \approx k_2 l$ and we can further replace $c_2 k_2$ by ω to give

$$\frac{I_t}{I_0} \approx \left(\frac{2z_1}{\rho l \omega}\right)^2. \tag{6.46}$$

This is just the result we would arrive at if we assumed the panel to behave as a simple mass load ρl per unit area, driven by the difference in sound pressure between its faces.

All this discussion about panels assumes, however, that their extent is infinite, whereas this can never be the case in practice—the panel must be somehow supported on a rigid frame. Once we recognize a structure of this sort, we realize that individual segments of the panel can resonate—a situation that was not possible in the truly infinite case. A detailed consideration of transmission through such a resonant panel would take us too far afield, but it is easy to see the general effect on the basis of the comment following Eq. (6.46). All we need to do to reach at least a semiquantitative result is to replace the mass load ρ/ω per unit area at the resonance frequency ω^* by the resonant load $\rho l \omega^*/Q$, where Q is the quality factor of the resonance. This artifice will work, however, only for the fundamental resonance of the panel, at which frequency it all moves in-phase, and even then it is only approximate because of the effect of mode shape. For higher resonances the result is more complex. It is clear that such resonances can greatly increase sound transmission through panels, since Q may well be as large as 10 for a simple undamped panel.

All this discussion applies only to cases where the extended surface between two media is flat and very large, both compared with the sound wavelength involved. If the surfaces of concern are all flat and of large extent in this sense, then the familiar rules of geometrical optics are an adequate approximation for the treatment of reflections. It is only for large areas, such as the walls or ceilings of concert halls, that this is of more than qualitative use in understanding behavior (Beranek, 1962; Rossing, 1982; Meyer, 1978).

At the other extreme, an object that is small compared with the wavelength of the sound wave involved will scatter the wave almost equally in all directions, the fractional intensity scattered being proportional to the sixth power of the size of the object. When the size of the object ranges from, for example, one-tenth of a wavelength up to 10 wavelengths, then scattering behavior is very complex, even for simply shaped objects (Morse, 1948, pp. 346–356; Morse and Ingard, 1968, pp. 400–449).

There is similar complexity in the "sound shadows" cast by objects. Objects that are very large compared with the sound wavelength create well-defined shadows, but this situation is rarely encountered in other than architectural acoustics. More usually, objects will be comparable in size to the wavelength involved, and diffraction around the edges into the shadow zone will blur its edges or even eliminate the shadow entirely at distances a few times the diameter of the object. Again, the discussion is complex even for a simple plane edge (Morse and Ingard, 1968, pp. 449–458). For the purposes of this book, a qualitative appreciation of the behavior will be adequate.

6.5. Absorption

Even in an unbounded uniform medium such as air, a plane sound wave is attenuated as it propagates because of losses of various kinds (Kinsler et al., 1982, Chapter 7). This occurs for two different reasons. The first is that, when a small element of the air is compressed in the wave, its shape changes since the compression is only along the direction of propagation. Any shape change is resisted by viscous forces and, though these are small in gases, they still dissipate some of the energy. The second reason has to do with thermal effects. When the gas is compressed in the wave, its temperature and density both rise to follow the compression. The process is complicated, however, by the fact that heat can be conducted from the warmer compressed parts of the gas to the cooler expanded parts only half a wavelength away. A further complication arises from the fact that molecules of oxygen and nitrogen are able to rotate and vibrate, and the compressional heat energy, initially confined to the translation velocity of these molecules, gradually becomes shared with the internal modes as well, reducing the temperature and increasing the density.

We can express all these processes formally by making the propagation constant k a complex quantity and writing

$$k \to \frac{\omega}{c} - j\alpha \tag{6.47}$$

so that the wave amplitude decays as $\exp(-\alpha x)$ for a plane wave and $(1/r)\exp(-\alpha r)$ for a spherical wave. The intensity in the two cases decays as $\exp(-2\alpha x)$ and $(1/r^2)\exp(-2\alpha r)$, respectively. The quantity α is called the attenuation coefficient, and its value, or more usually the value of 8686α, corresponding to the attenuation in decibels per kilometer, is available in standard tables (Evans and Bass, 1986).

If we consider just viscous and thermal-conduction relaxation, we can deduce the frequency dependence of the attenuation coefficient α by a simple argument. Consider first the viscous losses. For a given pressure amplitude, the energy loss per cycle is proportional to the magnitude of the shear shape change, which is inversely proportional to wavelength, so that the total energy loss per second is proportional to ω^2. In the case of thermal losses, the diffusion of heat energy is proportional to the second derivative of the temperature profile, and thus also to ω^2. These two mechanisms therefore give a simple ω^2 variation for the attenuations coefficient α, both contributing by comparable amounts.

For absolutely dry air, the relaxation time for transfer of energy to the molecular rotation modes is very short compared with the period of a sound wave, so that these modes stay close to equilibrium excitation and do not contribute to the attenuation. The relaxation time for the vibrational modes, in contrast, is very long, so that again they do not contribute to the attenuation. The behavior of such dry air is therefore fairly well described by the simple theory outlined above. Normal air, however, contains traces of other gases, particularly of water vapor, and these, through their collisions with oxygen and nitrogen molecules, effectively catalyze the transfer of energy to the vibration modes of these molecules and so greatly reduce the relaxation time so that it comes into the sonic or even ultrasonic range. This molecular relaxation greatly increases the attenuation of sound over a wide range of frequencies centered on that for which $\omega\tau \sim 1$. In air of very low relative humidity, α is increased over the classical value by as much as a factor 100 near the attenuation peak, whereas an increase of a factor 10 over the high-frequency part of the audible range is more typical of normal humidities. Because of this wide variation with relative humidity, it is not possible to give a simple formula for the attenuation of sound in air. A reasonable fit to the values in the tables for air at 50% relative humidity is, however, provided by the equations

$$\begin{aligned} \alpha &\approx 4 \times 10^{-7} f, \qquad 100\text{ Hz} < f < 1\text{ kHz} \\ \alpha &\approx 1 \times 10^{-10} f^2, \qquad 2\text{ kHz} < f < 100\text{ kHz} \end{aligned} \tag{6.48}$$

where f is the frequency in hertz. To an adequate approximation one can simply multiply these values by 10^4 to find the attenuation in dB/km.

For propagation of sound over large distances outdoors, attenuation of the higher frequencies by atmospheric absorption is very significant, and may be made even more so by the added effects of wind turbulence. For indoor applications in small rooms one can often ignore the propagation attenuation, though in large halls the high frequencies may be significantly attenuated by atmospheric absorption, which amounts to about 0.1 dB/m at 10 kHz.

More important, in most halls, is the absorption of sound upon reflection from the walls, ceilings, furnishings, and audience. If an impulsive sound is made in a hall, then the sound pressure level decays nearly linearly, corresponding to an exponential decay in sound pressure. The time T_{60} for the level to decay by 60 dB is known as the reverberation time. Details of the behavior are complicated, but to a first approximation T_{60} is given by the Sabine equation:

$$T_{60} = \frac{0.161V}{\sum_i \alpha_i S_i}, \tag{6.49}$$

where V is the hall volume (in cubic meters) and the S_i are areas (in square meters) of surface with absorption coefficients α_i. Satisfactory reverberation time is important for good music listening but depends on building size and musical style. For a small hall for chamber music, T_{60} may be as small as 0.7 s, for a large concert hall, it may be 1.7 to 2 s, and for a large cathedral suitable for nineteenth century organ music, as long as 10 s (Meyer, 1978; Beranek, 1962, pp. 555–569).

In musical instruments, we will often be concerned with sound waves confined in tubes or boxes or moving close to other surfaces. In such cases, there is attenuation caused by viscous forces in the shearing motion near the surface and by heat conduction from the wave to the solid. Both these effects are confined to a thin boundary layer next to the surface, the thickness of which decreases as the frequency increases. The viscous and thermal boundary layers have slightly different thicknesses, δ_v and δ_t, given by (Benade, 1968)

$$\delta_v = \left(\frac{\eta}{2\pi\rho}\right)^{1/2} f^{-1/2} \approx 1.6 \times 10^{-3} f^{-1/2},$$

and (6.50)

$$\delta_t = \left(\frac{\kappa}{2\pi\rho C_p}\right)^{1/2} f^{-1/2} \approx 1.9 \times 10^{-3} f^{-1/2},$$

where ρ is the density, η the viscosity, κ the thermal conductivity, and C_p the specific heat of air (per unit volume). For sound waves in musical instruments, these boundary layers are thus about 0.1 mm thick. We shall return to consider these surface losses in more detail in Chapter 8.

6.6. Normal Modes in Cavities

While many problems involve the propagation of acoustic waves in nearly unconfined space, many other practical problems involve the acoustics of enclosed spaces such as rooms, tanks, and other cavities. We defer discussion of wave propagation in pipes and horns—which generally have at least one open end—to Chapter 8, and consider here only completely enclosed volumes.

The mathematical principles involved are straightforward—we simply need to solve the three-dimensional version of the wave propagation equation (6.6) with appropriate boundary conditions on the walls of the enclosure. To be more explicit, the three-dimensional wave equation can be written in vector notation in the form

$$\frac{\partial^2 p}{\partial t^2} = c^2 \nabla^2 p \tag{6.51}$$

and we need simply to express the vector differential operator ∇^2 in coordinates related to the geometry of the enclosure so that the boundary conditions are simple. If the enclosure is rectangular, for example, then the wave equation (6.51) becomes

$$\frac{\partial^2 p}{\partial t^2} = c^2 \left(\frac{\partial^2 p}{\partial x^2} + \frac{\partial^2 p}{\partial y^2} + \frac{\partial^2 p}{\partial z^2} \right) \tag{6.52}$$

and we can write $p(x, y, z) = p_x(x)p_y(y)p_z(z)$. We return to this case in a moment.

The boundary conditions at the boundaries of the enclosure depend upon the physical nature of the walls. Suppose that the walls present an acoustic wave impedance z_W, then at the walls we must have

$$p = z_W u = \frac{j z_W}{\rho \omega} \frac{\partial p}{\partial n} \tag{6.53}$$

where the second form of writing comes from Eq. (6.4) and n is a coordinate normal to the surface of the wall. It is easy enough to deal with this general boundary condition if z_W is imaginary, so that the walls are either simply springy or simply slack and heavy. More generally the walls will have resistive losses so that z_W will contain a real part. The simplest case is that in which the walls are completely rigid, for then $z_W = \infty$ and Eq. (6.53) reduces to

$$\frac{\partial p}{\partial n} = 0. \tag{6.54}$$

It is usually an adequate approximation for real enclosures to adopt Eq. (6.54) as a boundary condition when determining the mode frequencies and to make allowance for any real part on z_W by adding damping to the modes.

To illustrate some aspects of the mode problem, suppose that we have a rectangular enclosure with sides of lengths a, b, and c. Solutions of Eq. (6.52) that satisfy the boundary condition (6.54) on all the walls have the form

$$p(x, y, z, t) = A \cos\left(\frac{l\pi x}{a}\right) \cos\left(\frac{m\pi y}{b}\right) \cos\left(\frac{n\pi z}{c}\right) \sin \omega t \tag{6.55}$$

where l, m, n are integers (including the possibility of zero) and the frequency

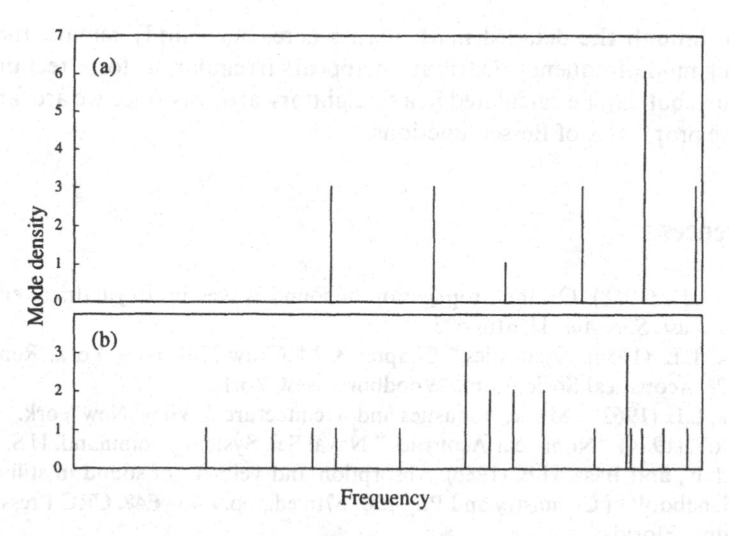

Fig. 6.3. Distribution of mode frequencies for two rectangular rooms of equal volume and dimension ratios (a) 2 : 2 : 2 and (b) 1 : 2 : 3. Where mode frequencies are coincident, the relevant line has been lengthened proportionally.

ω satisfies

$$\omega = \pi c \left(\frac{l^2}{a^2} + \frac{m^2}{b^2} + \frac{n^2}{c^2} \right)^{1/2}. \tag{6.56}$$

The frequencies given by this equation are the mode frequencies for the enclosure. Clearly their distribution depends upon the relative values of a, b, and c, and thus on the shape of the enclosure. Figure 6.3 shows such mode distributions for two representative rooms with equal volume but with dimensions in the ratio 2 : 2 : 2 and 1 : 2 : 3, respectively. It is one of the important parameters of good architectural acoustics to arrange the room shape so that the frequencies of the lower modes of the space are reasonably evenly distributed—any concentration of modes near some particular frequency can give a tonal character to the reverberation that is generally detrimental to music listening. Clearly the 2 : 2 : 2 room has a "peaky" response with many coincident resonances, whereas the 1 : 2 : 3 room has a much more even spread and is likely to be more satisfactory musically.

We can adopt the same approach to calculate the mode frequencies for other enclosures with simple geometry, in particular for a spherical cavity and for a cavity in the form of a circular cylinder. For the spherical case the wave equation can be separated in spherical polar coordinates (r, θ, ϕ) and the mode functions for the pressure can be written in terms of spherical Bessel functions and spherical harmonics. The cylindrical case can be separated in cylindrical polar coordinates (r, ϕ, z) and the pressure can be expressed in terms of ordinary Bessel functions and trigonometric (circular) functions. We shall not go through the detailed mathematics here, but simply remark that the resulting mode-frequency distribution appears irregular, as for a rectangular enclosure, but can be calculated in a straightforward way once we are familiar with the properties of Bessel functions.

References

Benade, A.H. (1968). On the propagation of sound waves in a cylindrical conduit. *J. Acoust. Soc. Am.* **44**, 616–623.

Beranek, L.L. (1954). "Acoustics," Chapter 3. McGraw-Hill, New York. Reprinted 1986, Acoustical Society Am., Woodbury, New York.

Beranek, L.L. (1962). "Music, Acoustics and Architecture." Wiley, New York.

Beyer, R.T. (1974). "Nonlinear Acoustics." Naval Sea Systems Command, U.S. Navy.

Evans, L.B., and Bass, H.E. (1986). Absorption and velocity of sound in still air. In "Handbook of Chemistry and Physics," 67th ed., pp. E45–E48. CRC Press, Boca Raton, Florida.

Fletcher, H., and Munson, W.A. (1933). Loudness, its definition, measurement and calculation. *J. Acoust. Soc. Am.* **5**, 82–108.

Fletcher, N.H. (1974). Adiabatic assumption for wave propagation. *Am. J. Phys.* **42**, 487–489.

Gulick, W.L. (1971). "Hearing: Physiology and Psychophysics." Oxford Univ. Press, London and New York.

Kinsler, L.E., Frey A.R., Coppens, A.B., and Sanders, J.V. (1982). "Fundamentals of Acoustics," 3rd ed., Wiley, New York.

Meyer, J. (1978). "Acoustics and the Performance of Music." Verlag Das Musikinstrument, Frankfurt am Main.

Morse, P.M. (1948). "Vibration and Sound." McGraw-Hill, New York. Reprinted 1981, Acoustical Society Am., Woodbury, New York.

Morse, P.M., and Feshbach, H. (1953). "Methods of Theoretical Physics," 2 vols. McGraw-Hill, New York.

Morse, P.M., and Ingard, K.U. (1968). "Theoretical Acoustics." McGraw-Hill, New York; reprinted 1986, Princeton Univ. Press, Princeton, New Jersey.

Rossing, T.D. (1982). "The Science of Sound," Chapter 23. Addison–Wesley, Reading, Massachusetts.

Stevens, S.S., and Davis, H. (1938). "Hearing: Its Psychology and Physiology." Wiley, New York. Reprinted 1983, American Institute of Physics, New York.

CHAPTER 7

Sound Radiation

In Chapter 6, we discussed some of the basic properties of sound waves and made a brief examination of the way sound waves are influenced by simple structures, such as tubes and cavities. In the present chapter, we take up the inverse problem and look at the way in which vibrating structures can generate sound waves. This is one of the most basic aspects of the acoustics of musical instruments—it is all very well to understand the way in which a solid body vibrates, but unless that vibration leads to a radiated sound wave, we do not have a musical instrument. We might, of course, simply take the fact of sound radiation for granted, and this is often done. This neglects, however, a great deal of interesting and important physics and keeps us from understanding much of the subtlety of musical instrument behavior.

Our plan, therefore, will be to look briefly at the properties of some of the simplest types of sources—monopoles, dipoles, and higher multipoles—to see the behavior we might expect. We then look at radiation from vibrating cylinders, since vibrating strings are so common in musical instruments, and then go on to the much more complicated problem of radiation from the motion of reasonably large and more-or-less flat bodies. These are, of course, the essential sound-radiating elements of all stringed and percussion instruments. The radiation of sound from the vibrating air columns of wind instruments presents a related but rather different set of problems that we defer for discussion in Chapter 8.

7.1. Simple Multipole Sources

The simplest possible source is the point source, which is the limit of a pulsating sphere as its radius tends to zero. Suppose the sphere has a small radius a and that the pulsating flow has a frequency ω and amplitude

$$Q = 4\pi a^2 u(a), \tag{7.1}$$

where $v(a)$ is the radial velocity amplitude at the surface. This object clearly generates a spherical wave, and, from our discussion in Chapter 6 and specifically from Eqs. (6.26) and (6.28), the pressure and velocity amplitudes in such a wave at radius r are given by

$$p(r) = (A/r)e^{-jkr}, \tag{7.2}$$

and

$$u(r) = \frac{A}{\rho c r}\left(1 - \frac{j}{kr}\right)e^{-jkr}. \tag{7.3}$$

Matching Eq. (7.3) to Eq. (7.1) on the surface of the sphere and assuming $ka \ll 1$, we find a value for A from which Eq. (7.2) gives

$$p(r) = \frac{j\omega\rho}{4\pi r}Qe^{-jkr}. \tag{7.4}$$

This result does not depend on the sphere radius a, provided $ka \ll 1$, and so it is the pressure wave generated by a point source of strength Q. Such a source is also called a monopole source. Its radiated power P is simply $\frac{1}{2}p^2/\rho c$ integrated over a spherical surface, whence

$$P = \frac{\omega^2\rho Q^2}{8\pi c}. \tag{7.5}$$

Note that for a given source strength Q the radiated power increases as the square of the frequency.

The next type of source to be considered is the dipole, which consists of two simple monopole sources of strengths $\pm Q$ separated by a distance dz, in the limit $dz \to 0$. A physical example of such a source is the limit of a small sphere oscillating to and fro. Referring to Fig. 7.1, if r_+ and r_- are, respectively, the distances from the positive and negative sources in the dipole to the observation point (r, θ), then the acoustic pressure at this point is, from Eq. (7.4),

$$p = \left(\frac{j\omega\rho}{4\pi}\right)\left(\frac{e^{-jkr_+}}{r_+} - \frac{e^{-jkr_-}}{r_-}\right)Q. \tag{7.6}$$

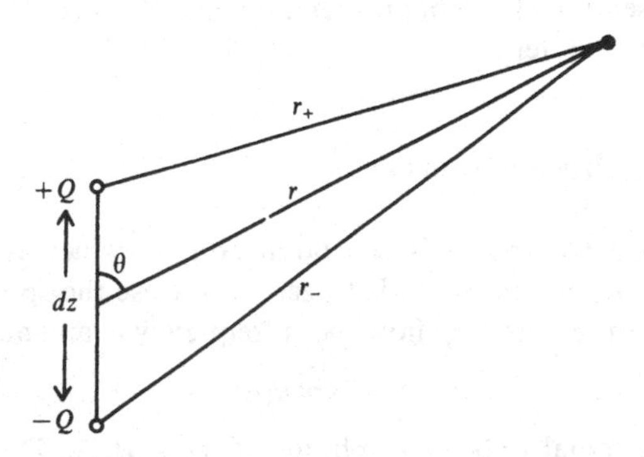

Fig. 7.1. A dipole source. In the limit $dz \to 0$, $Q \to \infty$, the dipole moment is $\mu = Q\,dz$.

As $dz \to 0$, this difference can be replaced by a differential, so that

$$p = \frac{j\omega\rho}{4\pi}\frac{\partial}{\partial z}\left(\frac{e^{-jkr}}{r}\right)Q\,dz$$
$$= \frac{\omega^2\rho}{4\pi cr}\left(1 + \frac{1}{jkr}\right)e^{-jkr}\mu\cos\theta, \tag{7.7}$$

where we have let $Q \to \infty$ as $dz \to 0$, so that the dipole moment $\mu = Q\,dz$ remains finite. The velocity field can be found by taking the gradient of p as usual. In the far field, where $kr \gg 1$, Eq. (7.7) can clearly be simplified by neglecting $1/jkr$ relative to 1. The radiated power P is simply

$$P = \frac{1}{2}\iint \frac{p^2}{\rho c} r^2 \sin\theta\, d\theta\, d\phi = \frac{\omega^4\rho\mu^2}{24\pi c^3}. \tag{7.8}$$

The power radiated by a dipole is thus a very strong function of frequency, and dipole sources are very inefficient radiators at low frequencies. From Eq. (7.7), the radiation is concentrated along the axis of the dipole.

The process of differentiating the field of a monopole to obtain that of a dipole, as expressed in Eq. (7.7), can be thought of as equivalent to reflecting the monopole source, and its radiation field, in a pseudomirror, which changes the sign of the source. The limit operation as we go from Eq. (7.6) to Eq. (7.7) is equivalent to moving the source closer and closer to the mirror plane.

The next step in complication is to reflect the dipole in another mirror plane, as shown in Fig. 7.2, to produce a quadrupole. This can be done in two ways to produce either an axial (longitudinal) quadrupole, in which the four simple poles lie on a straight line, or a plane (lateral) quadrupole, in which they lie at the corners of a square. The sign of the reflection is always chosen so that the quadrupole has no dipole moment. A physical example of a quadrupole source is a sphere, vibrating so that it becomes alternately a prolate and an oblate spheroid.

The pressure field for a quadrupole can be found by differentiating that for a dipole, Eq. (7.7), either with respect to z for an axial quadrupole or x for a plane quadrupole. To find the far-field radiation terms, the differentiation can be confined to the exponential factor. If the monopole source strength is Q,

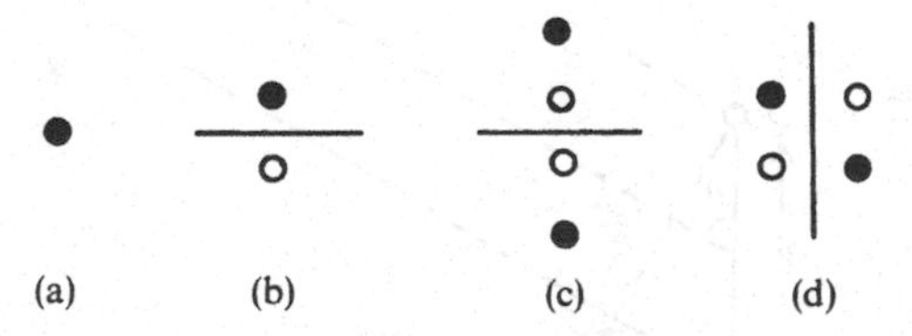

Fig. 7.2. Generation of two possible configurations of a quadrupole by successive reflection of a monopole, with a sign change on each reflection: (a) monopole, (b) dipole, (c) axial quadrupole, and (d) planar quadrupole.

then we clearly have the following sequence of results for the far-field pressure generated by monopole, dipole, and quadrupole sources, respectively:

$$p_m = jk\left(\frac{\rho c Q}{4\pi r}\right)e^{-jkr}, \tag{7.9}$$

$$p_d = -k^2\left(\frac{\rho c Q\,dz}{4\pi r}\right)\cos(r,z)e^{-jkr}, \tag{7.10}$$

and

$$p_q = -jk^3\left(\frac{\rho c Q\,dz\,dx}{4\pi r}\right)\cos(r,z)\cos(r,x)e^{-jkr}. \tag{7.11}$$

The notation is obvious and, for an axial quadrupole, x is replaced by z in Eq. (7.11). The important points to note are the increasingly complex angular behavior and more steeply decreasing radiation efficiency at low frequencies as we proceed through the series.

7.2. Pairs of Point Sources

To guide our later discussion, it is now helpful to examine briefly the radiation behavior of combinations of several point sources whose separation is not necessarily small compared with the sound wavelength. First, let us treat the case of two monopoles of strength Q separated by a distance d as shown in Fig. 7.3. The sources can be of either the same or opposite sign, and we seek the pressure p at a large distance $r \gg d$ in direction θ. For $r \sim d$, the expression is complicated, as in Eq. (7.6), but for $r \gg d$, we have

$$p \approx \left(\frac{j\omega\rho Q}{4\pi r}\right)e^{-jkr}(e^{1/2jkd\cos\theta} \pm e^{-1/2jkd\cos\theta}), \tag{7.12}$$

where the plus sign goes with like sources and the minus sign with sources of

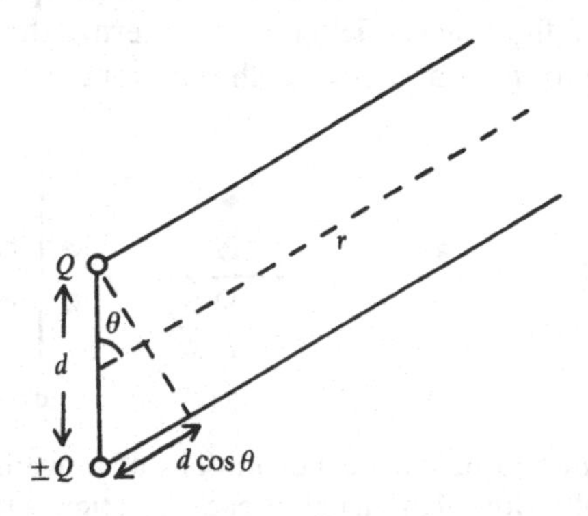

Fig. 7.3. Radiation from a pair of separated monopoles.

opposite phases. The absolute value of the square of p is

$$|p^2| = \left(\frac{\omega\rho Q}{2\pi r}\right)^2 \left[\cos^2\left(\frac{1}{2}kd\cos\theta\right);\ \sin^2\left(\frac{1}{2}kd\cos\theta\right)\right], \tag{7.13}$$

where $\cos^2$ goes with the plus and $\sin^2$ with the minus sign. The total radiated power is then

$$P = \frac{1}{2}\iint\left(\frac{|p^2|}{\rho c}\right)r^2\sin\theta\,d\theta\,d\phi = \frac{\omega^2\rho Q^2}{4\pi c}\left[1 \pm \frac{\sin kd}{kd}\right]. \tag{7.14}$$

The results in Eqs. (7.13) and (7.14) contain a great deal of information. From Eq. (7.13), the angular variation of the acoustic intensity $p^2/\rho c$ is very complex if kd is not small relative to unity, and there are many values of θ for which the intensity vanishes, whether the two sources are in phase or out of phase. From Eq. (7.14), however, the behavior of the total radiated power is much simpler. Comparing Eq. (7.14) with the result in Eq. (7.5) for a monopole source of strength Q, we see that if $kd \ll 1$ then the radiated power is either zero or four times that for a single source, corresponding to coherent superposition of the radiation from the two monopoles. On the other hand, if $kd \gg 1$, then P is just twice the value for a single source, irrespective of the phases of the two sources, corresponding to incoherent superposition of the individual radiations. The transition between these two forms of behavior is shown in Fig. 7.4. This important general result is true of the radiation from any two sources, irrespective of exact similarity. If the separation between the sources is greater than about half a wavelength ($kd > 3$), then the total radiated power is very nearly equal to the sum of the powers radiated by the two sources

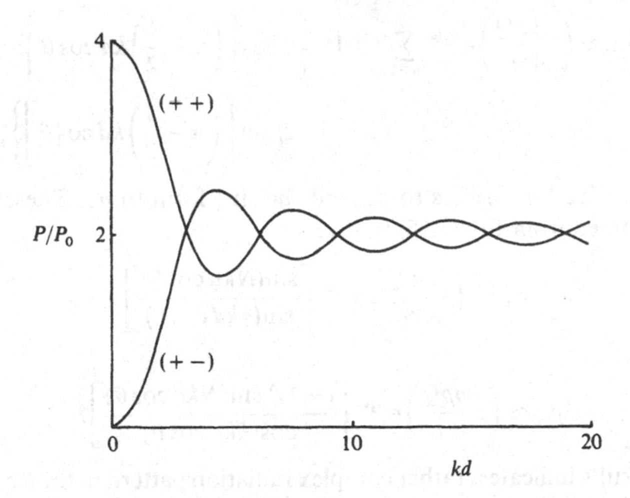

Fig. 7.4. Total power P radiated from a pair of monopoles of the same (+ +) or opposite (+ −) phase and separation d, as functions of the frequency parameter kd. The power radiated from an isolated monopole is P_0.

treated independently. A more detailed treatment along the same lines as that given above (Junger and Feit, 1986, Chapter 2) gives information about the angular variation of acoustic intensity and of the acoustic pressure in the near-field region where $r \sim d$, this pressure generally being higher than given by the far-field approximation [Eq. (7.12)] if used for the near field. These details need not concern us here.

7.3. Arrays of Point Sources

In some fields of acoustics, for example in sonar, we need to be concerned with arrays of point sources all related in phase. Something rather similar may apply to radiation from the open finger holes of a woodwind instrument. More usually, we will be concerned with radiation from an extended vibrating source, such as a drumhead or a piano soundboard, which is divided by nodal lines into areas vibrating in antiphase with their neighbors. It is instructive to look at the point-source approximations to these systems.

Suppose we have a line of $2N$ sources of strength Q, each separated from its neighbors by a distance d. If we choose the origin to be at the midpoint of the line of sources, then, by analogy with Eq. (7.12), the acoustic pressure p at a distance $r \gg 2Nd$ is

$$p_{\pm} \approx \left(\frac{j\omega\rho Q}{4\pi r}\right) e^{-jkr} \sum_{n=1}^{N} (\pm 1)^n [e^{(n-1/2)jkd\cos\theta} \pm e^{-(n-1/2)jkd\cos\theta}], \quad (7.15)$$

where the plus sign applies to a line of sources all with the same phase and the minus sign to a line in which the phase alternates between 0 and π. We rewrite Eq. (7.15) as

$$p_{\pm} \approx \left(\frac{j\omega\rho Q}{4\pi r}\right) e^{-jkr} \sum_{n=1}^{N} (\pm 1)^n \left\{2\cos\left[\left(n - \frac{1}{2}\right) kd\cos\theta\right]; \right.$$
$$\left. 2j\sin\left[\left(n - \frac{1}{2}\right) kd\cos\theta\right]\right\}, \quad (7.16)$$

where the cosine form refers to p_+ and the sine form to p_-. These series are readily summed from Eq. (7.15) to give

$$p_+ \approx \left(\frac{j\omega\rho Q}{4\pi r}\right) e^{-jkr} \left[\frac{\sin(Nkd\cos\theta)}{\sin(\frac{1}{2}kd\cos\theta)}\right], \quad (7.17)$$

and

$$p_- \approx \left(\frac{j\omega\rho Q}{4\pi r}\right) e^{-jkr} \left[\frac{(-1)^N \sin(Nkd\cos\theta)}{\cos(\frac{1}{2}kd\cos\theta)}\right]. \quad (7.18)$$

These results indicate a rather complex radiation pattern in the θ coordinate, but the important features are immediately clear. Let us look first at the case where the sources are all in phase, giving radiated acoustic pressure p_+. The term in square brackets in Eq. (7.17) is always large when $\theta = 90°$, the zero in the numerator being balanced by one in the denominator, and has at that angle the value $2N$. The radiation intensity at $\theta = 90°$ is thus equivalent to that from a source of strength $2NQ$, and, from the form of the bracket, the width of this radiation lobe in radians is approximately $2\pi/Nkd = \lambda/Nd$, where λ is the acoustic wavelength. If $kd < 2\pi$ or $\lambda > d$, then this is the only large maximum of the term in square brackets, which is of order unity at all other angles. The radiation pattern of the array at low frequencies is thus as shown in Fig.7.5(a).

If the frequency is higher, so that $\lambda < d$ or $kd > 2\pi$, then the denominator in the square brackets of Eq. (7.17) can vanish at other angles θ^* for which

$$\theta^* = \cos^{-1}\left(\frac{2n\pi}{kd}\right), \tag{7.19}$$

and this will have solutions for one or more values of the integer n. Each zero in the denominator is again balanced by a zero in the numerator, and the bracket again has the value $2N$. The radiation pattern now has the form shown in Fig. 7.5(b), with more lobes added at higher frequencies.

The total power radiated by the in-phase array can be found by integrating the intensity $p_+^2/\rho c$ over the surface of a large sphere surrounding the source. The behavior is broadly similar to that shown for the in-phase source pair in Fig. 7.4, except that the high-frequency power approaches NP_0 and the

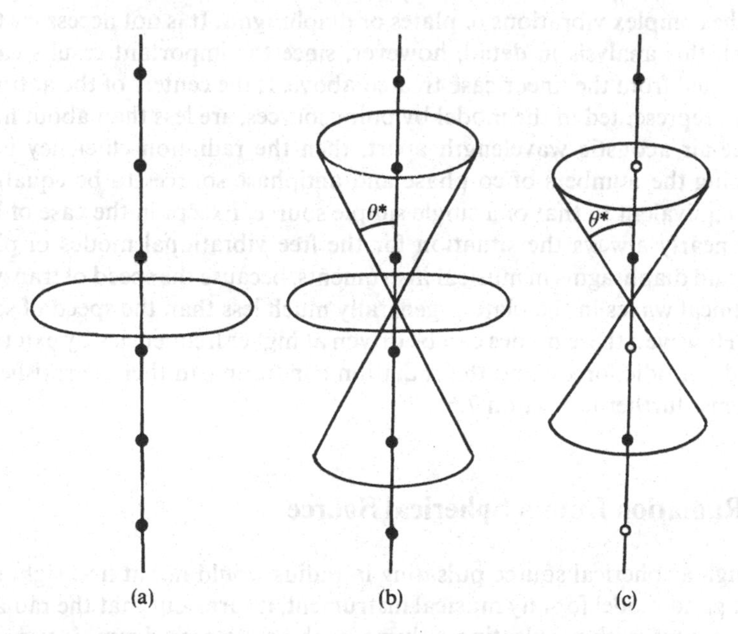

Fig. 7.5. Radiation intensity patterns for (a) a linear co-phase array of sources of separation d for $kd < 2\pi$ or $\lambda > d$; (b) a co-phase array for $kd > 2\pi$, $\lambda < d$; and (c) an antiphase array for $kd > \pi$, $\lambda < 2d$. There is very little radiation from an antiphase array with $kd < \pi$, $\lambda > 2d$.

low-frequency power approaches N^2P_0 when the wavelength is greater than the length of the entire array, P_0 being the power radiated by a simple source of strength Q.

The antiphase array, with radiated pressure p_-, behaves rather differently. If $\lambda > 2d$ so that $kd < \pi$, then there are no zeros in the denominator and the term in square brackets is of order unity at all angles θ. The radiated pressure is thus only comparable with that from a single source of strength Q and the radiation process is very inefficient. Most of the flow of the medium is simply a set of closed loops from one source to its antiphase neighbors and, though the acoustic pressure is actually large at distances less than about d from the source line, very little of this escapes as radiation.

At frequencies sufficiently high that $\lambda < 2d$ or $kd > \pi$, a frequency that is half that for an in-phase source, zeros occur in the denominator of Eq. (7.18) for angles θ^* given by

$$\theta^* = \cos^{-1}\left[\frac{(2n-1)\pi}{kd}\right] \tag{7.20}$$

for one or more positive integers n. The radiation pattern then has the form shown in Fig. 7.5(c), and the total radiated power becomes approximately NP_0.

We can extend these methods to the practically important case of a square-grid array of antiphase sources as a prototype for the radiation to be expected from the complex vibrations of plates or diaphragms. It is not necessary to go through this analysis in detail, however, since the important results can be appreciated from the linear case treated above. If the centers of the antiphase regions, represented in the model by point sources, are less than about half of the free-air acoustic wavelength apart, then the radiation efficiency is low (assuming the numbers of co-phase and antiphase sources to be equal) and about equivalent to that of a single simple source. Except in the case of bells, this is nearly always the situation for the free vibrational modes of plates, shells, and diaphragms in musical instruments, because the speed of transverse mechanical waves in the plate is generally much less than the speed of sound in air. However, these modes can be driven at higher frequencies by externally applied periodic forces, and the radiation condition can then be satisfied, as we discuss further in Section 7.6.

7.4. Radiation from a Spherical Source

Although a spherical source pulsating in radius would not at first sight seem to be a good model for any musical instrument, it turns out that the radiation from a source with a pulsating volume, such as a closed drum, is very little dependent on its shape provided its linear dimensions are small in comparison to the relevant sound wavelength in air. We will, however, look at radiation from a pulsating sphere rather more generally than this.

Suppose that the sphere has radius a and pulsates with surface velocity $u\exp(j\omega t)$. If we set this equal to the radial velocity at distance a from a simple source, as given by Eq. (7.3), then from Eq. (7.2) the acoustic pressure at distance r becomes

$$p(r) = \frac{j\rho cka^2u}{r}\left(\frac{1-jka}{1+k^2a^2}\right)e^{-jk(r-a)}. \tag{7.21}$$

There are obvious simplifications if $ka \ll 1$. The radiated power P is then

$$P = \frac{2\pi\rho ck^2a^4u^2}{1+k^2a^2} \tag{7.22}$$

$$\to 2\pi\rho ca^2u^2 = \frac{\rho cQ^2}{8\pi a^2} \quad \text{if} \quad ka \gg 1 \tag{7.23}$$

$$\to 2\pi\rho c(ka)^2a^2u^2 = \left(\frac{\rho cQ^2}{8\pi a^2}\right)(ka)^2 \quad \text{if} \quad ka \ll 1, \tag{7.24}$$

where Q is the volume flow amplitude of the source. For a given surface velocity u, the radiated power per unit area increases as $(ka)^2$ while $ka \ll 1$ but then saturates for $ka \gg 1$. Most musical instruments in this approximation

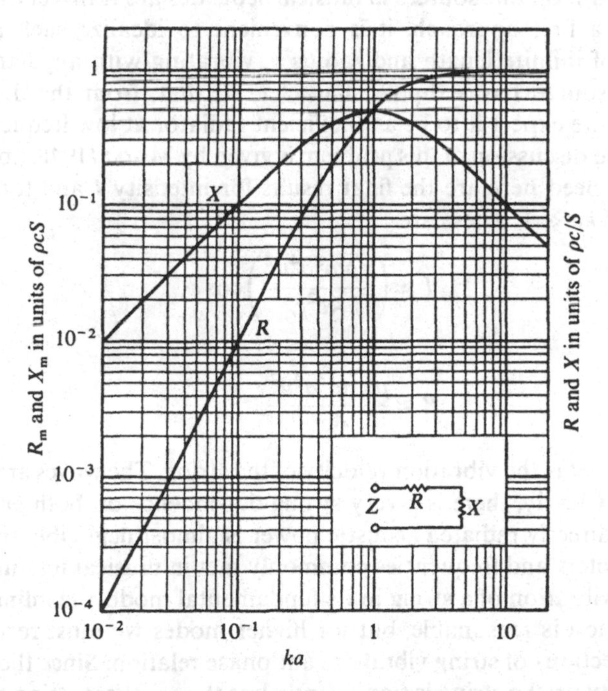

Fig. 7.6. Real and imaginary parts of the mechanical load $R_m + jX_m$ on the surface of a sphere of radius a pulsating with frequency ω and velocity amplitude u. R_m and X_m are given in units of ρcS, where $S = 4\pi a^2$. R and X are corresponding acoustic quantities in units of $\rho c/S$ (after Beranek, 1954).

operate in the region $ka < 1$, so that, other things being equal, there is usually an advantage in increasing the size of the flexible radiating enclosure.

It is also useful to know the mechanical load on the spherical surface. From Eq. (7.21), this is simply

$$F = 4\pi a^2 p(a) = \rho c S\left(\frac{k^2a^2 + jka}{1 + k^2a^2}\right)u = (R_m + jX_m)u, \tag{7.25}$$

where $S = 4\pi a^2$ is the area of the sphere. The real part $R_m u$ of this force, which increases as $(ka)^2$ until it saturates at ρcSu, represents the dissipative load of the radiation. The imaginary part $X_m u$, which increases as ka for $ka < 1$ and decreases as $(ka)^{-1}$ above $ka = 1$, represents the mass load of the co-moving air. Equation (7.25) is plotted in Fig. 7.6. The load on pulsating volume sources of all shapes behaves very similarly. Also shown in Fig. 7.6 are the acoustic quantities R and X; these are just $1/S^2$ times their mechanical counterparts.

7.5. Line Sources

The only common line sources in musical acoustics are transversely vibrating strings. As a first approach, it is convenient to idealize such a string as a cylinder of infinite length and radius a, vibrating with angular frequency ω. Such a source has a dipole character, so that, from the discussion of Section 7.1, we expect it to be an inefficient radiator at low frequencies.

A detailed discussion of this problem is given by Morse (1948, pp. 299–300). All that we need here are the final results for intensity I and total radiated power P for $ka \ll 1$. These are

$$I \approx \left(\frac{\pi\rho\omega^3a^4u^2}{4c^2r}\right)\cos\phi, \tag{7.26}$$

and

$$P \approx \frac{\rho\pi^2\omega^3a^4u^2}{4c^2}, \tag{7.27}$$

where $u \exp j\omega t$ is the vibration velocity of the string. The waves are, of course, cylindrical. Clearly, there is a very strong dependence on both ω and a and, in fact, the directly radiated acoustic power is almost negligibly small for the string diameters and frequencies commonly met in musical instruments.

For the vibration of a string in its fundamental mode, the infinite cylinder approximation is reasonable, but for higher modes we must recognize that adjoining sections of string vibrate in antiphase relation. Since the transverse wave velocity on the string is significantly less than c, these string sections are separated by less than half the sound wavelength in air, so there is an additional cause for cancellation of the radiated sound intensity, as discussed in Section 7.3.

It is therefore reasonable to neglect the contribution of direct radiation from vibrating strings to the sound of musical instruments. It is only when a vibrating cylinder has a quite large radius, as in the cylinders of tubular bells, that direct radiation becomes significant.

7.6. Radiation from a Plane Source in a Baffle

Few, if any, musical radiators consist of some sort of moving part set in an infinite plane baffle, but we examine the behavior of this system because it is the only case for which a simple general result emerges. Fortunately, it also happens that replacement of the plane baffle by an enclosure of finite size does not have a really major effect on the results, though the changes are significant.

Referring to Fig. 7.7, suppose that the area S on an otherwise rigid plane baffle is vibrating with a velocity distribution $u(\mathbf{r}')$ and frequency ω normal to the plane, all points being either in phase or in antiphase. The small element of area dS at $\mathbf{r}'$ then constitutes a simple source of volume strength $u(\mathbf{r}')\,dS$, which is doubled to twice this value by the presence of the plane which restricts its radiation to the half-space of solid angle 2π. The pressure dp produced by this element at a large distance $\mathbf{r}$ is

$$dp(\mathbf{r}) = \frac{j\omega\rho}{2\pi r} e^{-jk|\mathbf{r}-\mathbf{r}'|} u(\mathbf{r}')\,dS. \tag{7.28}$$

If we take $\mathbf{r}$ to be in the direction (θ, ϕ) and $\mathbf{r}'$ in the direction $(\pi/2, \phi')$, then we

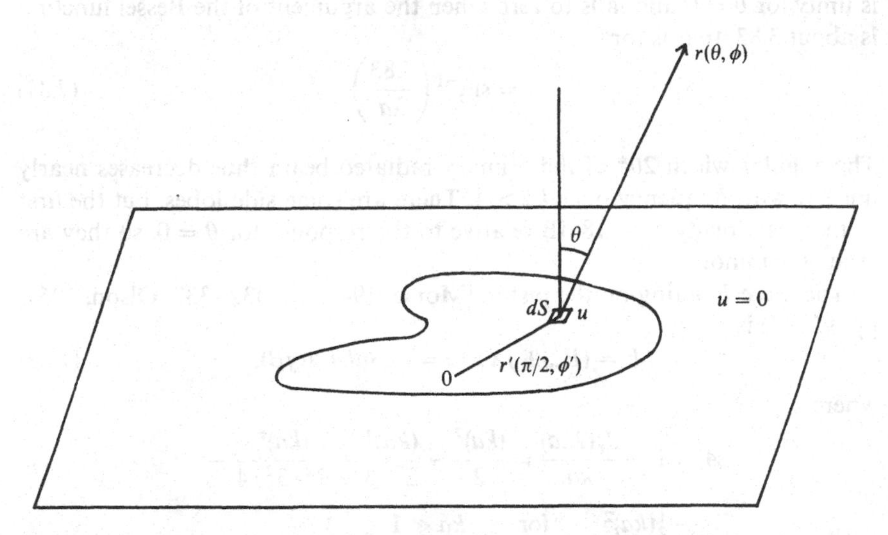

Fig. 7.7. A vibrating plane source set in an infinite plane baffle. Radiation pressure is evaluated at a point at a large distance r in the direction shown.

can integrate Eq. (7.28) over the whole surface of the plane, remembering that $u = 0$ outside S, to give

$$p(r, \theta, \phi) = \frac{j\omega\rho}{2\pi r} e^{-jkr} \int_S \int e^{jkr' \sin\theta \cos(\phi - \phi')} u(\mathbf{r}') r' \, d\phi' \, dr'. \tag{7.29}$$

The integral in Eq. (7.29) has the form of a spatial Fourier transform of the velocity distribution $u(\mathbf{r}')$. This is our general result, due in the first place to Lord Rayleigh.

It is now simply a matter of algebra to apply Eq. (7.29) to situations of interest. These include a rigid circular piston and a flexible circular piston (Morse, 1948, pp. 326–335) and both square and circular vibrators excited in patterns with nodal lines (Skudrzyk, 1968, pp. 373–429; Junger and Feit, 1986, Chapter 5). There is not space here to review this work in detail, but we shall select particular examples and relate the conclusion to the simplified treatments given in the earlier sections of this chapter.

The integral in Eq. (7.29) can be performed quite straightforwardly for the case of a circular piston of radius a with u constant across its surface. The result for the far field (Morse, 1948, pp. 327–328) is

$$p \approx \tfrac{1}{2} j\omega\rho u a^2 \left(\frac{e^{-jkr}}{r}\right) \left[\frac{2J_1(ka \sin\theta)}{ka \sin\theta}\right], \tag{7.30}$$

where J_1 is a Bessel function of order one. The factor in square brackets is nearly unity for all θ if $ka \ll 1$, so the radiation pattern in the half-space $0 \leq \theta < \pi/2$ is isotropic at low frequencies. For higher frequencies, the bracket is unity for $\theta = 0$ and falls to zero when the argument of the Bessel function is about 3.83, that is for

$$\theta^* = \sin^{-1}\left(\frac{3.83}{ka}\right). \tag{7.31}$$

The angular width $2\theta^*$ of the primary radiated beam thus decreases nearly linearly with frequency once $ka > 4$. There are some side lobes, but the first of these is already at -18 dB relative to the response for $\theta = 0$, so they are relatively minor.

The force F acting on the piston (Morse, 1948, pp. 332–333; Olson, 1957, pp. 92–93) is

$$F = (R_m + jX_m)u = \rho c S u(A + jB), \tag{7.32}$$

where

$$\begin{aligned} A = 1 - \frac{J_1(2ka)}{ka} &= \frac{(ka)^2}{2} - \frac{(ka)^4}{2^2 \cdot 3} + \frac{(ka)^6}{2^2 \cdot 3^2 \cdot 4} - \cdots \\ &\rightarrow \tfrac{1}{2}(ka)^2 \quad \text{for} \quad ka \ll 1 \\ &\rightarrow 1 \quad \text{for} \quad ka \gg 1, \end{aligned} \tag{7.33}$$

and

$$B = \frac{1}{\pi k^2 a^2}\left[\frac{(2ka)^3}{3} - \frac{(2ka)^5}{3^2 \cdot 5} + \frac{(2ka)^7}{3^2 \cdot 5^2 \cdot 7} - \cdots\right]$$

$$\to 8ka/3\pi \qquad \text{for} \qquad ka \ll 1$$

$$\to 2/\pi ka \qquad \text{for} \qquad ka \gg 1. \tag{7.34}$$

These functions, which also apply to a pipe with an infinite baffle, are shown in Fig. 8.7. For the moment, we simply note the close agreement between their asymptotic forms and the same quantities for a pulsating sphere of radius a as given in Eq. (7.25) and Fig. 7.6.

When we consider the radiation from a vibrating circular membrane or plate of the type discussed in Chapter 3, we realize that a rigid piston is not a good model for the motion. Indeed, for a membrane, the first mode has the form of a Bessel function $J_0(\alpha_0 r)$, and higher modes are of the form $J_n(\alpha_m r)\cos n\phi$, with the α_m determined by the condition that the functions vanish at the clamped edge. All except the fundamental J_0 mode have either nodal lines or nodal circles or both, and there is a good deal of cross flow and hence low radiation efficiency. All the axisymmetric modes $J_0(\alpha_m r)$, however, have a nonzero volume displacement and hence some monopole radiation component. The J_n modes with $n \neq 0$ have no monopole component, and their radiation is therefore much less efficient. No explicit tabulation of this behavior is readily available, but Morse (1948, pp. 329–332) details a related case for free-edge modes for which the radial slope of the displacement is required to vanish at the boundary—a condition appropriate to a flexible piston closing a flanged circular pipe.

Detailed discussion of radiation from flexible plane vibrators with nodal lines is algebraically complex but reflects the behavior we found for antiphase arrays of point sources, provided that allowance is made for the fact that a finite vibrator may have a net volume flow and hence a monopole radiation contribution. As has already been remarked, antiphase point-source arrays with spacing less than half a wavelength of sound in air are inefficient radiators, and their source strength is of the order of that of a single one of their component sources. Exactly the same result is found for continuous vibrators, with the effective source arising from noncancelling elements at the center and around the edges of the vibrator (Skudrzyk, 1968, pp. 419–429).

Only for the free vibrations of thick metal plates do we reach a situation in which the transverse wave speed in the plate exceeds the sound speed in air, so that high intensity radiated beams can be produced, as shown in Fig. 7.5(c). At 1kHz, this requires a steel plate about 10 mm in thickness (Skudrzyk, 1968, p. 378)—a situation often encountered in heavy machinery but scarcely applicable to musical instruments.

There is an interesting and important consequence of these conclusions that is investigated in some detail by Skudrzyk (1968, pp. 390–398). If a localized force drives an elastic plate, such as a piano soundboard, then it excites all vibrational modes to amplitudes that are dependent on the frequency separation between the exciting force frequency ω and the resonance frequency ω_n of the mode in question. If $\omega \gg \omega_n$, then the amplitude of mode n will be small, but, because its nodal lines are far apart, it will radiate efficiently at frequency ω. Conversely, a mode with $\omega_n \approx \omega$ may be strongly excited, but, because of the small distance between its nodal lines, it may radiate very poorly. The total power radiated by the forced plate must be found by summing the contributions from all the efficiently radiating modes as well as the smaller contributions from higher modes. This effect acts to smooth the frequency response of a large planar forced vibrator.

7.7. Unbaffled Radiators

Few if any musical instruments involve a large plane baffle, even one that extends for about a wavelength around the vibrating plate or membrane. In instruments like the timpani, the baffle is folded around so that it encloses one side of the membrane and converts it to a one-sided resonator; the body of the violin serves a somewhat similar function, though there is considerable vibration of the back as well. In instruments like the piano, both sides of the soundboard are able to radiate, but the case provides a measure of separation between them. Only in the case of cymbals, gongs, and bells is there no baffle at all.

For a half-enclosed radiator, like the membrane of the timpani, the enclosed air does, of course, have an effect on the vibration of the membrane. From the point of view of radiation, however, the source is still one-sided, and the major difference from the baffled case arises from the fact that radiation is into a whole space of solid angle 4π rather than into a half-space of solid angle 2π. For high frequencies, such that $ka > 4$, this effect is not large, for the energy of the radiation is concentrated into a broad beam along the direction of the axis ($\theta = 0$), and little of it passes into the halfplane $\theta > \pi/2$ anyway. For low frequencies, $ka < 1$, however, the radiation tends to become isotropic and to fill nearly uniformly the whole 4π solid angle. The vibrating membrane thus experiences a radiation resistance at low frequencies that is only half that for the baffled case, so that the power radiated is reduced by a factor 2, or 3 dB, for a given membrane velocity amplitude. More than this, since the radiation is into 4π rather than 2π, the intensity in any given direction is reduced by a further factor of 2, or 3 dB. Thus, for a given radiator velocity amplitude, the absence of a large baffle leaves the radiated intensity and power unchanged at high frequencies ($ka > 4$) but reduces the radiated intensity in the forward half-space by 6 dB and the total radiated power by 3 dB at low frequencies ($ka < 1$).

These phenomena can have a significant effect on the fullness of tone quality of an instrument in its low-frequency range, and it is usual to increase the bass intensity by providing a reflecting wall close behind the player. This recovers 3 dB of intensity in the bass, and a further 3 dB, relative to anechoic conditions, can be recovered from close floor reflections. Of course, these effects are subjectively assumed when listening to normal playing—it is only in anechoic conditions that the loss of intensity at low frequencies becomes noticeable.

Instruments such as the piano have a case structure that goes some way toward separating the two sides of the soundboard acoustically. This is, of course, desirable since they vibrate in antiphase. We have seen, however, that antiphase source distributions cancel each other only if their separation is less than about half a wavelength. A semiquantitative application of this principle gives a distance of about 2 m between the top and bottom of the soundboard of a piano and hence suggests that such cancellations should not occur above about 70 Hz. The different geometries of the lid and the floor in any case ensure that cancellation is only partial. Cancellations do, of course, occur between neighboring antiphase regions on the same side of the soundboard.

Finally, let us look briefly at radiation from cymbals and bells. For a nearly planar radiator, such as a cymbal or tam-tam, cancellation of radiation from opposite surfaces may clearly be very significant. An unbaffled plane piston radiator with both sides exposed acts as a dipole source at low frequencies and, to a good approximation, the mechanical radiation resistance presented to each side is (Olson, 1957, pp. 98–99)

$$\begin{aligned} R_m &\approx 3 \times 10^{-2} \rho c S (ka)^4, \qquad ka < 2, \\ &\approx \rho c S, \qquad ka > 3. \end{aligned} \tag{7.35}$$

Much of the low-frequency radiation is therefore suppressed. The initial amplitude of the lower modes is, however, usually high, so that some of their sound is heard. Much of the effect of such a gong depends on the shimmer of high frequency modes, which have comparable radiation efficiency despite near cancellation of radiation from neighboring antiphase areas. Details of the residual radiation from noncancelling areas around the edge of such a gong are discussed by Skudrzyk (1968, pp. 419–429). This overall cancellation has another effect, of course, and that is to reduce the radiation damping of the oscillation and prolong the decay of the sound.

Radiation from curved shells, such as found in bells, is a very complex subject (Junger and Feit, 1986, Chapter 7). There is significant cancellation of radiation from neighboring antiphase regions, but their different geometrical environments inhibit cancellation between the interior and the exterior of the shell. Indeed, modes that would seem at first to be of high order, such as the quadrupole mode associated with the circular to elliptical distortion of a bell, can lead to a change in interior volume and therefore an inside-to-outside dipole source.

7.8. Radiation from Large Plates

It is often important in practical situations to have some appreciation of the radiation properties of extended vibrating objects. The general subject is very difficult, but we can attain some insights by considering very simple situations. Suppose, for example, that we have an infinite plane plate upon which is propagating a plane transverse wave of angular frequency ω. The speed $v_P(\omega)$ of this wave is determined by the thickness and elastic properties of the plate, and also by the frequency ω, since a stiff plate is a dispersive medium for transverse waves, as described by Eq. (3.13). If we suppose the plate to lie in the plane $z = 0$ and the plate wave to propagate in the x direction, then we can represent the displacement velocity of the plate surface by the equation

$$u(x, 0) = u_P e^{-jk_P x} e^{j\omega t} \tag{7.36}$$

where the wave number on the surface of the plate is given by

$$k_P = \omega / v_P. \tag{7.37}$$

In the air above the plate there is no variation of physical quantities in the y direction, and the acoustic pressure $p(x, z)$ and acoustic particle velocity $u(x, z)$ both satisfy the wave equation in the form

$$\frac{\partial^2 p}{\partial x^2} + \frac{\partial^2 p}{\partial z^2} = \frac{1}{c^2} \frac{\partial^2 p}{\partial t^2} \tag{7.38}$$

which has solutions of the form

$$p(x, z) = \rho c u(x, z) = A e^{-j(k_x x + k_z z)} \tag{7.39}$$

where k_x and k_z are the components of the wave vector $\mathbf{k}$ in directions respectively parallel to and normal to the plane of the plate and there is a time variation $e^{j\omega t}$ implied. Because the acoustic particle velocity in the air must match the normal velocity of the plate in the plane $z = 0$, we must have

$$k_x = k_P; \qquad A = u_P \tag{7.40}$$

so that, if we substitute this into Eq. (7.38), we find that

$$k_z^2 = \frac{\omega^2}{c^2} - k_P^2 = \omega^2 \left(\frac{1}{c^2} - \frac{1}{v_P^2} \right). \tag{7.41}$$

Equation (7.41) has very important implications. If the velocity v_P of the wave on the plate is less than the velocity c of sound in air, or equivalently if the wavelength of the wave on the plate is smaller than the wavelength in air, then k_z is imaginary. The acoustic disturbance is then exponentially attenuated in the z direction rather than propagating as a wave. The whole motion of the air is thus confined to the immediate vicinity of the plate surface and there is no acoustic radiation. If, however, $v_P > c$, then an acoustic wave is radiated in a direction making an angle θ with the surface of the plate, given by

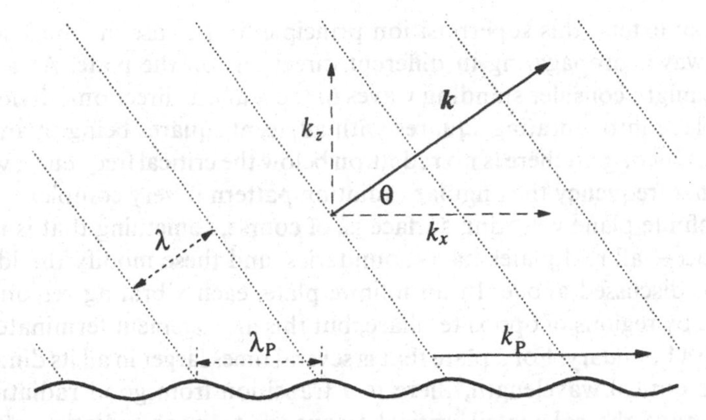

Fig. 7.8. Relation between a plate wave with propagation number k_P and the acoustic wave with propagation vector k that it radiates, for the case $\lambda_P > \lambda$.

$$\tan\theta = \frac{k_z}{k_x} = \left(\frac{v_P^2}{c^2} - 1\right)^{1/2}. \tag{7.42}$$

The physical reason for this behavior is clear from Fig. 7.8, which illustrates the relation between the plate and air waves. If $v_P > c$, the wavelength λ_P of the plate wave is greater than the wavelength λ of the air wave, and the diagram fits together with the propagation direction θ given by Eq. (7.42). If $v_P < c$, however, then $\lambda_P < \lambda$ and there is no way that the acoustic wave can be matched to the plate wave.

Since the wave speed v_P on the plate increases with increasing frequency, there is a particular critical frequency for any plate above which radiation can occur. If we have a complex wave propagating on a plate, this means that only the high-frequency components of the wave will be radiated, and their radiation directions will be spread out in angle rather like white light passing through a prism or reflected from a diffraction grating. It is clear that there is a close analogy between this behavior and that already noted for two simple anti-phase sources in Section 7.2 and for linear arrays of antiphase sources as discussed in Section 7.3.

If, instead of a propagating wave on the plate, we have a standing wave, then the situation is rather similar. The standing wave has a spatial variation in the x direction given by

$$\cos k_P x = \frac{1}{2}(e^{jk_P x} + e^{-jk_P x}) \tag{7.43}$$

which is just two similar waves propagating in opposite directions. Each behaves as described above, and the resulting radiation pattern, if $v_P > c$, is a superposition of two plane waves at angles $\pm\theta$ to the plane. These interfere to give a plane wave propagating in the z direction but with an amplitude variation $\cos k_P x$ in the x direction.

We can extend this superposition principle to the case in which we have several waves propagating in different directions on the plate. As a special case, we might consider standing waves in the x and y directions, dividing the whole plane into vibrating squares with adjacent squares being in antiphase relation. Once again there is no radiation below the critical frequency, whereas above that frequency the angular radiation pattern is very complex.

An infinite plane vibrating surface is, of course, something that is not met in practice—all real plates have boundaries, and these modify the idealized behavior discussed above. In an infinite plate, each vibrating region is surrounded by regions of opposite phase, but this arrangement terminates when we reach a boundary. For a plate that is several times larger in all its dimensions than the critical wavelength, there is a transition from good radiation efficiency above the calculated critical frequency to poor radiation efficiency below, but this transition is not completely sharp and there is still some radiation at lower frequencies. As the frequency is decreased below the critical frequency, lower and lower normal modes are progressively excited until the plate is vibrating essentially in its fundamental mode, giving approximately a finite-dipole source, the radiated intensity from which varies as ω^4 for constant panel velocity amplitude, compared with the frequency-independent radiation characteristic above the critical frequency.

References

Beranek, L.L. (1954). "Acoustics." McGraw-Hill, New York; reprinted 1986, Acoustical Soc. Am., Woodbury, New York.

Junger, M.C., and Feit, D. (1986). "Sound, Structures, and Their Interaction," 2nd ed., MIT Press, Cambridge, Massachusetts.

Morse, P.M. (1948). "Vibration and Sound." McGraw-Hill, New York; reprinted 1981, Acoustical Soc. Am., Woodbury, New York.

Olson, H.F. (1957). "Acoustical Engineering." Van Nostrand-Reinhold, Princeton, New Jersey.

Skudrzyk, E. (1968). "Simple and Complex Vibratory Systems." Penn. State Univ. Press, University Park, Pennsylvania.

CHAPTER 8

Pipes and Horns

The wave propagation phenomena in fluids that we have examined in previous chapters have referred to waves in infinite or semi-infinite spaces generated by the vibrational motion of some small object or surface in that space. We now turn to the very different problem of studying the sound field inside the tube of a wind instrument. Ultimately, we shall join together the two discussions by considering the sound radiated from the open end or finger holes of the instrument, but for the moment our concern is with the internal field. We begin with the very simplest cases and then add complications until we have a reasonably complete representation of an actual instrument. At this stage, we will find it necessary to make a digression, for a wind instrument is not excited by a simple source, such as a loudspeaker, but is coupled to a complex pressure-controlled or velocity-controlled generator—the reed or air jet—and we must understand the functioning of this before we can proceed. Finally, we go on to treat the strongly coupled pipe and generator system that makes up the instrument as played.

8.1. Infinite Cylindrical Pipes

The simplest possible system of enclosure is an infinite cylindrical pipe or tube with its axis parallel to the direction of propagation of a plane wave in the medium (Morse and Ingard, 1968). If the walls of the pipe are rigid, perfectly smooth, and thermally insulating, then the presence of the tube wall has no effect on wave propagation. A pressure wave propagating in the x direction has the form

$$p(x, t) = p \exp[j(-kx + \omega t)], \tag{8.1}$$

and the resultant acoustic volume flow is, as we saw in Chapter 6,

$$U(x, t) = \left(\frac{Sp}{\rho c}\right) \exp[j(-kx + \omega t)], \tag{8.2}$$

where ω is the angular frequency, k is the angular wave number $k = 2\pi/\lambda = \omega/c$, and S is the cross-sectional area of the pipe. As usual, ρ is the density of and

c the velocity of sound in air. The acoustic impedance of the pipe at any point x is

$$Z_0(x) = \frac{p(x,t)}{U(x,t)} = \frac{\rho c}{S}. \tag{8.3}$$

To treat this problem in more detail, we must solve the wave equation directly in cylindrical polar coordinates (r, ϕ, x). If a is the radius of the pipe and its surface is again taken to be perfectly rigid, then the boundary condition is

$$\frac{\partial p}{\partial r} = 0 \quad \text{at} \quad r = a, \tag{8.4}$$

which implies that there is no net force and therefore no flow normal to the wall. The wave equation in cylindrical coordinates is

$$\frac{1}{r}\frac{\partial}{\partial r}\left(r\frac{\partial p}{\partial r}\right) + \frac{1}{r^2}\frac{\partial^2 p}{\partial \phi^2} + \frac{\partial^2 p}{\partial x^2} = \frac{1}{c^2}\frac{\partial^2 p}{\partial t^2}, \tag{8.5}$$

and this has solutions of the form

$$p_{mn}(r, \phi, x) = p_{\sin}^{\cos}(m\phi)J_m\left(\frac{\pi q_{mn} r}{a}\right)\exp[j(-k_{mn}x + \omega t)], \tag{8.6}$$

where J_m is a Bessel function and q_{mn} is defined by the boundary condition [Eq. (8.4)], so that the derivative $J_m'(\pi q_{mn})$ is zero. The (m, n) mode thus has an (r, ϕ) pattern for the acoustic pressure p with n nodal circles and m nodal diameters, both m and n running through the integers from zero. In the full three-dimensional picture, these become nodal cylinders parallel to the axis and nodal planes through the axis, respectively.

In Fig. 8.1, the pressure and flow velocity patterns for the lowest three modes of the pipe, omitting the simple plane-wave mode, are shown. The

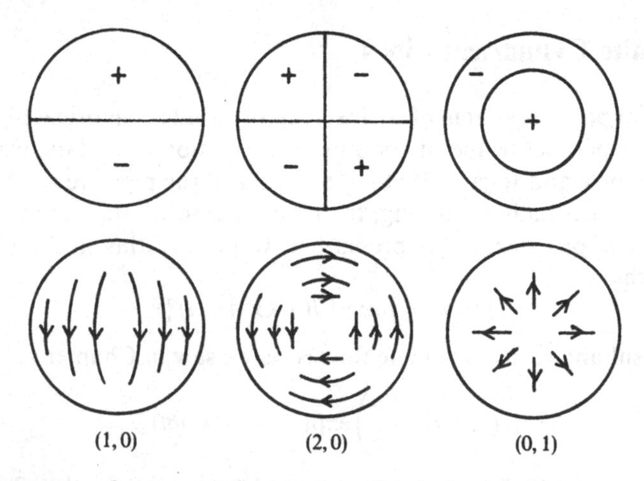

Fig. 8.1. Pressure and transverse flow patterns for the lowest three transverse modes of a cylindrical pipe. The plane-wave mode is not shown.

pressure patterns have nodal lines as already observed, and there are similar nodal diameters in the transverse flow patterns. Nodal circles for pressure occur for modes of the type $(0, n)$, which have n such nodal circles within the boundary. A general mode (m, n) has both nodal lines and circles in the pressure.

The propagation wave vector k_{mn} for mode (m, n) is obtained by substituting Eq. (8.6) into Eq. (8.5), whence

$$k_{mn}^2 = \left(\frac{\omega}{c}\right)^2 - \left(\frac{\pi q_{mn}}{a}\right)^2. \tag{8.7}$$

Thus, while the plane-wave mode with $m = n = 0$ will always propagate with $k = k_{00} = \omega/c$, this is not necessarily true for higher modes. In order for a higher mode (m, n) to propagate, the frequency must exceed the cutoff value;

$$\omega_c = \frac{\pi q_{mn} c}{a}. \tag{8.8}$$

For frequencies less than ω_c, k_{mn} is imaginary and Eq. (8.6) shows that the mode is attenuated exponentially with distance. The attenuation is quite rapid for modes well below cutoff, and the amplitude falls by a factor e, or about 10 dB, within a distance less than the pipe radius.

The first higher mode to propagate is the antisymmetric $(1, 0)$ mode, which has a single nodal plane, above a cutoff frequency $\omega_c = 1.84\ c/a$. Next is the $(2, 0)$ mode, with two nodal planes, for $\omega > 3.05\ c/a$, and then the lowest nonplanar axial mode $(0, 1)$, for $\omega > 3.80\ c/a$. Propagating higher modes are thus possible only when the pipe is greater in diameter than about two-thirds of the free-space acoustic wavelength. The nonpropagating higher modes are necessary to explain certain features of the acoustic flow near wall irregularities, such as finger holes or mouthpieces. Indeed, it is possible to match any disturbance distributed over an opening or a vibrating surface in a duct with an appropriate linear combination of duct modes. The plane wave component of this combination will always propagate along the duct away from the disturbance, but this will not be true for modes with q_{mn} values that are too large. The propagating wave will thus be a low-pass filtered version of the disturbance while the nonpropagating modes will simply modify the flow in the near neighborhood of the source.

It is helpful to sketch the three-dimensional acoustic flow streamlines associated with a few of these modes for both propagating and nonpropagating cases. This can be done from the form of the pressure pattern given by Eq. (8.6) together with the relation

$$u = \frac{j}{\rho\omega} \nabla p \tag{8.9}$$

for the flow velocity u in a mode excited at frequency ω. Figure 8.2 shows this for the $(1, 0)$ and $(0, 1)$ modes. In the case of the propagating modes, the flow pattern itself moves down the pipe with the characteristic phase velocity of the mode—nearly the normal sound velocity c, except very close to cutoff when the phase velocity is higher than c.

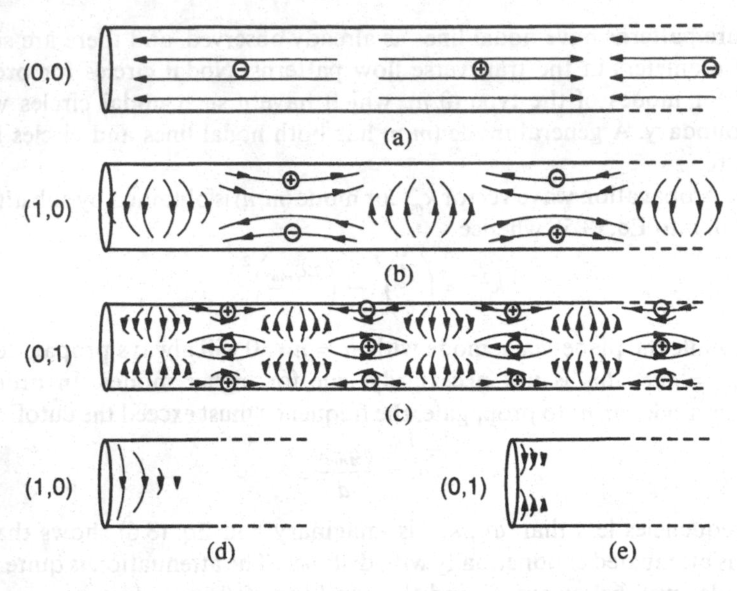

Fig. 8.2. Acoustic flow patterns and pressure maxima and minima for higher modes in a cylindrical duct. (a)–(c) are modes propagating to the right at a frequency a little above cut-off; (d) and (e) are evanescent modes below cutoff.

It is not important to go into detail about the impedance behavior of these higher modes, since this depends greatly upon the geometry with which they are driven, the net acoustic flow along the pipe axis being zero except for the plane (0, 0) mode. The impedance is always a real function multiplied by ω/k_{mn}, so it is real for ω above cutoff, becomes infinite at cutoff, and is imaginary below cutoff.

8.2. Wall Losses

So far in our discussion, we have assumed a rigid wall without introducing any other disturbance. In a practical case this can never be achieved, though in musical instruments the walls are at least rigid enough that their mechanical vibrations can be neglected—we return to the subtleties of this statement later. More important, however, are viscous and thermal effects from which no real walls or real fluids are immune.

Detailed consideration of these effects is complicated (Benade, 1968), but the basic phenomena and final results are easily discussed. The walls contribute a viscous drag to the otherwise masslike impedance associated with acceleration of the air in the pipe. The relative magnitude of the drag depends upon the thickness of the viscous boundary layer, itself proportional to the square root of the viscosity η divided by the angular frequency ω, in relation to the pipe radius a. A convenient parameter to use is the ratio of pipe radius to the boundary layer thickness:

$$r_v = \left(\frac{\omega\rho}{\eta}\right)^{1/2} a. \tag{8.10}$$

Similarly, thermal exchange between the air and the walls adds a lossy resistance to the otherwise compliant compressibility of the air, and the relative magnitude of this loss depends on the ratio of the pipe radius a to the thermal boundary layer thickness, as expressed by the parameter

$$r_t = \left(\frac{\omega\rho C_p}{\kappa}\right)^{1/2} a, \tag{8.11}$$

where C_p is the specific heat of air at constant pressure and κ is its thermal conductivity. Near 300 K (27°C), we can insert numerical values to give (Benade, 1968)

$$r_v \approx 632.8 a f^{1/2}(1 - 0.0029\,\Delta T), \tag{8.12}$$

and

$$r_t \approx 532.2 a f^{1/2}(1 - 0.0031\,\Delta T), \tag{8.13}$$

where a is the tube radius in meters, f is the frequency in hertz, and ΔT is the temperature deviation from 300 K.

It is clear that the effect of these loss terms will be to change the characteristic impedance Z_0 of the pipe from its ideal real value $\rho c/S$ to a complex quantity. This, in turn, will make the wave number k complex and lead to attenuation of the propagating wave as it passes along the pipe.

The real and imaginary parts of the characteristic impedance Z_0, as fractions of its ideal value $\rho c/S$, are shown in Figs. 8.3 and 8.4, both as functions of r_v. The correction to Z_0 begins to be appreciable for $r_v < 10$, while for $r_v < 1$ the real and imaginary parts of Z_0 are nearly equal and vary as r_v^{-1}.

It is convenient to rewrite the wave vector k as the complex number $\omega/v - j\alpha$, where α is now the attenuation coefficient per unit length of path

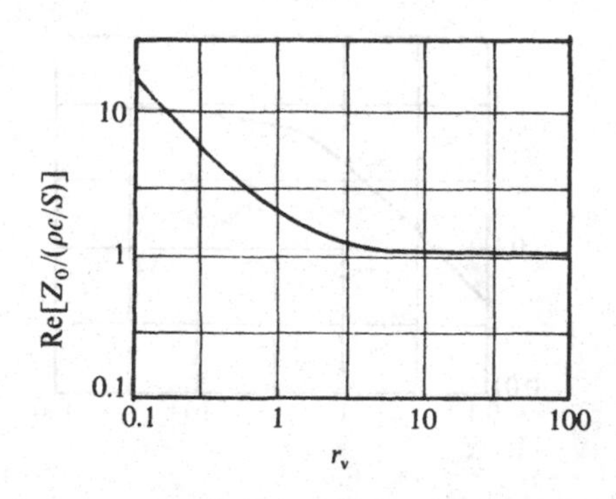

Fig. 8.3. Real part of the characteristic impedance Z_0, in units of $\rho c/S$, as a function of the parameter r_v (after Benade, 1968).

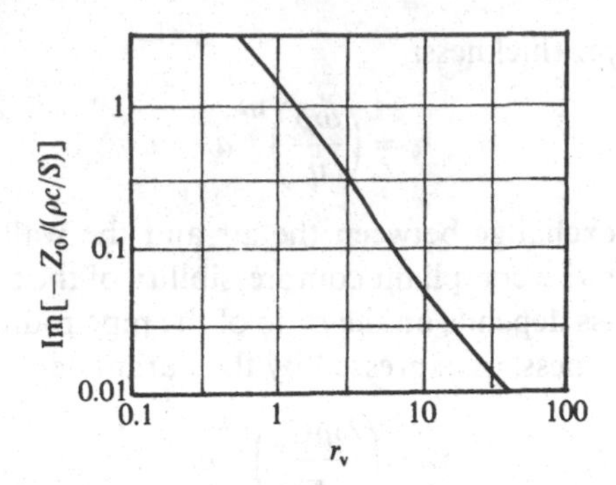

Fig. 8.4. Imaginary part of the characteristic impedance Z_0, in units of $\rho c/S$, as a function of the parameter r_v (after Benade, 1968).

and v is the phase velocity. We can then most usefully plot the phase velocity v, measured in units of the free-air sound velocity c, and the attenuation coefficient α, divided by f, both as functions of r_v. This is done in Figs. 8.5 and 8.6. The phase velocity v is significantly less than c for pipes so narrow that $r_v < 10$, while the attenuation coefficient exceeds λ^{-1} if $r_v < 6$. Since the phase velocity and attenuation coefficient for relatively wide tubes are both of fundamental significance for the physics of musical instruments, it is useful to restate Benade's (1968) versions of Rayleigh's (1894) approximate formulas, which are good for $r_v > 10$ and useful down to about $r_v = 3$. They are

$$v \approx c\left[1 - \frac{1}{r_v\sqrt{2}} - \frac{(\gamma - 1)}{r_t\sqrt{2}}\right] \approx c\left[1 - \frac{1.65 \times 10^{-3}}{af^{1/2}}\right], \tag{8.14}$$

and

$$\alpha \approx \frac{\omega}{c}\left[\frac{1}{r_v\sqrt{2}} + \frac{(\gamma - 1)}{r_t\sqrt{2}}\right] \approx \frac{3 \times 10^{-5} f^{1/2}}{a}, \tag{8.15}$$

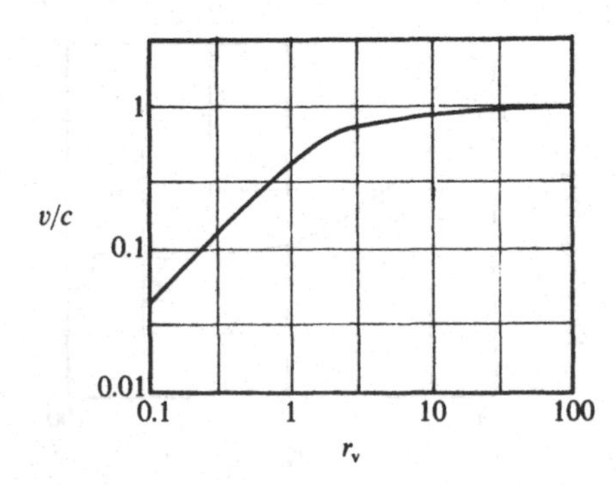

Fig. 8.5. The phase velocity v, relative to the free-air sound velocity c, as a function of the parameter r_v (after Benade, 1968).

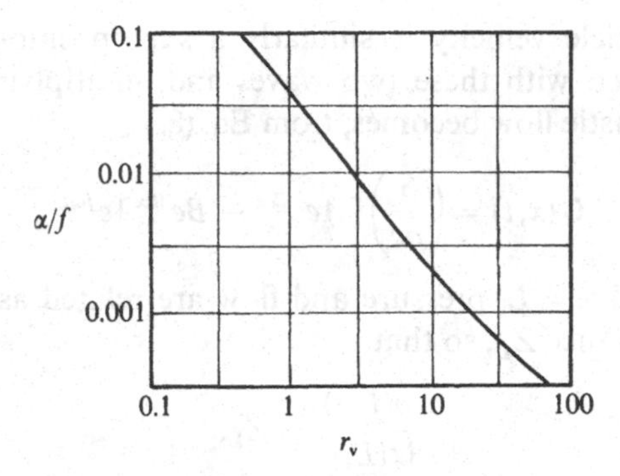

Fig. 8.6. The attenuation coefficient α in (meters)$^{-1}$ at frequency f, plotted as α/f, as a function of the parameter r_v (after Benade, 1968).

where α is given in (meters)$^{-1}$ if a is in meters. Here, γ is the ratio of specific heats C_p/C_v, which for air is approximately 1.40.

In most of the more practical discussions that follow, we will find it adequate simply to use a complex form for k, with real and imaginary parts derived from Eqs. (8.14) and (8.15). The fact that Z_0 has a small imaginary part is not generally significant for the main pipes of musical instruments. For a few discussions, such as those related to the smaller tubes of finger holes, the more general results shown in the figures may be necessary.

8.3. Finite Cylindrical Pipes

All of the pipes with which we deal in musical instruments are obviously of finite length, so we must allow for the reflection of a wave from the remote end, whether it is open or closed. Because we are concerned with pipes as closely coupled driven systems, rather than as passive resonators, we shall proceed by calculating the input impedance for a finite length of pipe terminated by a finite load impedance Z_L, rather than examining doubly open or closed pipes in isolation. The terminating impedance Z_L will generally represent an open or a closed end, but it is not restricted to these cases. The development here is essentially the same as that set out in Chapter 2 for a string stretched between nonrigid bridges but, since the results are central to our discussion of pipes and horns, we start again from the beginning.

Suppose the pipe extends from $x = 0$ to $x = L$, and that it is terminated at $x = L$ by the impedance Z_L. The pressure in the pipe is a superposition of two waves, moving to the right and left, respectively, with amplitudes A and B, taken as complex quantities so that they can include a phase factor. Thus, at the point x,

$$p(x, t) = [Ae^{-jkx} + Be^{jkx}]e^{j\omega t}. \tag{8.16}$$

The acoustic particle velocity is similarly a superposition of the particle velocities associated with these two waves and, multiplying by pipe cross section S, the acoustic flow becomes, from Eq. (8.3),

$$U(x,t) = \left(\frac{S}{\rho c}\right)[Ae^{-jkx} - Be^{jkx}]e^{j\omega t}. \tag{8.17}$$

At the remote end $x = L$, pressure and flow are related as required by the terminating impedance Z_L, so that

$$\frac{p(L,t)}{U(L,t)} = Z_L, \tag{8.18}$$

and this equation is enough to determine the complex ratio B/A. If we write for the characteristic impedance of the pipe

$$Z_0 = \rho c/S \tag{8.19}$$

as in Eq. (8.3), then

$$\frac{B}{A} = e^{-2jkL}\left[\frac{(Z_L - Z_0)}{(Z_L + Z_0)}\right], \tag{8.20}$$

and the power reflected from Z_L has a ratio to incident power of

$$R = \left|\frac{B}{A}\right|^2 = \left|\frac{Z_L - Z_0}{Z_L + Z_0}\right|^2. \tag{8.21}$$

Clearly, there is no reflection if $Z_L = Z_0$ and complete reflection if $Z_L = 0$ or ∞. Since Z_0 is real for a lossless tube, there is also perfect reflection if Z_L is purely imaginary; however, if Z_L has a real part that is nonzero, then there will always be some reflection loss.

The quantity in which we are interested now is the input impedance Z_{IN} at the point $x = 0$. From Eqs. (8.16)–(8.19), this is

$$Z_{IN} = Z_0\left[\frac{A+B}{A-B}\right], \tag{8.22}$$

or from Eq. (8.20),

$$Z_{IN} = Z_0\left[\frac{Z_L \cos kL + jZ_0 \sin kL}{jZ_L \sin kL + Z_0 \cos kL}\right]. \tag{8.23}$$

Two important idealized cases are readily derived. The first corresponds to a pipe rigidly stopped at $x = L$ so that $Z_L = \infty$. For such a pipe,

$$Z_{IN}^{(ST)} = -jZ_0 \cot kL. \tag{8.24}$$

For the converse case of an ideally open pipe with $Z_L = 0$, which is not physically realizable exactly, as we see below,

$$Z_{IN}^{(OP)} = jZ_0 \tan kL. \tag{8.25}$$

The familiar resonance frequencies for open and stopped pipes arise from applying the idealized condition that the input end at $x = 0$ is also open, so that resonances occur if $Z_{IN} = 0$. For a stopped pipe, this requires that $\cot kL = 0$, giving

$$\omega^{(ST)} = \frac{(2n-1)\pi c}{2L}, \tag{8.26}$$

corresponding to an odd number of quarter wavelengths in the pipe length, while for an ideally open pipe, $\tan kL = 0$, giving

$$\omega^{(OP)} = \frac{n\pi c}{L}, \tag{8.27}$$

corresponding to an even number of quarter wavelengths, or any number of half wavelengths, in the pipe length.

While Eq. (8.24) applies quite correctly to a physically stopped pipe, the treatment of a physically open pipe is more difficult since, while $Z_L \ll Z_0$, it is not a sufficient approximation to set it to zero. It is relatively straightforward to calculate the radiation load Z_L on a pipe that terminates in a plane flange of size much larger than a wavelength (and therefore effectively infinite). The formal treatment of Rayleigh (1894) (Olson, 1957) makes the assumption that the wavefront exactly at the open end of the pipe is quite planar, normally a very good approximation, and gives the result

$$Z^{(F)} = R + jX, \tag{8.28}$$

where, as discussed for Eqs. (7.32)–(7.34),

$$R = Z_0\left[\frac{(ka)^2}{2} - \frac{(ka)^4}{2^2\cdot 3} + \frac{(ka)^6}{2^2\cdot 3^2\cdot 4} - \cdots\right], \tag{8.29}$$

$$X = \frac{Z_0}{\pi k^2 a^2}\left[\frac{(2ka)^3}{3} - \frac{(2ka)^5}{3^2\cdot 5} + \frac{(2ka)^7}{3^2\cdot 5^2\cdot 7} - \cdots\right], \tag{8.30}$$

and a is the radius of the pipe.

The behavior of R and X as functions of frequency, or more usefully as functions of the dimensionless quantity ka, is shown in Fig. 8.7. If $ka \ll 1$, then $|Z^{(F)}| \ll Z_0$ and most of the wave energy is reflected from the open end. If $ka > 2$, however, then $Z^{(F)} \approx Z_0$ and most of the wave energy is transmitted out of the end of the pipe into the surrounding air.

In musical instruments, the fundamental, at least, has $ka \ll 1$, though this is not necessarily true for all the prominent partials in the sound. It is therefore useful to examine the behavior of the pipe in this low-frequency limit. From Eqs. (8.29) and (8.30), $X \gg R$ if $ka \ll 1$, so that

$$Z^{(F)} \approx jZ_0 k\left(\frac{8a}{3\pi}\right). \tag{8.31}$$

By comparison with Eq. (8.25), since $ka \ll 1$, this is just the impedance of

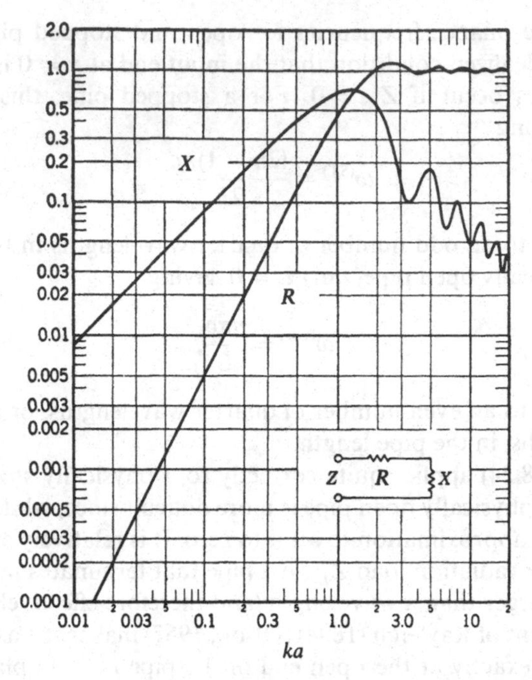

Fig. 8.7. The acoustic resistance R and the acoustic reactance X, both in units of $\rho c/\pi a^2$, for a circular piston (or open pipe) of radius a set in an infinite plane baffle, as functions of the frequency parameter ka (after Beranek, 1954).

an ideally open short pipe of length

$$\Delta^{(F)} = \frac{8a}{3\pi} \approx 0.85a. \tag{8.32}$$

It is thus a good approximation in this frequency range to replace the real flanged pipe by an ideally open pipe of length $L + \Delta^{(F)}$, and to neglect the radiation loss. From Fig. 8.7, it is clear that the end correction $\Delta^{(F)}$, which is proportional to X/ka, decreases slightly as $ka \to 1$ and continues to decrease more rapidly as ka increases past this value.

A real pipe, of course, is not generally flanged, and we need to know the behavior of Z_L in this case. The calculation (Levine and Schwinger, 1948) is very difficult, but the result, as shown in Fig. 8.8, is very similar to that for a flanged pipe. The main difference is that for $ka \ll 1$ R is reduced by about a factor 0.5 and X by a factor 0.7 because the wave outside the pipe has freedom to expand into a solid angle of nearly 4π rather than just 2π. The calculated end correction at low frequencies is now

$$\Delta^{(0)} \approx 0.61a. \tag{8.33}$$

The calculated variation of this end correction with the frequency parameter ka is shown in Fig. 8.9.

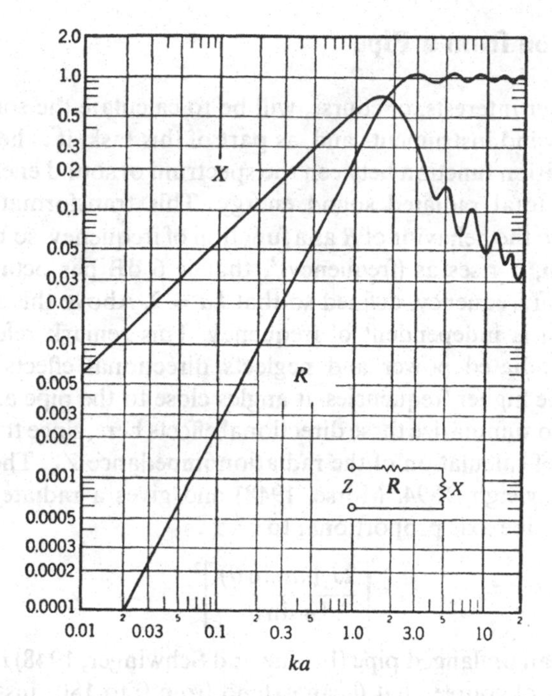

Fig. 8.8. The acoustic resistance R and the acoustic reactance X, both in units of $\rho c/\pi a^2$, for the open end of a circular cylindrical pipe of radius a, as functions of the frequency parameter ka (after Beranek, 1954).

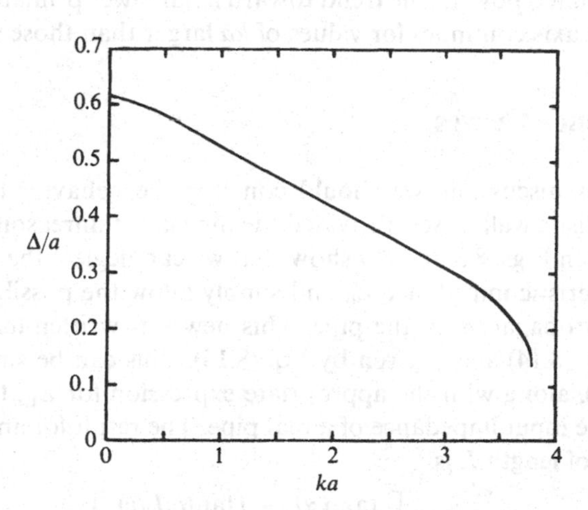

Fig. 8.9. The calculated end correction Δ for a cylindrical pipe of radius a, plotted as Δ/a, as a function of the frequency parameter ka, (after Levine and Schwinger, 1948).

8.4. Radiation from a Pipe

One of our later interests, of course, will be to calculate the sound radiation from musical wind instruments and, as part of this task, it is helpful to know the transformation function between the spectrum of sound energy within the pipe and the total radiated sound energy. This transformation is simply proportional to the behavior of R as a function of frequency, so that, to a good approximation, it rises as (frequency)^2, that is 6 dB per octave, below the reflection cutoff frequency, defined so that $ka = 2$. Above this frequency, the transformation is independent of frequency. This remark refers, of course, to the total radiated power and neglects directional effects that tend to concentrate the higher frequencies at angles close to the pipe axis.

It is useful to summarize these directional effects here, since they are derived in the course of calculation of the radiation impedance Z_L. The flanged case is simplest (Rayleigh, 1894; Morse, 1948) and gives a radiated intensity at angle θ to the pipe axis proportional to

$$\left[\frac{2J_1(ka\sin\theta)}{ka\sin\theta}\right]^2. \tag{8.34}$$

The result for an unflanged pipe (Levine and Schwinger, 1948) is qualitatively similar except, of course, that θ can extend from 0 to 180° instead of just to 90°. The angular intensity distribution for this case is shown in Fig. 8.10 for several values of ka, the results being normalized to the power radiated along the axis (Beranek, 1954). The directional index (DI) is the intensity level on the axis compared to the intensity level produced by an isotropic source with the same total radiated power. The trend toward a narrower primary beam angle along the pipe axis continues for values of ka larger than those shown.

8.5. Impedance Curves

Finally, in this discussion, we should consider the behavior of pipes with physically realistic wall losses. Provided the pipe is not unreasonably narrow, say $r_v > 10$, then Figs. 8.3 and 8.4 show that we can neglect the small change in the characteristic impedance Z_0 and simply allow the possibility that k is complex for propagation in the pipe. This new k is written $(\omega/v - j\alpha)$ with v given by Eq. (8.14) and α given by Eq. (8.15). This can be simply inserted into Eq. (8.23), along with the appropriate expression for Z_L, to deduce the behavior of the input impedance of a real pipe. The result for an ideally open pipe ($Z_L = 0$) of length L is

$$Z_{IN} = Z_0\left[\frac{\tanh\alpha L + j\tan(\omega L/v)}{1 + j\tanh\alpha L\tan(\omega L/v)}\right]. \tag{8.35}$$

This expression has maxima and minima at the maxima and minima, respec-

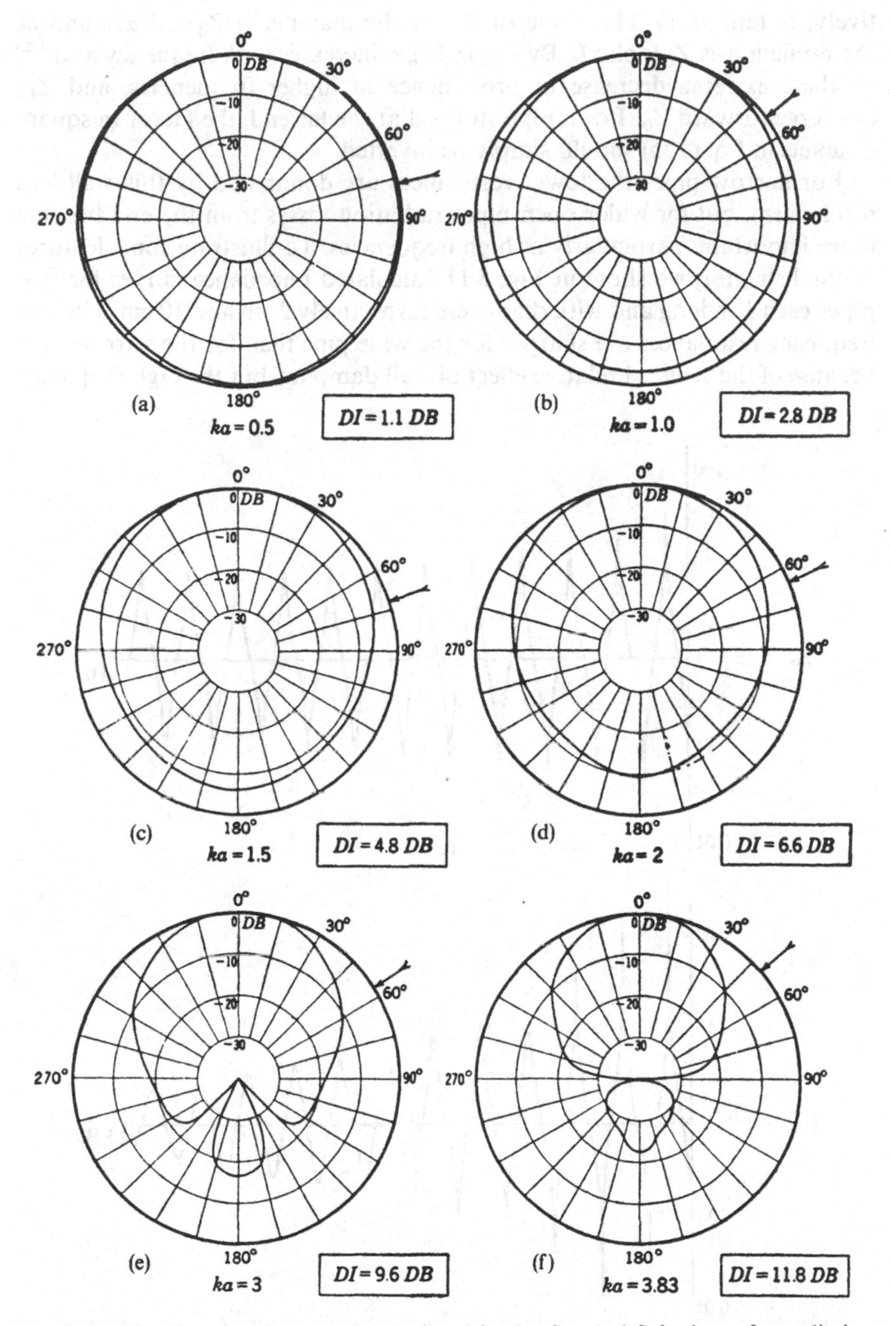

Fig. 8.10. The directional patterns calculated by Levine and Schwinger for radiation from an unbaffled circular pipe of radius a. The radial scale is in each case 40 dB and the directional index has the calculated value shown (after Beranek, 1954).

tively, of $\tan(\omega L/v)$. The value of Z_{IN} at the maxima is $Z_0 \coth \alpha L$, and at the minima it is $Z_0 \tanh \alpha L$. By Eq. (8.15), α increases with frequency as $\omega^{1/2}$, so these extrema decrease in prominence at higher frequencies, and Z_{IN} converges toward Z_0. For a pipe stopped at the far end, the factor in square brackets in Eq. (8.35) should simply be inverted.

For narrow pipes the lower resonances are dominated by this wall-loss mechanism, but for wider open pipes radiation losses from the end become more important, particularly at high frequencies. To illustrate some features of the behavior, we show in Fig. 8.11 calculated impedance curves for two pipes each 1 m long and with diameters respectively 2 cm and 10 cm. The low frequency resonances are sharper for the wide pipe than for the narrow pipe because of the reduced relative effect of wall damping, but the high frequency

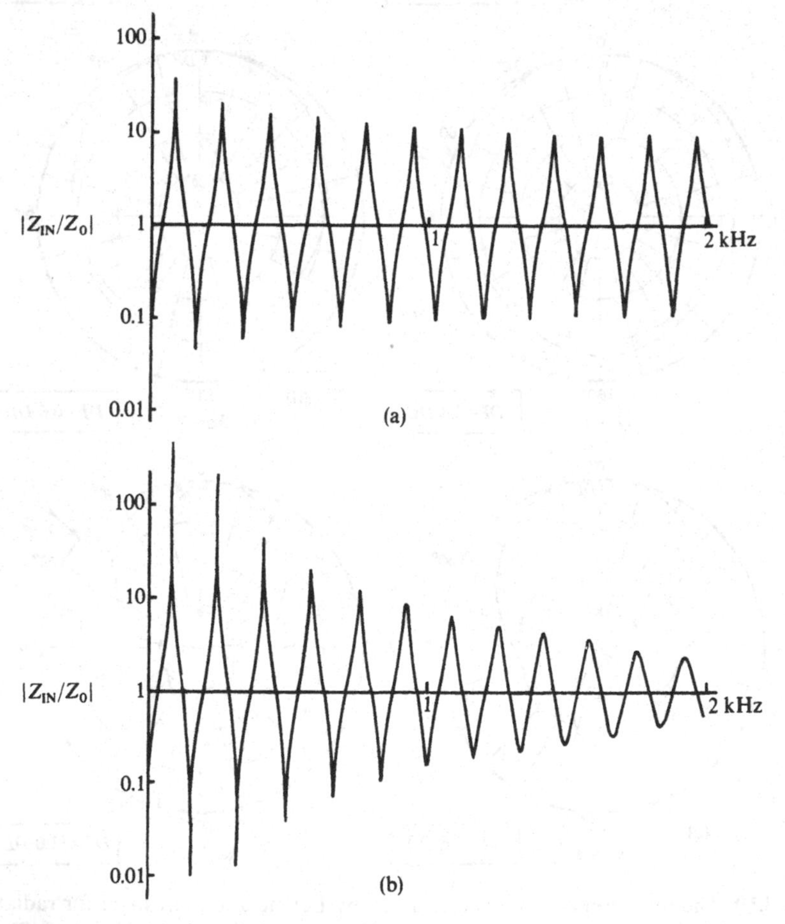

Fig. 8.11. Magnitude of the acoustic input impedance Z_{IN}, in terms of the characteristic impedance Z_0, for open cylindrical pipes of length 1 m and diameters of (a) 2 cm and (b) 10 cm.

resonances of the wide pipe are washed out by the effects of radiation damping. We can see that all the impedance maxima and minima have frequencies that are nearly harmonically related, that is as the ratio of two small integers. In fact, because the end correction decreases with increasing frequency, the frequencies of these extrema are all slightly stretched, and this effect is more pronounced for the wide than for the narrow pipe.

It is worthwhile to note incidentally that, because these input impedance curves have been plotted on a logarithmic scale, the corresponding admittance curves can be obtained simply by turning the impedance curve upside down. We will see later that sometimes we will be required to think of admittance maxima and sometimes of impedance maxima, depending upon the way in which the pipe is used.

When we consider musical instruments in detail, we find that several of them rely upon cylindrical pipes as their sound generators. The most obvious of these is the pipe organ, in which most of the pipes are cylindrical (a few are conical). Tone quality of air jet driven pipes is varied by the use of closed and open tubes, by differences in relative diameters, and by differences in the sort of termination at the open end—some are simple open ends, some have slots, some have bells, and some have narrow chimneys. These variations can all be treated on the basis of the above discussion supplemented by a separate consideration of the form of Z_L produced by the termination. We will return to consider these matters later on.

8.6. Horns

Following this introductory discussion of cylindrical pipes, we are now ready to begin a treatment of sound propagation in horns, a horn being defined quite generally as a closed-sided conduit, the length of which is usually large compared with its lateral dimensions. In fact, we shall only treat explicitly horns that are straight and have circular cross section, but much of the discussion is really more general than this.

Formulation of the wave propagation problem in an infinitely long horn simply requires solution of the wave equation

$$\nabla^2 p = \frac{1}{c^2}\frac{\partial^2 p}{\partial t^2}, \tag{8.36}$$

subject to the condition that $\mathbf{n}\cdot\nabla p = 0$ on the boundaries, $\mathbf{n}$ being a unit vector normal to the boundary at the point considered. More simply, we suppose the wave to have a frequency ω so that Eq. (8.36) reduces to the Helmholtz equation

$$\nabla^2 p + k^2 p = 0, \tag{8.37}$$

where $k = \omega/c$. Solution of this equation is simple provided that we can choose a coordinate system in which one coordinate surface coincides with the walls of the horn and in which Eq. (8.37) is separable. Unfortunately, the Helmholtz equation is separable only in coordinates that are confocal quadric surfaces or their degenerate forms (Morse and Feshbach, 1953). There are 11 varieties of these coordinate systems, but only a few of them are reasonable candidates for horns. These are rectangular coordinates (a pipe of rectangular cross section), circular cylinder coordinates, elliptic cylinder coordinates, spherical coordinates (a conical horn), parabolic coordinates, and oblate spheroidal coordinates, as shown in Fig. 8.12. Of these, we have already dealt with the circular cylinder case, and the rectangular and elliptic cylinder versions differ from it only in cross-sectional geometry and hence in their higher modes. The parabolic horn is not musically practical since it cannot be made to join smoothly onto a mouthpiece, so we are left with the conical horn and the horn derived from oblate spheroidal coordinates, which will prove to be of only passing interest.

We deal with the oblate spheroidal case first, because it illustrates some of the difficulties we will have to face later. The hornlike family of surfaces consists of hyperboloids of revolution of one sheet, as shown in Fig. 8.12. At large distances, these approach conical shapes, but near the origin they become almost cylindrical. Indeed, one could join a simple cylinder parallel to the axis in the lower half plane to a hyperboloid horn in the upper half plane without any discontinuity in slope of the walls. The important thing to notice, however, is the shape of the wavefronts as shown by the orthogonal set of coordinate surfaces. These are clearly curved and indeed they are oblately spheroidal, being nearly plane near the origin and nearly spherical at large

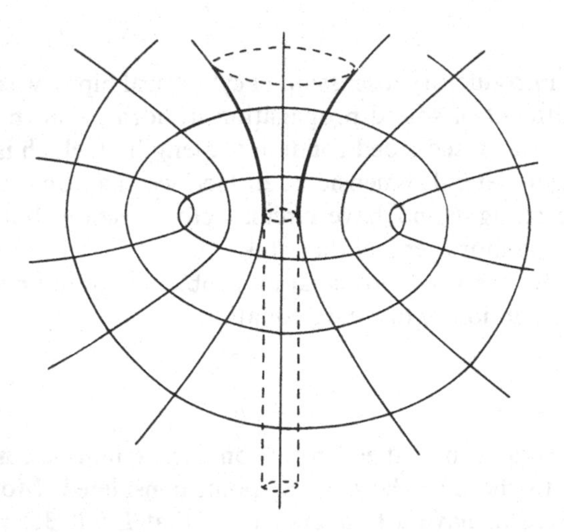

Fig. 8.12. The oblate spheroidal coordinate system in which the wave equation is separable. If the hyperboloid of revolution (shown in heavy outline) is taken as the horn, then the oblate spheroidal surfaces orthogonal to this and lying within it are the wave fronts. Note that such a hyperboloid horn can be smoothly joined to a cylindrical pipe of appropriate diameter, as shown.

distances. Waves can propagate in this way as a single mode, like the plane waves in a cylinder. Such behavior is possible only for separable coordinate systems. For nonseparable systems that we may try to separate approximately, there will always be an admixture of higher modes. Horn systems resembling a cylinder joined to a narrow-angle hyperboloid horn as described above are in fact used in many brass instruments, though not because of any consideration of separability of the wave equation. Indeed, once the length of the horn is made finite, we produce an unresolvable inseparability near the open end so that there is no real practical design assistance derived from near separability inside the horn.

Rather than setting out the exact solution for a hyperboloid or a conical horn in detail, let us now go straight to the approximate solution for propagation in an infinite horn of rather general shape. We assume that we have some good approximation to the shapes of the wavefronts—something more or less spherical and, since the wave fronts must be orthogonal to the horn walls, centered approximately at the apex of the cone that is locally tangent to the horn walls, as shown in Fig. 8.13. This description will be exact for a conical horn, but only an approximation for other shapes. If $S(x)$ is the area of this wavefront in the horn at position x, defined by its intersection with the axis, then, during an acoustic displacement ξ, the fractional change in the volume of air in the horn at position x is $(1/S)\,\partial(S\xi)/\partial x$. This contrasts with the simpler expression $\partial\xi/\partial x$ for a plane wave in unconfined space. Proceeding now as for the plane wave case, we find a wave equation of the form

$$\frac{1}{S}\frac{\partial}{\partial x}\left(S\frac{\partial p}{\partial x}\right)=\frac{1}{c^2}\frac{\partial^2 p}{\partial t^2}, \tag{8.38}$$

which is known as the Webster equation (Webster, 1919; Eisner, 1967), although

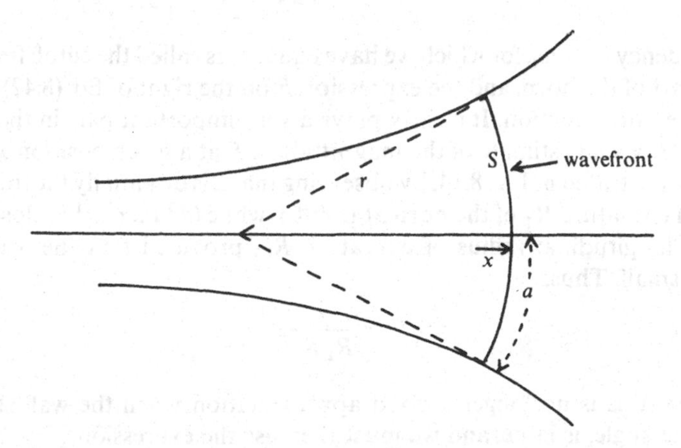

Fig. 8.13. In a horn, the wavefront has approximately the form of a spherical cap of area S and effective radius r based upon the local tangent cone and cutting the axis at a point with coordinate x.

its origins date back to the time of Bernoulli. Actually, in Webster's case, the curvature of the wave fronts was neglected so that x was taken as the geometrical distance along the horn axis and S as the geometrical cross section at position x. This plane-wave approximation is good for horns that are not rapidly flaring, but breaks down for a horn with large flare, as we see later.

We have assumed that p is constant across the wave front in the horn, which is equivalent to assuming separability. This is not a bad approximation for horns that do not flare too rapidly, but we must not expect too much of it in extreme cases. In this spirit, we now make the transformation

$$p = \psi S^{-1/2} \tag{8.39}$$

in the reasonable expectation that, with the even spreading of wave energy across the wavefront, ψ should be essentially constant in magnitude, independent of x. If we also assume that p varies with angular frequency ω and write S in terms of a local equivalent radius a so that

$$S = \pi a^2, \tag{8.40}$$

then Eq. (8.38) becomes

$$\frac{\partial^2 \psi}{\partial x^2} + \left(k^2 - \frac{1}{a}\frac{\partial^2 a}{\partial x^2}\right)\psi = 0, \tag{8.41}$$

where $k = \omega/c$. This form of the equation, attributable to Benade and Jansson (1974), serves as a good basis for discussion of the behavior of horns.

The first thing to notice about Eq. (8.41) is that the wave function ψ, and hence the original pressure wave p, is propagating or nonpropagating according as

$$k^2 \equiv \left(\frac{\omega}{c}\right)^2 \gtrless \frac{1}{a}\frac{d^2 a}{dx^2} \equiv F. \tag{8.42}$$

The frequency $\omega = \omega_0$ for which we have equality is called the cutoff frequency at this part of the horn, and the expression F on the right of Eq. (8.42) may be called the horn function. It clearly plays a very important part in the theory of horns. A visual estimate of the magnitude of F at a given position x can be made, as illustrated in Fig. 8.14, by observing that a is essentially the transverse radius of curvature R_T of the horn at point x while $(d^2a/dx^2)^{-1}$ is close to the external longitudinal radius of curvature R_L, provided that the wall slope da/dx is small. Thus,

$$F \approx \frac{1}{R_L R_T}. \tag{8.43}$$

Of course, this is no longer a good approximation when the wall slope, or local cone angle, is large, and we must then use the expression

$$F = \frac{1}{a}\frac{d^2 a}{dx^2}, \tag{8.44}$$

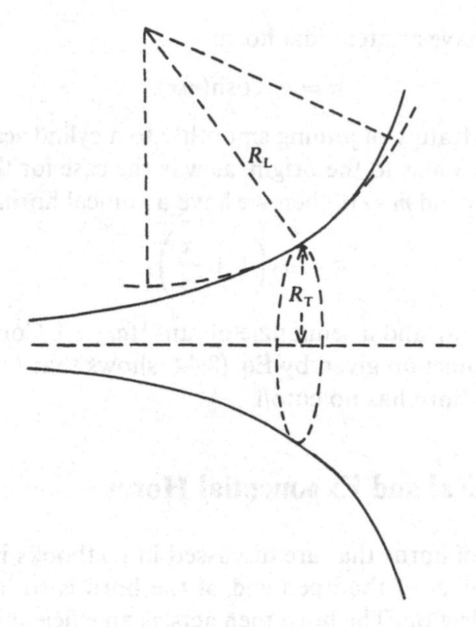

Fig. 8.14. The geometry of a horn at any one place is characterized by the external longitudinal radius of curvature R_L and the internal transverse radius of curvature R_T.

with a interpreted as the equivalent internal radius measured along the wavefront as discussed previously.

Of particular theoretical simplicity is the class of horns called Salmon horns (Salmon, 1946a, b), for which the horn function F, and therefore the cutoff frequency ω_0, is constant along the whole length of the horn (Morse, 1948). Clearly, from Eq. (8.44), this implies

$$a = Ae^{mx} + Be^{-mx}, \tag{8.45}$$

where $F = m^2$ and m is called the flare constant. It is more convenient to rewrite Eq. (8.45) as

$$a = a_0[\cosh(mx) + T\sinh(mx)], \tag{8.46}$$

where T is an alternative parameter. The pressure wave in the horn then has the form

$$p = \left(\frac{p_0}{a}\right)e^{j\omega t}e^{-j\sqrt{k^2-m^2}\,x} \tag{8.47}$$

and is nonpropagating if $k < m$. These expressions should strictly all be interpreted in terms of curved wavefront coordinates, as in Fig. 8.13, but it is usual to neglect this refinement and simply use the plane wave approximation.

The family of horns described by Eq. (8.46) has several important degenerate forms. If $T = 1$, then we have an exponential horn:

$$a = a_0\exp(mx). \tag{8.48}$$

If $T = 0$, then we have a catenoidal horn:

$$a = a_0 \cosh(mx), \tag{8.49}$$

which has the nice feature of joining smoothly to a cylindrical pipe extending along the negative x axis to the origin, as was the case for the hyperboloidal horn. If $T = 1/mx_0$ and $m \to 0$, then we have a conical horn:

$$a = a_0\left(1 + \frac{x}{x_0}\right), \tag{8.50}$$

with its vertex at $-x_0$ and a semiangle of $\tan^{-1}(a_0/x_0)$. Consideration of the value of the horn function given by Eq. (8.44) shows that $F = 0$ for this case, so that the conical horn has no cutoff.

8.7. Finite Conical and Exponential Horns

Many of the uses of horns that are discussed in textbooks involve situations in which the diameter of the open end of the horn is so large that there is no appreciable reflection. The horn then acts as an efficient impedance transformer between a small diaphragm piston in the throat and a free spherical wave outside the mouth. Exponential and catenoidal horns have near-unity efficiency, as defined by Morse (1948), above their cutoff frequencies, while the efficiency of a conical horn never becomes zero but rises gradually with increasing frequency until it reaches unity. We shall not discuss these situations further—those interested should consult Morse (1948) or Olson (1957). It is, however, worthwhile to quote results analogous to Eq. (8.23) for the throat impedance of a truncated conical or exponential horn terminated by a given mouth impedance Z_L, which is typically the radiation impedance, though this requires some modification in careful work because of the curvature of the wavefronts (Fletcher and Thwaites, 1988).

For a conical horn with a throat of area S_1 located at position x_1, a mouth of area S_2 at position x_2 and length $L = x_2 - x_1$, we find (Olson, 1957)

$$Z_{IN} = \frac{\rho c}{S_1}\left\{\frac{jZ_L[\sin k(L-\theta_2)/\sin k\theta_2]+(\rho c/S_2)\sin kL}{Z_L[\sin k(L+\theta_1-\theta_2)/\sin k\theta_1 \sin k\theta_2]-(j\rho c/S_2)[\sin k(L+\theta_1)/\sin k\theta_1]}\right\}, \tag{8.51}$$

where $k\theta_1 = \tan^{-1} kx_1$ and $k\theta_2 = \tan^{-1} kx_2$, both x_1 and x_2 being measured along the axis from the position of the conical apex. Similarly, for an exponential horn of the form of Eq. (8.48) and length L,

$$Z_{IN} = \frac{\rho c}{S_1}\left[\frac{Z_L \cos(bL + \theta) + j(\rho c/S_2)\sin bL}{jZ_L \sin bL + (\rho c/S_2)\cos(bL - \theta)}\right], \tag{8.52}$$

where $b^2 = k^2 - m^2$ and $\theta = \tan^{-1}(m/b)$. It is not simple to allow for wall effects in these expressions, since the imaginary part of k varies with position in the horn. For a horn with a wide mouth and not too narrow a throat, radiation effects may dominate so that k can be taken as real. This is, however, not a valid approximation in musical instruments, which use long horns of quite small diameter. We shall see that more complex calculations are necessary in such cases.

The expression [Eq. (8.51)] for the input impedance of a conical horn, measured at the end that is at a distance x_1 from the apex, deserves some further discussion. In the first place, we should note that it is applicable for the impedance at either the wide or the narrow end of a conical pipe. For a flaring cone, $x_2 > x_1$ and $L > 0$, while for a tapering cone, $x_2 < x_1$ and $L < 0$.

In the second place, we should examine several special cases of open and stopped cones, making the approximation that $Z_L = 0$ at an open end and $Z_L = \infty$ at a closed end. For a cone of length L with an ideally open end $Z_L = 0$, Eq. (8.51) gives, for either the large or the small end of a cone, the formal result

$$Z_{IN} = j\left(\frac{\rho c}{S_1}\right)\frac{\sin kL \sin k\theta_1}{\sin k(L + \theta_1)}. \tag{8.53}$$

This does not imply that the input impedance is the same from both ends, since, as noted above, the sign of L and the magnitude of θ_1 are different in the two cases.

Zeros in Z_{IN} occur at frequencies for which $\sin kL = 0$, so that these frequencies are the same in each case and exactly the same as those for a cylindrical pipe with the same length L. To allow for the finite reactance associated with the radiation impedance Z_L, it is approximately correct, for a narrow cone, to simply add an appropriate end correction equal to 0.6 times the open end radius to the geometrical length L, as discussed in relation to Eq. (8.33).

The infinities in Z_{IN} occur, however, at frequencies that differ between the two cases and are not simply midway between those of the zeros, as was the case with a cylindrical pipe. Rather, the condition for an infinity in Z_{IN} is, from Eq. (8.53),

$$\sin k(L + \theta_1) = 0, \tag{8.54}$$

or equivalently,

$$kL = n\pi - \tan^{-1} kx_1. \tag{8.55}$$

For a cylinder, $x_1 \to \infty$ so that $\tan^{-1} kx_1 = \pi/2$, as we already know. For a cone measured at its narrow end, L is positive and, since $\tan^{-1} kx_1 < \pi/2$, the frequencies of the infinities in Z_{IN} are higher than those for a cylinder of the same length. The converse is true for a tapering cone. If the cone is nearly complete, so that $kx_1 \ll 1$, then $\tan^{-1} kx_1 \approx kx_1$ and, since $L = x_2 - x_1$, Eq. (8.55) becomes $kx_2 \approx n\pi$, so that the frequencies of the infinities in impedance approach those of an open cylinder of length $x_2/2$. Figure 8.15 shows an input impedance curve for an incomplete cone calculated for the narrow end.

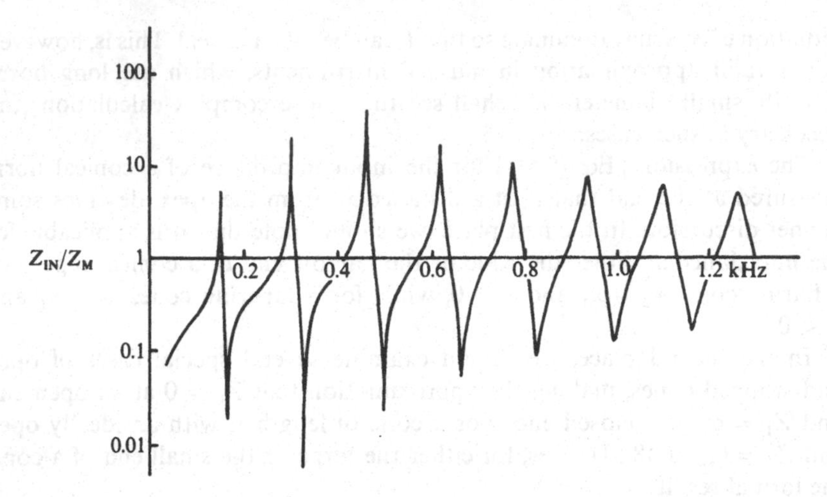

Fig. 8.15. The input impedance Z_{IN} of a conical horn of length 1 m, throat diameter 1 cm, and mouth diameter 10 cm, relative to the impedance $Z_M = (\rho c/S_1)$, as a function of frequency.

Cones that are stopped at the remote end are of less musical interest. If $Z_L = \infty$, then, from Eq. (8.51),

$$Z_{IN} = j\left(\frac{\rho c}{S_1}\right)\frac{\sin k(L-\theta_2)\sin k\theta_1}{\sin k(L+\theta_1-\theta_2)}. \tag{8.56}$$

The zeros are given by $k(L-\theta_2) = n\pi$, which, writing $L = x_2 - x_1$, becomes

$$(kx_2 - \tan^{-1}kx_2) - kx_1 = n\pi. \tag{8.57}$$

Thus, if we are considering a tapering cone and the distance x_2 from the stopped end to the imaginary apex is small enough that kx_2 is rather less than unity, the bracketed terms nearly cancel and the cone behaves approximately as though it is of length x_1 and complete to its vertex. No such simplification occurs for a flaring cone or for the infinities in the impedance, for which the condition is that $k(L+\theta_1-\theta_2) = n\pi$.

An extensive discussion of the conical horn in musical acoustics has been given by Ayers et al. (1985). Using straightforward physics, this paper treats both conical frusta and complete cones and points out a number of misconceptions, or at least erroneous expositions, in the standard physics literature.

The particular theoretical attraction of the family of horns defined by Eq. (8.46) is that they have a constant cutoff frequency along their length. Unfortunately, it is only in the cases of cylindrical and conical horns that this property can be combined with the musical requirement that the modes for a finite horn, defined by the condition of minimum or maximum input impedance, should have harmonically related frequencies. For this reason, there is little to be gained here by a more detailed discussion of exponential horns based upon Eq. (8.52). Instead, we shall go on to discuss more general types of horns.

8.8. Bessel Horns

One particular family of horns that is worthy of attention because of its formal simplicity and that provides a good approximation to musically useful horns (Benade, 1959; Benade and Jansson, 1974) is the Bessel horn family, defined by

$$S = Bx^{-2\varepsilon} \quad \text{or} \quad a = bx^{-\varepsilon}, \tag{8.58}$$

where x is the geometrical distance measured from a reference point $x = 0$. If $\varepsilon = 0$, the horn is cylindrical, and if $\varepsilon = -1$ it is conical, so that these two degenerate cases are included in the picture. More usefully for our present discussion, we suppose ε to be positive, in which case the horn has a rapid flare at the origin $x = 0$, which thus represents the open mouth of the horn as shown in Fig. 8.16. The particular analytical simplicity of the Bessel horns arises from the fact that, in the plane-wave approximation, the wave equation, Eq. (8.38), in the form Eq. (8.41) has, for the case of an ideally open horn mouth at $x = 0$, the standing wave solution (Jahnke and Emde, 1938)

$$\psi = x^{1/2} J_{\varepsilon+1/2}(kx), \tag{8.59}$$

where J is a Bessel function, hence the name of the horn family. From Eq. (8.39), the pressure standing wave has the form

$$p(x) = Ax^{\varepsilon+1/2} J_{\varepsilon+1/2}(kx). \tag{8.60}$$

The existence of this analytical solution is a great help for semiquantitative discussion of the behavior of this family of horns, whose shape can be made to vary very considerably by choice of the parameter ε. If we are considering

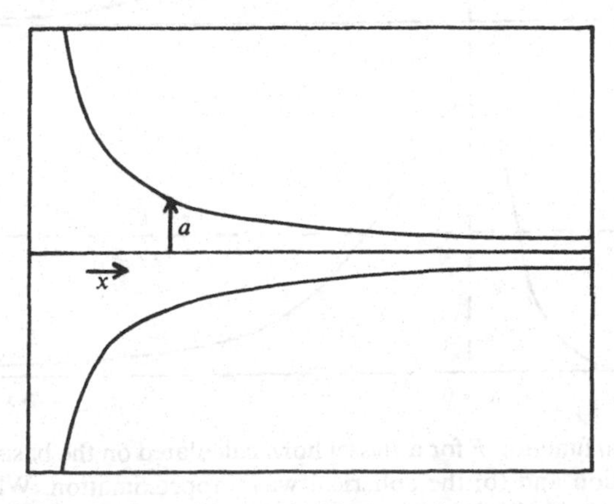

Fig. 8.16. The form of a Besssel horn with parameter $\varepsilon > 0$.

a horn composed of segments of different Bessel horns joined end to end, then Eq. (8.60) must be supplemented in each segment by a similar term involving the Neumann function $N_{\varepsilon+1/2}(kx)$. We can then calculate the behavior, in the plane-wave approximation, of composite horns having sections of Bessel, exponential, conical, and cylindrical geometry joined end to end (Pyle, 1975). We will not follow such a course in detail but, instead, we will examine briefly the behavior of waves near the mouth of a Bessel horn to show some of the complications involved.

As we saw in Eqs. (8.41)–(8.44), the propagation of a wave in a flaring horn is governed by the value of the horn function F at the point concerned. If F is greater than k^2, the wave is attenuated exponentially with distance and a reflection occurs rather than propagation. For a horn mouth of Bessel type, F is easily calculated in the plane-wave approximation, in which wave coordinate x is replaced by the axial geometrical coordinate, and has the form shown in Fig. 8.17(a). F goes to infinity at the mouth of the horn, so that waves of all frequencies are reflected, and the reflection point for waves of low frequency (small k) is further inside the mouth than for those of high frequency.

Close to the open mouth, however, the plane-wave approximation is clearly inadequate, and the spherical approximation gives a much better picture

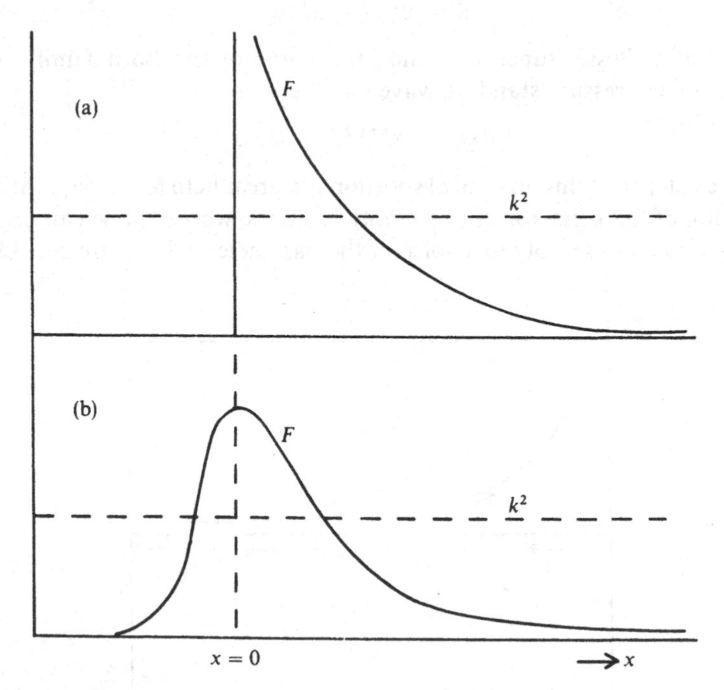

Fig. 8.17. The horn function F for a Bessel horn calculated on the basis of (a) the plane wave approximation and (b) the spherical wave approximation. When $k^2 < F$, the wave is attenuated instead of being propagated (after Benade and Jansson, 1974).

(Benade and Jansson, 1974). The function F in this approximation is shown in Fig. 8.17(b). Its form is similar to the plane-wave version, but F never becomes infinite, so that there is an upper cutoff frequency above which waves can propagate freely from the mouth without reflection. An infinite F, as in the plane-wave approximation, would in fact confine all sound energy within the horn because of the infinite barrier height; the more realistic curve given by the spherical approximation allows some wave energy to leak out at any frequency by tunneling through the barrier.

When we consider the losses in a real horn in detail, not only must we supplement the standing wave solution [Eq. (8.60)] with extra terms in $N_{\varepsilon+1/2}(kx)$, which combine to represent the small fraction of energy contained in propagating waves lost through the mouth of the horn, but we must also take account of wall losses by adding a small imaginary part $-j\alpha$ to k. For a horn more complicated in profile than a simple cylinder, α depends on the local horn radius and therefore varies from place to place along the horn. The calculations are then quite involved (Kergomard, 1981). Fortunately, we can ignore these complications in our present discussion though they must be taken into account in any really accurate computations.

Because the horns of real musical instruments do not conform exactly to any of the standard types we have considered, we will not go into further detailed discussion of them at this stage, but defer this until we come to describe the instruments themselves in a later chapter.

8.9. Compound Horns

As we see in Chapter 14, most brass wind instruments actually have horn profiles that are nearly cylindrical for about half their length, starting from the mouthpiece, and then expand to an open flaring bell. In modern instruments, the profile of this flaring section is well approximated by a Bessel horn of the form of Eq. (8.58), while for older instruments and some of the more mellow modern instruments, much of the expanding section is nearly conical.

It is not worthwhile to model such compound horns in detail, since real instruments do not conform precisely to any such oversimplified prescription. The complications of mode tuning are illustrated, however, by consideration of the frequencies of the input impedance maxima—which are the sounding modes—for a compound horn consisting of a cylindrical and a conical section smoothly joined together. Part of the complication is produced by the acoustical mismatch at the joint, but similar mode behavior would be found for other profiles.

The input impedance Z_c for a conical horn of length L_1 is given by Eq. (8.51). We can simplify our discussion by taking the radiation impedance Z_L at the open mouth to be zero, giving, as another form of Eq. (8.53),

$$Z_c = \frac{j\rho c}{S_1}\left(\cot kL_1 + \frac{1}{kx_1}\right)^{-1}, \tag{8.61}$$

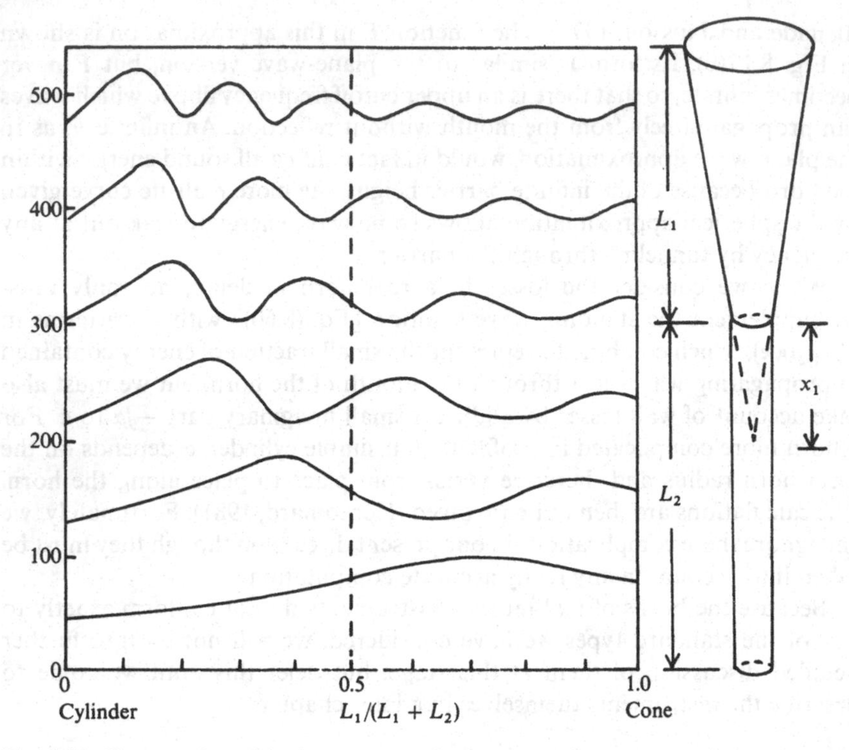

Fig. 8.18. Frequencies of the input impedance maxima for a compound horn, of the shape shown, as a function of the fraction of horn length that is conical.

where S_1 is the throat area and x_1 is the distance from the throat to the vertex of the cone, as shown in Fig. 8.18. We now take Z_c as the terminating impedance for the cylindrical section of length L_2 and matching area S_1. From Eq. (8.23), the input impedance to the compound horn then becomes infinite if

$$jZ_c \sin kL_2 + \left(\frac{\rho c}{S_1}\right) \cos kL_2 = 0, \tag{8.62}$$

and, with Eq. (8.61), this immediately leads to the condition

$$\tan kL_2 - \cot kL_1 - \left(\frac{1}{kx_1}\right) = 0. \tag{8.63}$$

Solution of this equation for $k = \omega/c$ then gives the frequencies of the impedance maxima. The same calculation for a cylinder joined to an exponential horn would have used Eq. (8.52) and led to the formally similar result

$$\tan kL_2 - \left(\frac{b}{k}\right) \cot bL_1 - \left(\frac{m}{k}\right) = 0, \tag{8.64}$$

where $b = (k^2 - m^2)^{1/2}$.

The behavior of the frequencies of the first few modes of a cylinder–cone compound horn of roughly trumpet dimensions is shown in Fig. 8.18. The total length $L_1 + L_2$ and the mouth and throat areas are all kept constant, while the fraction of the length that is conical is varied. Clearly, the frequency variation of individual modes is complicated, but note that the modes of a horn with half its length conical are nearly harmonic. Compound horns with other profiles show rather similar behavior.

For a complete understanding of horn acoustics, we must, of course, include the effect of the nonzero radiation impedance at the open mouth, take account of wall losses if the horn is narrow, and calculate a complete impedance curve. The mathematical apparatus for all this is contained in our discussion.

8.10. Perturbations

As a final part of this chapter, let us consider the effect of a small perturbation in the shape of some idealized horn. This is important for several reasons. The first of these is that if the effect of such perturbations is understood then the instrument designer can use them to adjust the horn shape slightly in order to properly align or displace horn resonances in which he or she is interested. The second is that in instruments with finger holes in the side of the horn perturbations are unavoidable, and it is important to understand and control their effects.

Suppose that we know a standing wave solution $p_0(x, t)$ for a horn of profile $S_0(x)$ that corresponds to a mode of frequency ω_0, as described by Eq. (8.38), with appropriate terminating impedances at each end of the horn. Now, let the bore of the horn be altered by a small amount so that the new cross section becomes

$$S(x, t) = S_0(x, t) + \delta S(x, t). \tag{8.65}$$

This perturbation will change the resonance frequency ω_0 to a new value ω, which we write

$$\omega = \omega_0 + \delta\omega, \tag{8.66}$$

and the pressure distribution will become

$$p(x, t) = \beta p_0(x, t) + p_1(x, t), \tag{8.67}$$

where $\beta \approx 1$ and p_1 is functionally orthogonal to p_0. If we substitute Eqs. (8.65), (8.66), and (8.67) into Eq. (8.38) and also use the unperturbed version of Eq. (8.38) with S_0, ω_0, and p_0, then we can collect all the terms of first order in the perturbations to give

$$\frac{1}{S_0}\frac{d\delta S}{dx}\frac{dp_0}{dx} - \frac{\delta S}{S_0^2}\frac{dS_0}{dx}\frac{dp_0}{dx} = -\frac{2\omega_0\,\delta\omega}{c^2}p_0 + \text{terms in } p_1. \tag{8.68}$$

Now, we multiply by $S_0 p_0$ and integrate over the whole length of the horn.

The terms in p_1 vanish since they are orthogonal to p_0, and we find

$$\delta\omega = -\frac{c^2}{2\omega_0 N}\int S_0 \frac{d}{dx}\left(\frac{\delta S}{S_0}\right) p_0 \frac{dp_0}{dx}\, dx, \tag{8.69}$$

where

$$N = \int S_0 p_0^2\, dx. \tag{8.70}$$

Clearly, these two equations allow us to calculate the shift $\delta\omega$ in the resonance frequency produced by the bore perturbation $\delta S(x)$.

It is easiest to evaluate the effect of such perturbations by considering what happens when the bore is enlarged by a small amount Δ at a position x_0. To do this, we write

$$\delta S(y) = \Delta\, \delta(x - x_0), \tag{8.71}$$

where $\delta(x - x_0)$ is a Dirac delta function. Substituting this into Eq. (8.69) we find, since $d\delta(x - x_0)/dx$ under an integral yields the negative of the derivative of the integrand at x_0, that

$$\delta\omega = \frac{c^2 \Delta}{2\omega_0 N}\left[\frac{d}{dx}\left(p_0 \frac{dp_0}{dx}\right)\right]_{x=x_0}. \tag{8.72}$$

To see what this means, suppose that in some region of the horn the pressure pattern has a spatial variation like $\sin kx$. Then the bracket in Eq. (8.72) behaves like $\cos 2kx_0$. Thus, when the perturbation is near a maximum in the pressure variation ($\sin kx_0 \approx 1$), $\cos 2kx_0 \approx -1$ and $\delta\omega$ has its maximum negative value. Conversely, if $\sin kx_0 \approx 0$, then $\cos 2kx_0 \approx +1$, so that near a maximum in the velocity $\delta\omega$ has its maximum positive value. Both these remarks assume that Δ is positive, so that the bore is being enlarged by the perturbation. Opposite conclusions apply for a constriction in the bore. Since different modes have their pressure maxima at different places in the horn, it is possible to change the relative frequencies of selected modes and so effect musically desirable changes in the behavior of the horn.

A few examples make this point clear. Suppose we consider the modes of a cylindrical pipe open at the end $x = 0$ and closed at the other end, $x = L$. The nth normal mode then has a pressure pattern like $\sin[(2n - 1)\pi x/2L]$. If the pipe diameter is enlarged near the open end ($x = 0$) then, by Eq. (8.72), $\delta\omega$ is positive for all n, and the frequencies of all modes are raised. Conversely, if the diameter is enlarged near the closed end, the frequencies of all modes are lowered.

More interesting is the case in which the bore is enlarged at a point one-third of the length away from the closed end, that is at $x = 2L/3$. The bracket in Eq. (8.72) then behaves like $\cos(2\pi/3) = -0.5$ for the first mode, $n = 1$, and like $\cos 2\pi = +1$ for the second mode, $n = 2$. Thus, the frequency of the first mode is lowered while that of the second mode is raised.

8.11. Numerical Calculations

With the ready availability of computers, it is often practically convenient, though generally less instructive, to calculate the behavior of horns numerically. This can be done conveniently only in the plane-wave approximation or the local spherical-wave approximation, so that its accurate use is limited to horns with small flare, but this is adequate for many wind instruments.

The bare bones of the procedure have already been set out, but its use in this way has not been made explicit. The basis of the method is the recognition that an arbitrary horn can always be represented to very good accuracy by a succession of small conical sections joined end to end. In the limit in which the length of the sections becomes infinitesimal, the representation is exact. Now, in the spherical-wave approximation, the input impedance of a section of conical horn of length L is related to its terminating impedance Z_L by Eq. (8.51). Indeed, if the length L is small so that the cross section is nearly constant, the propagation constant k in Eq. (8.51) can be made complex to allow for wall losses according to Eqs. (8.14), (8.15), and (8.35). To make a numerical calculation for an arbitrary horn therefore, we simply start from the open end, with Z_L the radiation impedance shown in Fig. 8.8. We then use Eq. (8.51) successively for short distances back along the horn until the input throat is reached. A modification of the program readily allows the pressure distribution along the horn to be calculated at the same time. In Chapters 14–16, we will refer to calculations for particular musical instruments that have been carried out in this way.

For the plane-wave version of this calculation, which trades off a simpler calculation at each step against an increased number of steps, we can approximate the horn by a series of very short cylindrical sections. Equation (8.23) replaces the more complicated conical form of Eq. (8.51), but each cylindrical section must be made very short in order to give a reasonable approximation for a flaring horn.

8.12. The Time Domain

Nearly all of our discussion has been carried on in the frequency domain—we have examined the propagation in a horn of sinusoidal waves of steady frequency ω. While this is generally the most convenient framework in which to study the physics of musical instruments, it is sometimes helpful to revert to the time domain and examine the buildup and propagation of pressure disturbances along the horn and their reflection from its open end. This is clearly a good way to treat the initial transients of musical sounds, and the time-domain method can also be used for steady tones. Formally, treatment of a problem in the time domain or the frequency domain must give identical results, but in practice we are forced to make approximations in our analysis in order to get a reasonable answer, and the nature of these approximations can be quite different in the two cases, so that one can usually be employed more easily than the other (Schumacher, 1981; McIntyre et al., 1983).

In Sections 8.3 and 8.7, we gave explicit expressions for the input impedance $Z(\omega)$ of horns of various profiles, while Figs. 8.11 and 8.15 illustrated the behavior of $Z(\omega)$ relative to the characteristic impedance $Z_0 = \rho c/S_1$ of an infinite cylindrical tube having the same area S_1 as the throat of the horn. In general, $Z(\omega)$ displays a long series of more or less sharp impedance peaks at frequencies ω_n that are, for useful musical instruments, related moderately closely to some harmonic series $n\omega_0$.

If now we seek the pressure waveform $p(t)$ observed in the mouth of the horn when the acoustic flow into it is $U(t)$, we can proceed in two ways. Either we express $U(t)$ as a Fourier integral,

$$U(t) = \int_{-\infty}^{\infty} U(\omega)e^{j\omega t}\,d\omega, \tag{8.73}$$

and use the definition of $Z(\omega)$ to write

$$p(t) = \int_{-\infty}^{\infty} Z(\omega)U(\omega)e^{j\omega t}\,d\omega \tag{8.74}$$

or else we define an impulse response function $G(t - t')$, which gives the pressure response at time t to a unit impulse of flow at time t', and then write directly the convolution integral

$$p(t) = \int_{-\infty}^{t} G(t - t')U(t')\,dt'. \tag{8.75}$$

As we remarked before, these results are formally equivalent, and, indeed, the impulse response function $G(t)$ can be shown to be simply the Fourier transform of the impedance function $Z(\omega)$.

As Schumacher (1981) has pointed out, the problem with using Eq. (8.75) as a computational formula from which to derive $p(t)$ arises from the fact that $G(t - t')$ has a considerable extension in time; an acoustic flow pulse injected into the horn reflects from its two ends for some tens of periods before its amplitude is reduced enough for it to become negligible. This follows at once from the sharply peaked nature of $Z(\omega)$, which is an equivalent feature in the frequency domain.

A useful way out of this computational problem is to note that for a short time after the injection of a flow pulse and before any reflections have returned to the throat from discontinuities along the horn the input impedance of a cylindrical horn is the simple resistive quantity Z_0. This suggests that we should write

$$G(t) = Z_0\,\delta(t) + \tilde{G}(t), \tag{8.76}$$

where $\delta(t)$ is a Dirac delta function. It is clear then that $\tilde{G}(t) \approx 0$ for t less than the wave transit time τ from the throat of the horn to the first significant reflecting discontinuity—often the open mouth—and back. In fact $\tilde{G}(t)$ is the impulse response of the horn if it is assumed that its throat is blocked by a nonreflecting termination. $\tilde{G}(t)$ shows the effect of reflections from all irregularities along the horn as well as the reflection from its open mouth.

Schumacher (1981) has shown that $\tilde{G}(t)$ can be usefully expressed as the Fourier transform of the reflection coefficient $r(\omega)$ of the horn as seen from its throat, as defined by

$$r(\omega) = \frac{Z(\omega) - Z_0}{Z(\omega) + Z_0}, \tag{8.77}$$

and from this he derives the result

$$p(t) = Z_0 U(t) + \int_0^\infty r(t')[Z_0 U(t - t') + p(t - t')]\, dt', \tag{8.78}$$

where $r(t)$ is the Fourier transform of $r(\omega)$. It turns out that $r(t)$ is nearly zero for t less than the wave transit time τ and that $r(t)$ has a significantly smaller extension in time than the original impulse function $G(t)$. It is therefore relatively straightforward to use Eq. (8.78) as an integral equation from which to calculate numerically the transient behavior of a tube-loaded acoustic generator.

As an example, let us apply Eqs. (8.77) and (8.78) to the case of a uniform cylindrical tube open at its far end. Neglecting radiation corrections at the open mouth, we have, from Eq. (8.25),

$$Z = jZ_0 \tan kL, \tag{8.79}$$

where L is the length of the tube and $k = \omega/c$. Then, from Eq. (8.77),

$$r(\omega) = -e^{-2j\omega L/c}. \tag{8.80}$$

Taking the Fourier transform,

$$r(t) = -\int e^{j\omega t} e^{-j\omega L/c}\, d\omega = -\delta(t - \tau), \tag{8.81}$$

where

$$\tau = 2L/c. \tag{8.82}$$

Substituting into Eq. (8.78), we find

$$p(t) = Z_0 U(t) - Z_0 U(t - \tau) - p(t - \tau), \tag{8.83}$$

and applying Eq. (8.78) again to $p(t - \tau)$,

$$p(t) = p(t - 2\tau) + Z_0[U(t) - 2U(t - \tau) + U(t - 2\tau)]. \tag{8.84}$$

For a lip-driven or reed-driven instrument, $Z_0 U(t)$ is always much smaller than $p(t)$, since the excitation mechanism is pressure controlled. We can therefore neglect the U terms to give

$$p(t) = p(t - 2\tau), \tag{8.85}$$

so that the pipe acts as a quarter-wave resonator with frequency $1/2\tau = c/4L$.

The behavior of a nearly complete conical horn is rather more complex and really requires a different derivation, but the pressure wave reflects to the throat with a time delay τ given by Eq. (8.82) as before. When we consider Eq. (8.83), however, we find from Eqs. (6.26) and (6.27) that the terms $Z_0 U$ behave like $1/r^2$ near the origin and so dominate the terms in p, which behave like $1/r$. We can therefore ignore the p terms in Eq. (8.83) and conclude that, in the steady state,

$$U(t) = U(t - \tau), \tag{8.86}$$

so that the horn acts as a half-wave resonator with frequency $1/\tau = c/2L$.

A detailed discussion of the impulse response of a conical horn has been given by Ayers et al. (1985) and includes the more complex case when the conical frustum is not nearly complete.

For a more general type of horn, such as is found in brass wind instruments, there is usually an initial cylindrical section which then flares to conical or Bessel form near the mouth. We expect the reflection behavior to be intermediate between that of a cylinder and a cone of the same length, but the reflected pulse will be considerably distorted by dispersion effects. Details can be found either by direct measurement or by taking the Fourier transform of the input impedance.

References

Ayers, R.D., Eliason, L.J., and Mahgerefteh, D. (1985). The conical bore in musical acoustics. *Am. J. Phys.* **53**, 528–537.

Benade, A.H. (1959). On woodwind instrument bores. *J. Acoust. Soc. Am.* **31**, 137–146.

Benade, A.H. (1968). On the propagation of sound waves in a cylindrical conduit. *J. Acoust. Soc. Am.* **44**, 616–623.

Benade, A.H., and Jansson, E.V. (1974). On plane and spherical waves in horns with nonuniform flare. *Acustica* **31**, 80–98.

Beranek, L.L. (1954). "Acoustics," pp. 91–115. McGraw-Hill, New York; reprinted 1986, Acoustical Society Am., Woodbury, New York.

Eisner, E. (1967). Complete solutions of the "Webster" horn equation. *J. Acoust. Soc. Am.* **41**, 1126–1146.

Fletcher, N.H., and Thwaites, S. (1988). Response of obliquely truncated simple horns: Idealized models for vertebrate pinnae. *Acustica* **65**, 194–204.

Jahnke, E. and Emde, F. (1938). "Tables of Functions," p. 146. Teubner, Leipzig, Reprinted 1945, Dover, New York.

Kergomard, J. (1981). Ondes quasi-stationnaires dans les pavillons avec partis visco-thermiques aux parois: Calcul de l'impedance. *Acustica* **48**, 31–43.

Levine, H., and Schwinger, J. (1948). On the radiation of sound from an unflanged pipe. *Phys. Rev.* **73**, 383–406.

McIntyre, M.E., Schumacher, R.T., and Woodhouse, J. (1983). On the oscillation of musical instruments. *J. Acoust. Soc. Am.* **74**, 1325–1345.

Morse, P.M. (1948). "Vibration and Sound," 2nd ed pp. 233–288. McGraw-Hill, New York; reprinted 1976, Acoustical Society of Am., Woodbury, New York.

Morse, P.M., and Feshbach, H. (1953). "Methods of Mathematical Physics," Vol. 1, pp. 494–523, 655–666. McGraw-Hill, New York.

Morse, P.M., and Ingard, K.U. (1968). "Theoretical Acoustics," pp. 467–553. McGraw-Hill, New York. Reprinted 1986, Princeton Univ. Press, Princeton, New Jersey.

Olson, H.F. (1957). "Acoustical Engineering," pp. 88–123. Van Nostrand-Reinhold, Princeton, New Jersey.

Pyle, R.W. (1975). Effective length of horns. *J. Acoust. Soc. Am.* **57**, 1309–1317.

Rayleigh, Lord (1894). "The Theory of Sound," 2 vols. Macmillan, London. Reprinted 1945. Dover, New York.

Salmon, V. (1946a). Generalized plane wave horn theory. *J. Acoust. Soc. Am.* **17**, 199–211.

Salmon, V. (1946b). A new family of horns. *J. Acoust. Soc. Am.* **17**, 212–218.

Schumacher, R.T. (1981). Ab initio calculations of the oscillations of a clarinet. *Acustica* **48**, 71–85.

Webster, A.G. (1919). Acoustical impedance, and the theory of horns and of the phonograph. *Proc. Nat. Acad. Sci. (US)* **5**, 275–282.

CHAPTER 9

Acoustic Systems

The preceding chapters have introduced us to the physics and mathematics of a wide range of acoustic components, and the explicit formulae we have derived allow us to calculate behavior of simple systems in a good amount of detail. Very often, however, we shall wish to understand and even to calculate the behavior of more complex systems combining a number of acoustic components to make resonators, mufflers, auditory systems, loudspeakers, and microphones. While a study of the detailed complexity of such systems would take us too far from the purpose of this small book, it is very useful to know the basic principles by which their behavior can be calculated.

The basic approach is closely similar to that of electrical network theory, with acoustic components replacing electric components. Close analogies between the acoustical and electrical situations can be drawn, but it is generally adequate to take a more relaxed approach and to point out the details only when this helps understanding.

9.1. Low-Frequency Components and Systems

The basic acoustic quantities with which we shall be concerned are acoustic pressure p and acoustic volume flow U, both considered to be quantities oscillating in time with angular frequency ω. In analogy with electrical quantities we can think of the acoustic pressure as being equivalent to electric potential and acoustic volume flow to electric current. Just as in the electrical case, there is a difference between network analysis at low frequencies where electrical components are connected by simple wires, and the high frequency case in which transmission lines and waveguides must be used. Let us consider the low-frequency case first.

Suppose we have a short pipe of length l and cross-section S, and that p is the pressure difference between its ends, considered as an acoustic quantity with time variation $e^{j\omega t}$. The air in the pipe behaves as a simple mass of magnitude ρlS and the force acting upon it is pS. Its velocity is U/S, and the relation between acceleration and force then gives

$$pS = \rho l S \frac{d}{dt}\left(\frac{U}{S}\right) \quad \text{or} \quad p = \left(\frac{\rho l}{S}\right)\frac{dU}{dt} = j\left(\frac{\rho l}{S}\right)\omega U \tag{9.1}$$

The analogy between this equation and the electrical relation between the voltage V across the inductance L and the current i through it

$$V = L\frac{di}{dt} = j\omega L i \tag{9.2}$$

is immediately apparent. The quantity

$$Z_{\text{pipe}} = j\omega(\rho l/S) \tag{9.3}$$

is an acoustic impedance, as discussed previously in Section 8.18, and is analogous to the electrical inductance. It is usually called an acoustic inertance.

In just the same way, the pressure inside a cavity of volume V is related to the acoustic current into the cavity by the equation

$$p = \frac{\gamma p_{\text{a}}}{V}\int U\,dt \tag{9.4}$$

where p_{a} is the steady atmospheric pressure and γ is the ratio of specific heats for air, so that γp_{a} is the bulk elastic modulus for the air in the cavity, as discussed in Section 6.1. Since the integral is just the total flow into the cavity, and is analogous to the electric charge, the quantity $V/\gamma p_{\text{a}}$ is analogous to the electrical capacitance, and is generally referred to as the acoustic compliance. A more convenient expression for it is, from Eq. (6.11), $V/\rho c^2$ and the acoustic impedance of the cavity is

$$Z_{\text{cav}} = -j\rho c^2/V\omega. \tag{9.5}$$

The third acoustic component we need is the analog of an electrical resistance, and is generally realized physically by a disk of permeable fabric or felt. The acoustic resistivity, in this case, results from the viscous drag associated with motion of air through the narrow passages in the resistive material. The flow through such an acoustic resistance is simply proportional to the pressure across it, and can be related to its thickness l, area S, and acoustic resistivity r_{a} by $Z_{\text{res}} = R = r_{\text{a}} l/S$.

Finally we need two other simple components representing a rigid stopper and a simple opening. The impedance of the stopper is simply infinite, as in an electrical open circuit, while to a first approximation the impedance of an opening is zero, corresponding to an electrical short circuit. Although the first idealization is always appropriate, the second requires modification in careful work to take account of the radiation impedance $Z_{\text{rad}} = R + jX$ as discussed in Section 8.3 and given explicitly in Eqs. (8.29) and (8.30). For the open end of an unbaffled pipe of radius a and area $S = \pi a^2$,

$$Z_{\text{rad}} \approx 0.16\frac{\rho\omega^2}{c} + 0.6 j\omega\frac{\rho a}{S} \tag{9.6}$$

provided $\omega < c/a$.

To complete our treatment of acoustic components at low frequencies we need two sorts of acoustic generators—an acoustic pressure generator and an acoustic current generator. The first is simply a pressure source with zero internal impedance—it can supply any amount of acoustic current and always maintains its driving pressure. The second is an acoustic flow generator with infinite internal impedance—it always supplies the same acoustic current, whatever driving pressure it requires. Such idealized sources do not exist in reality, of course, but free sound fields come close to being pressure sources and mechanically driven pistons are close to being acoustic current sources.

The free-field pressure source is particularly important, so we look briefly at the way in which it differs from a true acoustic pressure source. This differs somewhat from case to case, but in fact the internal impedance to be associated with a free field at low frequencies is just the radiation impedance of the aperture upon which it acts. While we are dealing with low frequencies, this impedance is small and, from Eq. (9.6), largely inertive—the resistive component is very small. At high frequencies we will see that things become a little more complicated.

The basic principles by which components are assembled into networks are straightforward, and are most easily presented through the examples that follow. Ambiguities that appear to arise about constructing the network can be resolved by considering, in a physical way, the flow of acoustic current through the system and the way in which pressure drops may occur across the circuit elements. The behavior of the network can then be described in terms of circulating currents U_i flowing through its interconnected circuits. These currents can generally be drawn in more than one way, but alternative descriptions lead to the same results. We need only follow a few general rules: (a) each current must flow around a complete circuit and must not cross over itself; (b) each circuit element must have at least one current flowing through it; (c) the number of currents should be as small as possible. The direction associated with each current is not important, because this is sorted out when solving the network equations, but it is generally best to draw currents that pass through pressure or current generators in the "natural" direction for the generator to avoid possible errors of sign.

We have already met the Helmholtz resonator in Section 1.6.2 as an example of an "air spring," and it is now appropriate to examine it in a little more detail as our first example of a network problem. The resonator consists simply of an enclosure of volume V vented by a pipe of length l and cross-section S, as shown in Fig. 9.1a. The network describing this system and its coupling to external acoustic radiation is shown in Fig. 9.1b—notice that we have included the radiation impedance. There is only a single circuit in this network, and we can easily write an equation to describe it, by setting the driving pressure in the circuit equal to the sum of the pressure drops across the individual circuit elements, to give

$$p = (Z_{\text{rad}} + Z_{\text{pipe}} + Z_{\text{cav}})U. \tag{9.7}$$

In a very careful calculation we might want to include the resistive losses

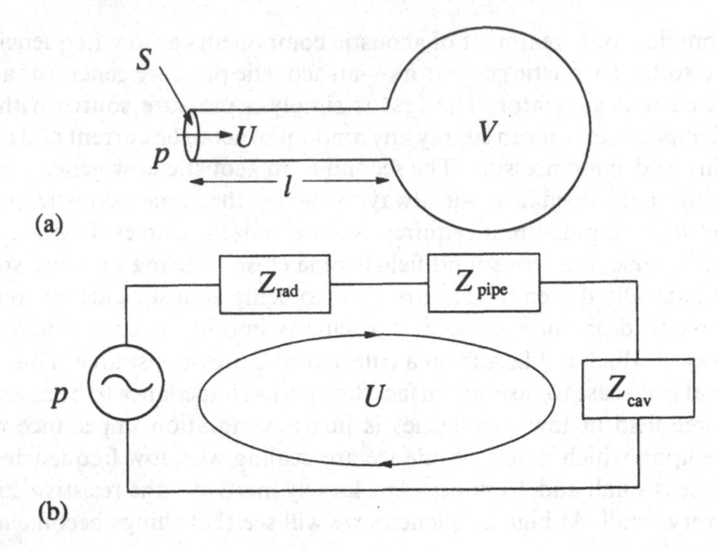

Fig. 9.1. (a) A simple Helmholtz resonator driven by an external sound field; and (b) the analog network describing its behavior.

associated with viscous drag on the walls of the pipe and thermal and viscous losses to the walls of the cavity, but here we simply ignore these in comparison with the radiation loss—an approximation that is justified if the pipe is short and wide. We can now use the explicit expressions given above to write

$$U \approx \left\{0.16\frac{\rho\omega^2}{c} + j\left[\frac{\omega\rho(l + 0.6a)}{S} - \frac{\rho c^2}{V\omega}\right]\right\}^{-1} p. \tag{9.8}$$

This is just a simple damped resonator equation, resonance with maximum acoustic flow occurring at the frequency

$$\omega^* \approx c\left[\frac{S}{V(l + 0.6a)}\right]^{1/2}. \tag{9.9}$$

The resonance Q value is determined in this case by the radiation resistance, or more generally by the sum of this resistance and the resistive losses inside the resonator. This is only a first-order treatment of the Helmholtz resonator, and we shall return to look at it more carefully in the next section after we have considered high-frequency networks.

It is interesting to consider the related situation in which the resonator is excited, not by an external sound field, but rather by a simple piston volume source located within the cavity, as in Fig. 9.2a. In this case the network appears as shown in Fig. 9.2b and has two circuits. We do not know the pressure p_0 produced by the flow generator, just the flow U_0, and there is no pressure generator in the second mesh, so that the equations for the two meshes can be written

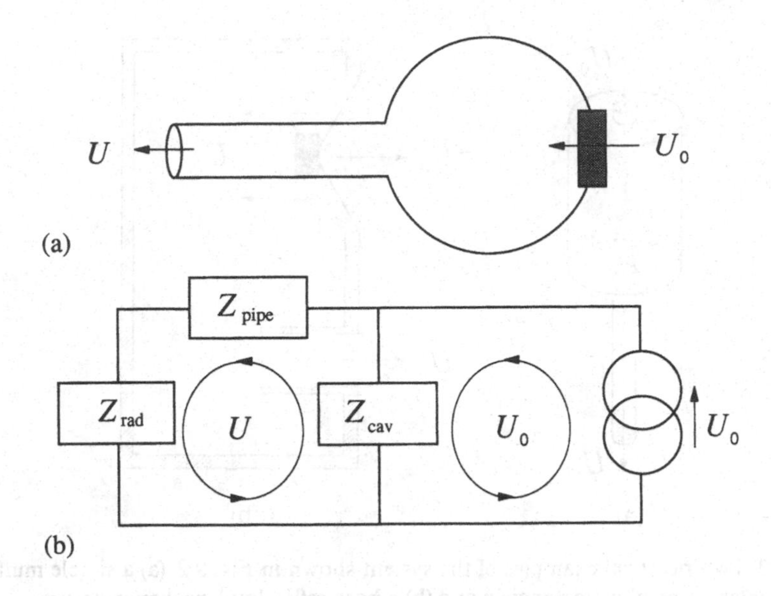

Fig. 9.2. (a) A Helmholtz resonator driven by a volume source within its cavity; and (b) the analog network describing its behavior.

$$\begin{aligned} p_0 &= Z_{\text{cav}}(U_0 - U) \\ 0 &= Z_{\text{cav}}(U - U_0) + (Z_{\text{pipe}} + Z_{\text{rad}})U. \end{aligned} \tag{9.10}$$

The first of these equations is not needed, and the second gives the current U through the open neck of the resonator as

$$U = \frac{Z_{\text{cav}} U_0}{Z_{\text{rad}} + Z_{\text{pipe}} + Z_{\text{cav}}}. \tag{9.11}$$

The resonance frequency is thus given again by Eq. (9.9). If we are interested in the pressure in the cavity, rather than the flow through the pipe, then this can easily be found by substituting U from Eq. (9.11) in the first of Eqs. (9.10). Both U and p have maximum amplitude at the same frequency, but they differ in phase by an amount that depends upon the magnitude of the resistive loss term.

Two practical applications of the system shown in Fig. 9.2 should be noted. The first is a simple exhaust muffler such as shown in Fig. 9.3a. The engine connected to the muffler produces a pulsating gas flow which can be regarded, to a first approximation, as a steady flow and an oscillating component of frequency ω. The design minimizes the amplitude of the flow out the exhaust pipe at frequency ω by arranging that the resonance frequency ω^* given by Eq. (9.9) is much lower than the engine frequency ω. From Eq. (9.11), under these conditions, the exhaust flow $U(\omega)$ varies as ω^{-2} and is significantly less than the engine flow $U_0(\omega)$. The muffler efficiency increases as ω^* is made smaller, which requires a large cavity and a long tail-pipe of small diameter.

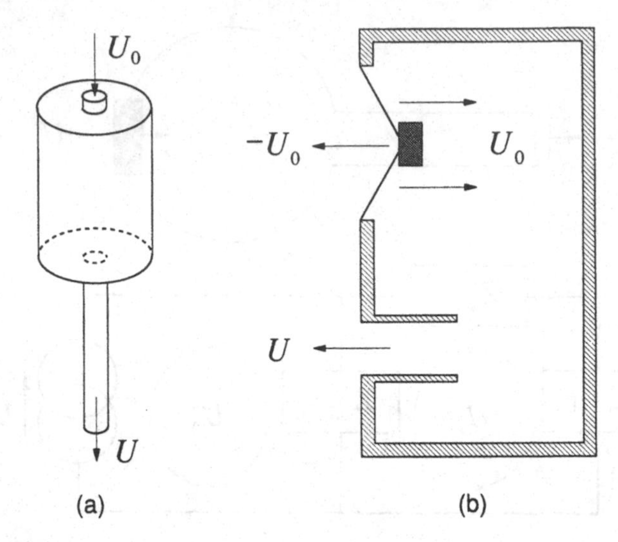

Fig. 9.3. Two practical examples of the system shown in Fig. 9.2: (a) a simple muffler for an internal-combustion engine; and (b) a bass-reflex loudspeaker enclosure.

The first requirement is inconvenient and the second impedes the steady component of the flow, so that design compromises are required.

The second application is to the so-called "bass-reflex" loudspeaker design shown in Fig. 9.3b. The rear of the loudspeaker cone provides a volume drive $U_0(\omega)$ and the acoustic flow through the vent port is $U(\omega)$, as given by Eq. (9.11), which has a resonance maximum at frequency ω^* and a falling characteristic above and below this frequency. This acoustic output is supplemented by the direct volume flow $-U_0(\omega)$ from the front surface of the speaker cone. The direct speaker output falls dramatically at frequencies below the natural resonance frequency of the loudspeaker cone suspension but, if the vented-box resonance is arranged to have a frequency somewhat below this cone resonance, phase relations are such that the vent flow reinforces the direct flow at all frequencies above the box resonance, giving improved bass response. In practice the enclosure is lined with heavy felt to broaden the box resonance. This can be modelled as a resistance in series with the impedance Z_{cav} in Fig. 9.2b. The coupling between the cone and the enclosure actually shifts the two resonance peaks apart a little in frequency, in a way that can be calculated.

It is perhaps worth noting that in the case of Fig. 9.1 we had a series-resonant circuit driven by a low-impedance source, and the resonance condition corresponded to a minimum in the input impedance of the resonator as seen by the source. In the case of Fig. 9.2 the resonator appeared as a parallel-resonant circuit driven by a high-impedance generator, and the resonance condition corresponded to a maximum in the impedance presented to

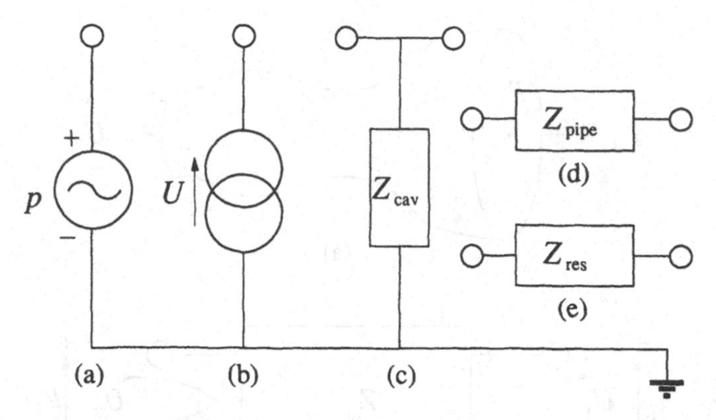

Fig. 9.4. Analog circuit elements for (a) a pressure generator, (b) a flow generator, (c) a cavity, (d) a pipe, and (e) a resistance, drawn in relation to a convenient "ground" potential.

the generator. Power transfer is always greatest when the impedance of the circuit is matched to the impedance of the generator—specifically the resistive parts of the impedances should be as nearly equal as possible, and the reactive parts should be equal in magnitude but opposite in sign.

Sometimes it is difficult to decide just how the analogs for components should be assembled into a network—whether, for example, the network should be series or parallel, as in the cases above. A simple procedure helps resolve this difficulty by introducing a baseline or "ground" lead to which some of the components are always connected. In particular, one terminal of a pressure or flow generator must always be connected to ground, and the same must be true of one terminal of a cavity element. The elements corresponding to pipes or resistive loads, however, can have both terminals above ground potential. This is illustrated in Fig. 9.4, where we have also modified the drawing of the cavity element to show explicitly the possibility that it might be connected to more than one other element—to an inlet and an outlet tube, for example. If one of the connections is unused in this case, it can simply be left open-circuited (not connected to anything), which is equivalent to closing the port with a rigid plate.

Before going on to consider high-frequency systems it is as well to note the limitations on the low-frequency analysis outlined above. Essentially what has been assumed is that all the elements involved are small in dimensions compared with the wavelength of the sound involved. This means that in the case of cavities the pressure is uniform throughout, and that in the case of pipes the flow into one end is exactly equal to the flow out the other. Because sound wavelengths of concern in air typically range from a few meters to a few centimeters, this means that we are justified in applying the analysis above only to rather low frequencies or to rather small systems.

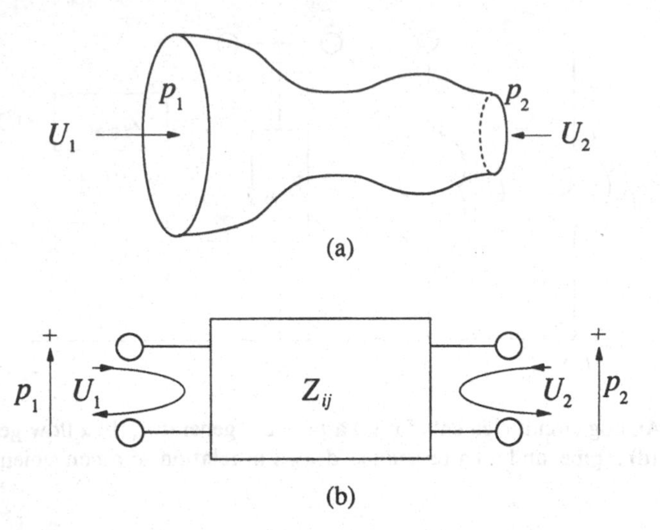

Fig. 9.5. Definition of (a) the acoustic quantities and (b) the network quantities to be associated with an extended acoustic element.

9.2. High-Frequency Components and Systems

Once the sound wavelength is comparable with the dimensions of an acoustic system, we must take a rather different approach, though clearly the results must reduce to those we have already derived if we let the frequency become sufficiently low. In this section we consider only what we might call one-dimensional components, implying by that phrase that only one of the component dimensions is comparable to the wavelength and the other dimensions are small. This essentially limits the discussion to pipes and horns of arbitrary length but of relatively small transverse dimensions. We will say a little about more general systems at the end.

With this limitation on the components we are considering, the important thing is that they all have two ends, and that we can no longer make the assumption that the acoustic flow out of one end is exactly equal to the flow in at the other. We therefore need more variables to describe each component, and a convenient scheme is shown in Fig. 9.5a. We choose the relevant physical quantities to be the acoustic pressures p_1 and p_2 at the two ends and the acoustic flows U_1 and U_2 in those two ends. Defining the flow directions in this way has the advantage of displaying the appropriate symmetry for simple components such as pipes of uniform cross-section. Figure 9.5b shows the analog network element, which is usually referred to as a two-port element for obvious reasons. Because we are dealing only with linear systems, we can write simple relationships between the acoustic quantities for the element in the form

$$\begin{aligned} p_1 &= Z_{11}U_1 + Z_{12}U_2 \\ p_2 &= Z_{21}U_1 + Z_{22}U_2 \end{aligned} \tag{9.12}$$

where the coefficients Z_{ij} are generally complex. There is a very general theorem, called the reciprocity theorem, which shows that, for any system not involving magnetic fields, $Z_{21} = Z_{12}$, and we shall use this result often. There is no similar relation between Z_{11} and Z_{22} unless it is imposed by the physical symmetry of the component—clearly this equality holds in the case of a uniform pipe, but not in the case of a flared horn.

Before we can use these ideas to analyze a system, we must have a way of determining the values of the impedance coefficients Z_{ij} for that system. Generally this involves determining the analytical form for waves propagating through the element, calculating the pressures and flows for two simple cases such as for a stopped and an ideally open end, and comparing the results with Eqs. (9.12). In this way we find, for the case of a uniform pipe of length l and cross-section S, the explicit forms

$$\begin{aligned} Z_{11} &= Z_{22} = -jZ_0 \cot kl \\ Z_{21} &= Z_{12} = -jZ_0 \operatorname{cosec} kl \end{aligned} \tag{9.13}$$

where $Z_0 = \rho c/S$ is called the characteristic impedance of the pipe, and $k = \omega/c$ as usual. We can, if we wish, include viscous and thermal losses to the walls of the pipe, as discussed in Section 8.2, by taking k to be the complex quantity $\omega/c - j\alpha$, where α is the attenuation coefficient for wave propagation in the pipe.

From Eqs. (9.12) and (9.13) we can easily calculate the input impedance Z_{in} for open and stopped pipes, simply by setting $p_2 = 0$ or $U_2 = 0$, respectively. After a little algebra we find

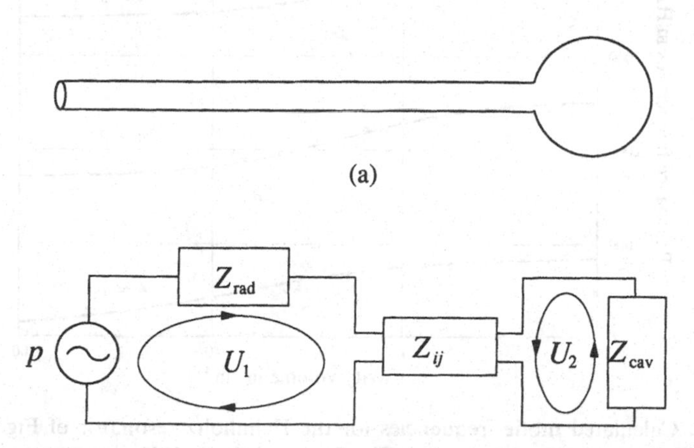

Fig. 9.6. (a) A Helmholtz resonator with a long neck and a small cavity; and (b) the associated network diagram.

$$Z_{\text{in}}^{\text{open}} = jZ_0 \tan kl \rightarrow \frac{j\omega l}{S} \tag{9.14}$$

$$Z_{\text{in}}^{\text{stopped}} = -jZ_0 \cot kl \rightarrow -\frac{j\rho c^2}{\omega l S}. \tag{9.15}$$

The final form of writing is in each case the low-frequency result for $kl \ll 1$, and these agree exactly with Eqs. (9.3) and (9.5) as required.

We can now carry out a more sophisticated examination of the Helmholtz resonator, allowing the possibility that the length l of the neck may become comparable with, or even considerably greater than, the sound wavelength. We must suppose, however, that the neck is rather narrow and that the diameter of the cavity is always a good deal less than the wavelength. The physical arrangement is shown in Fig. 9.6, together with the associated network. There are two circuits to the network, and we can easily write down the equations describing them as

$$\begin{aligned} p &= (Z_{\text{rad}} + Z_{11})U_1 + Z_{12}U_2 \\ 0 &= Z_{21}U_1 + (Z_{\text{cav}} + Z_{22})U_2 \end{aligned} \tag{9.16}$$

which can easily be solved to give

$$U_1 = \frac{(Z_{\text{cav}} + Z_{22})p}{(Z_{\text{rad}} + Z_{11})(Z_{\text{cav}} + Z_{22}) - Z_{12}Z_{21}}. \tag{9.17}$$

The resonance frequencies, of which there are now many, are given by the minima of the denominator.

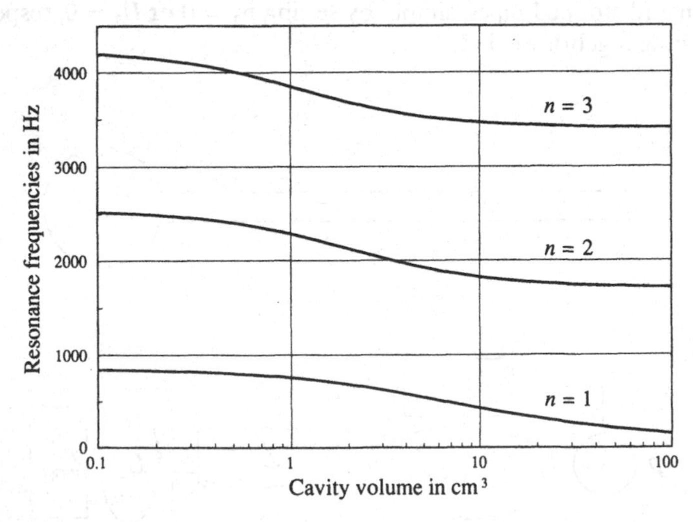

Fig. 9.7. Calculated mode frequencies for the Helmholtz resonator of Fig. 9.6 as a function of the volume of the cavity. The first mode has a behavior corresponding to that of a simple Helmholtz resonator.

If we take V to be the volume of the cavity, so that $Z_{cav} = -j\rho c^2/V\omega$, ignore the small radiation impedance Z_{rad}, and use the expressions for the Z_{ij} given in Eq. (9.13) then we find the resonance condition to be

$$\frac{\omega V}{Sc}\tan\frac{\omega l}{c} = 1. \tag{9.18}$$

These frequencies are plotted as functions to the cavity volume V in Fig. 9.7 for a particular case. The frequency of the lowest mode agrees with the value calculated from the simple low-frequency expression (9.9), except for omission of the radiation correction, but there are also important higher modes that start out, for very small cavity volume, at the frequencies for a pipe stopped at its far end, as given by Eq. (9.15), and end up, for large cavity volume, at the values for an open pipe as given by Eq. (9.14). We obviously get a great deal more information by using a more detailed approach.

Other problems can be handled in a very similar manner, a good case being that of an exhaust muffler consisting of a number of cylindrical expansion chambers connected by lengths of narrower pipe. Each section is represented by a two-port element with S and l chosen appropriately, and the transfer characteristic U_{out}/U_{in} can be calculated in a straightforward manner.

To obtain realistic bandwidths and input impedances for systems of the type we have been describing, it is necessary of course to include losses other than just the radiation losses at the open ports. The major dissipation mechanism will, in fact, generally be viscous and thermal losses to the walls of the tubes in the system. This can be dealt with by replacing the propagation constant $k = \omega/c$ in formulae such as Eq. (9.13) by the complex quantity $\omega/c - j\alpha$, where α is the attenuation coefficient in the pipe as given by Eq. (8.15). We must then carry through the evaluation of formal results such as Eq. (9.17) by complex arithmetic. This does not involve much extra labor if an appropriate computer program is written. Of course Eq. (8.15) applies to pipes with smooth rigid walls—we must allow for extra attenuation if the pipe is lined with some sort of absorbing material. Cavities in the system may be similarly lossy, particularly if they are lined with porous material, as is often the case in duct silencers. This is handled in a rather similar way by adding a resistive loss in series with the acoustic compliance within the element representing the cavity.

9.3. Finite Horns

We should devote special attention to the behavior of horns, since they are of particular importance in many applications of acoustics. The method of analysis is just the same as that used above for pipes, but we need explicit expressions for the impedance coefficients Z_{ij} for the horn profile and length involved. These are rather more complicated than in the tube case. The discussion below is an abbreviated form of that given by Fletcher and Thwaites (1988) and Fletcher (1992).

For a conical horn it is best to relate the geometry to that of a complete horn truncated at a distance x_1 from its apex to make port 1 of area S_1, and at a distance x_2 to make port 2 of area S_2. The horn length is $l = x_2 - x_1$, which is positive if port 1 is taken to be the narrow throat. The impedance coefficients for such a horn are given by

$$\begin{aligned} Z_{11} &= -\frac{j\rho c}{S_1}\left[\frac{\sin(kl+\theta_2)\sin\theta_1}{\sin(kl+\theta_2-\theta_1)}\right] \\ Z_{22} &= \frac{j\rho c}{S_2}\left[\frac{\sin(kl-\theta_1)\sin\theta_2}{\sin(kl+\theta_2-\theta_1)}\right] \\ Z_{21} = Z_{12} &= -\frac{j\rho c}{(S_1S_2)^{1/2}}\left[\frac{\sin\theta_1\sin\theta_2}{\sin(kl+\theta_2-\theta_1)}\right] \end{aligned} \tag{9.19}$$

where $k = \omega/c$, $\theta_1 = \tan^{-1} kx_1$ and $\theta_2 = \tan^{-1} kx_2$.

For an exponential horn with throat area S_1, mouth area S_2 and length l, the flare constant is defined to be $m = (1/2l)\ln(S_2/S_1)$. Above the cutoff frequency for which $k = \omega/c = m$, the impedance coefficients are given by

$$\begin{aligned} Z_{11} &= -\frac{j\rho c}{S_1}\left[\frac{\cos(\kappa l-\theta)}{\sin\kappa l}\right] \\ Z_{22} &= -\frac{j\rho c}{S_2}\left[\frac{\cos(\kappa l+\theta)}{\sin\kappa l}\right] \\ Z_{21} = Z_{12} &= -\frac{j\rho c}{(S_1S_2)^{1/2}}\left[\frac{\cos\theta}{\sin\kappa l}\right] \end{aligned} \tag{9.20}$$

where $\kappa = (k^2 - m^2)^{1/2}$ and $\theta = \tan^{-1}(m/\kappa)$. These formulae need to be transformed to a somewhat different form at frequencies below cutoff, since κ then becomes imaginary.

Because the mouth of a horn is often not small compared with the wavelength of the sound involved, this provides an added complication when we come to calculate horn response for two reasons. The first is that, although the equivalent sound pressure generator for a free field acting on a small aperture is just the free-field acoustic pressure p in series with the radiation impedance Z_{rad} of the aperture, the case of an aperture that is large compared with the wavelength demands an equivalent pressure generator of magnitude $2p$ in series with the radiation resistance, which is essentially just $\rho c/S$. This transition from p to $2p$ for the equivalent pressure generator occurs gradually over a frequency range near $ka = 1$, where a is the radius of the aperture. Let us denote this pressure magnification factor by $E(ka)$.

The second complication arises from the fact that the wavefront at the horn mouth is not plane, as expected in a one-dimensional approximation, but curved. This curvature introduces a further factor $F_\alpha(ka)$ into the effective magnitude of the free-field pressure generator for on-axis response, given by

$$F_\alpha(ka) = \frac{\sin[(ka/2)\tan(\alpha/2)]}{(ka/2)\tan(\alpha/2)} \tag{9.21}$$

where α is the semiangle of a cone that would be tangent to the surface of the horn at its mouth and a is the mouth radius. $F_\alpha(ka)$ is approximately unity for $ka < 1$, but decreases rapidly at higher frequencies, reflecting the fact that the free plane wave is unable to couple efficiently to the curved wavefront propagating in the horn. The matter is even more complicated if we consider off-axis response—again not envisioned in a simple one-dimensional treatment. If a plane wave falls on the horn mouth from a direction making an angle θ with the horn axis, then the effective driving pressure is reduced by a further factor

$$G_\theta(ka) \approx \frac{2J_1(ka\sin\theta)}{ka\sin\theta} \tag{9.22}$$

which defines the beam-width of the horn, J_1 being a Bessel function. This factor has a complicated angular dependence, as described in Section 8.4 and Fig. 8.10, but is unity on the axis where $\theta = 0$ and declines more or less rapidly with increasing θ for $ka > 1$.

A horn must be treated as a two-port element and some sort of connection, even if a free field, must be prescribed at both its mouth and its throat before its acoustic behavior can be calculated. A particularly instructive calculation relates the pressure gain between the free field and the horn throat for the particular case in which the horn throat is rigidly blocked. We assume that the mouth is port 1 and set $U_1 = 0$. The driving pressure at the mouth, assuming the source to be on-axis, is $pF_\alpha(ka)$, where p is the free-field pressure at the position of the mouth with the horn absent, and p_T is the pressure in the horn throat. The network equations are then

$$\begin{aligned} p_M &= Z_{12}U_2 \\ pE(ka)F_\alpha(ka) &= (Z_{22} + Z_{rad})U_2 \end{aligned} \tag{9.23}$$

which can be simply solved to give

$$p_T = \left[\frac{Z_{21}}{Z_{rad} + Z_{22}}\right] E(ka)F_\alpha(ka)p. \tag{9.24}$$

The results of evaluating this expression for three horns with the same throat and mouth diameters but with different flare profiles are shown in Fig. 9.8. There is clearly a trade-off between maximum gain and bandwidth, with the exponential horn giving the narrowest bandwidth and the highest gain.

We should emphasize that the curves in Fig. 9.8 apply only to the case in which the horn throat is rigidly blocked, which is not a practically useful arrangement. More generally we would expect the throat to be connected to some sort of transducer—either a human ear or a microphone—with a finite

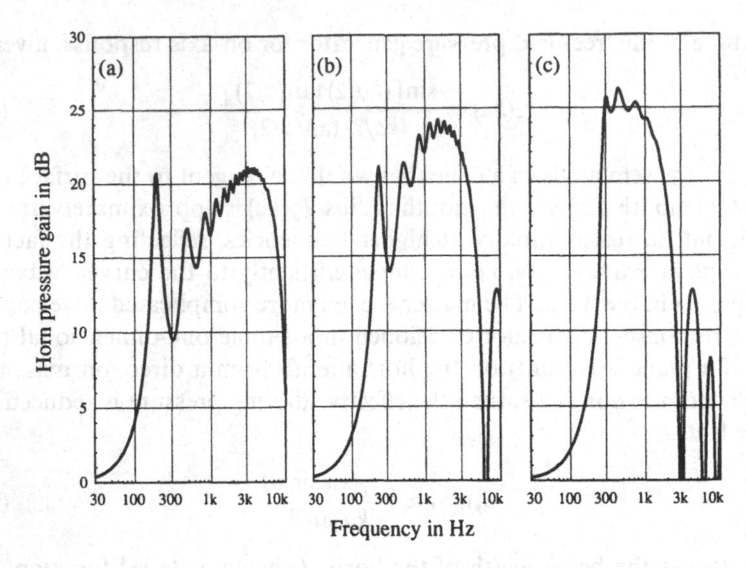

Fig. 9.8. Pressure response of (a) parabolic, (b) conical, and (c) exponential horns of the same length (50 cm), mouth diameter (50 cm), and throat diameter (5 cm), with the throat rigidly stopped. (Rescaled from Fletcher 1992.)

acoustic impedance. The response of such a system is easily calculated, along the lines indicated above, and is always lower than that for the rigidly blocked case. If the transducer has resonances, then the response function may look very different from those displayed in Fig. 9.8.

Finally we should point out that we can calculate, in a very similar manner, the pressure produced at a distant point when an acoustic volume source is located in the otherwise blocked horn throat. The reciprocity theorem guarantees, however, that the pressure response produced at this distant point by a small source of unit flux located in the horn throat will be identical to the pressure produced in the horn throat by the same source located at the distant point under consideration.

9.4. Coupled Mechanical Components

All the systems we have discussed above have been purely acoustic, in the sense that the only motion was due to acoustic waves in air. We can often separate off the acoustic part of a system in this way, regarding any associated mechanical vibrator as simply a source of acoustic flow or acoustic pressure. Sometimes, however, this approach is too simplistic—for example, when analyzing the operation of a microphone or an auditory system, in which the motion of a taut diaphragm is the penultimate output, the ultimate one being an electrical signal.

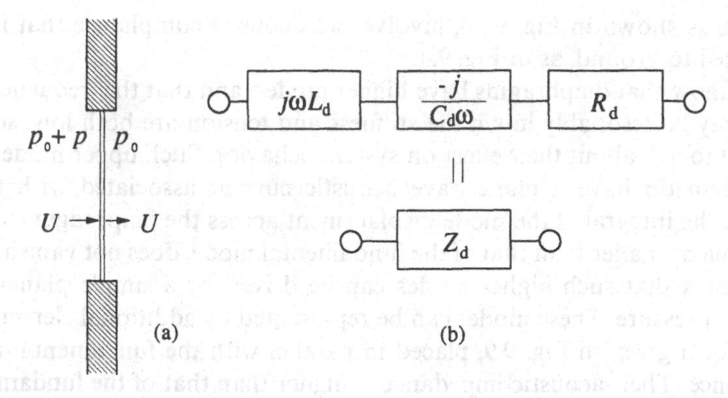

Fig. 9.9. (a) A simple elastic diaphragm, regarded as an acoustic element; (b) the acoustic analog circuit for the diaphragm element, broken into components and combined.

There are many possible mechanical elements that might be coupled to an acoustic system, and here we consider just one as an example—the elastic diaphragm shown in Fig. 9.9a. Clearly such an element can be integrated fairly simply into the analysis if we can assign to it an acoustic impedance, so the essence of our discussion is to establish how to do this. Suppose that the diaphragm has area S and thickness d, and that it is made from material with density ρ_d. The restoring force associated with diaphragm displacement may be either a tension or a stiffness, or a combination of both, and we need not be definite about this to a first approximation. The important thing is that the restoring force establishes a resonance frequency ω_d for the first mode of the diaphragm. This first mode is all that we are really concerned with, but we shall return to the higher modes later. Finally, we recognize that the diaphragm will have some internal damping associated with its motion which is most easily quantified by giving the Q value of the diaphragm resonance in vacuum, which we shall denote by Q_d.

To calculate the acoustic impedance of the diaphragm we must simply describe its motion using the acoustic quantities p and U. To an adequate approximation we can replace the flexible diaphragm by a simple piston of the same mass and area, tethered by a spring of the right stiffness and damping to give the defined resonance behavior. The velocity of the piston is then U/S and the force exerted on it by a pressure difference p between its faces is pS. Its oscillatory behavior is then described if we suppose it to have an acoustic inertance $L_d = \rho_d Sd$, an acoustic compliance $C_d = 1/\omega_d^2 L_d$, and an acoustic series resistance $L_d\omega_d/Q_d$. The formal expression for the acoustic impedance is then

$$Z_d = R + j\omega L_d - \frac{j}{C_d\omega}. \tag{9.25}$$

This is one case—almost the only case—in which the analog circuit for the element, as shown in Fig. 9.9b, involves an acoustic compliance that is not connected to ground, as in Fig. 9.4.

We know that diaphragms have higher modes, and that the frequencies of these may be resonably low if the stiffness and tension are both low, so it is relevant to ask about their effect on system behavior. Such upper modes of a diaphragm do have a plane-wave acoustic current associated with them, because the integral of the mode displacement across the diaphragm surface, while much smaller than that of the fundamental mode, does not vanish. This also means that such higher modes can be driven by a simple plane-wave acoustic pressure. These modes can be represented by additional elements Z_d of the form given in Fig. 9.9, placed in parallel with the fundamental-mode impedance. Their acoustic impedance is higher than that of the fundamental mode, however, by a factor approximately equal to the inverse square of the ratio of the integral of the areal mode displacements for the two cases. Generally we can neglect the second and higher modes in most practical applications.

9.5. Multi-Port Systems

Although individual components such as horns or pipes are two-port entities, with two separate connections to their environment, all the systems we have discussed so far, with the exception of the bass-reflex loudspeaker enclosure shown in Fig. 9.3, have been single-port systems in which there is only a single aperture communicating with the outside environment. It is appropriate now to give brief attention to multi-port systems, since some of them have important practical applications.

The principles of analysis for multi-port systems are just the same as those for systems with only a single port. The only important point to note is that, in the case of a passive system in a sound field, there will generally be a phase difference between the acoustic pressures at the different ports, depending upon the direction of incidence of the sound wave, and this can have important consequences. In a reciprocal sense, if the system is excited internally, then radiation from its separated ports will interfere to produce a characteristic radiation pattern.

The auditory systems of many animals have a multi-port nature—the same is true topologically of the human auditory system, but the Eustachian tubes connecting the open nostrils to the middle ear cavity are so narrow that the nostrils can generally be disregarded as input ports. One biologically useful feature of two-port systems is they allow the animal to determine the direction from which a sound is coming. The technical analog is the direction microphone, and it will illustrate the principles involved to analyze this in a little detail.

In an ordinary omnidirectional pressure microphone, sound pressure is allowed access to only one side of a flexible diaphragm, the other side being

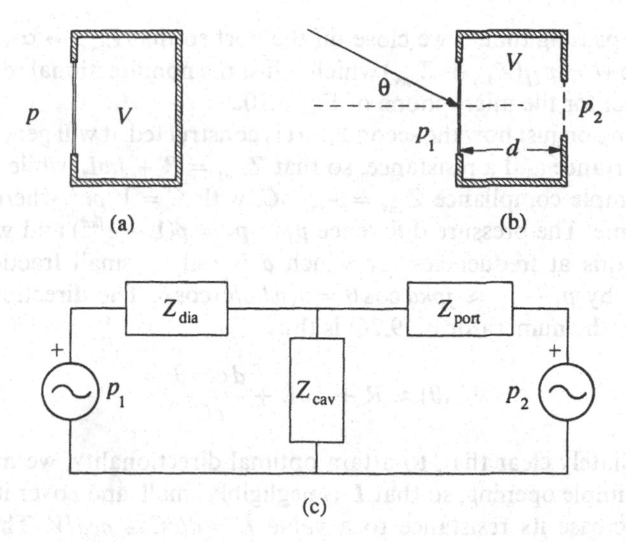

Fig. 9.10. (a) A simple omni-directional "pressure" microphone; (b) a simple cardioid response "directional" microphone; and (c) the network for the microphone in (b).

shielded by a closed cavity as shown in Fig. 9.10a. The diaphragm vibrates under the influence of the acoustic pressure, and its motion is detected by electrical (condenser microphone) or electromagnetic (dynamic microphone) means. The directional response pattern is uniform at low frequencies but directionality develops at frequencies high enough that the wavelength is comparable with the diaphragm diameter. A simple directional microphone can be made by setting a second port into the back or around the sides of the backing cavity, as shown in Fig. 9.10b. If the sound comes from a direction making an angle θ with the microphone axis, then the path difference between the sound at the two ports is approximately $d\cos\theta$ and the phase difference $kd\cos\theta$, where $k = \omega/c$. For careful design work this phase difference involves diffraction effects around the microphone casing and must be calculated more accurately, but this simple approximation illustrates the principle.

The network for the directional microphone is shown in Fig. 9.10c. In the interest of simplicity we have omitted radiation impedances at the two ports, since these make little difference if the diaphragm is fairly robust. We can write down and solve the network equations to find the acoustic current U through the diaphragm, since this is directly proportional to the diaphragm motion. The result is

$$U = \frac{Z_{\text{port}} p_1 + Z_{\text{cav}}(p_1 - p_2)}{Z_{\text{port}} Z_{\text{dia}} + Z_{\text{cav}} Z_{\text{port}} + Z_{\text{cav}} Z_{\text{dia}}} \tag{9.26}$$

where the impedances and pressures are as shown in the diagram. The denominator of this expression essentially determines the frequency response, and has no directional element—all the directionality is contained in the numerator.

We note in passing that, if we close off the port so that $Z_{port} \to \infty$, Eq. (9.26) simplifies to $U = p_1/(Z_{dia} + Z_{cav})$ which is just the nondirectional response we would expect for the microphone of Fig. 9.10a.

Depending on just how the second port is constructed, it will generally have both an inertance and a resistance, so that $Z_{port} = R + j\omega L$, while the cavity will be a simple compliance $Z_{cav} = -j/\omega C$, with $C = V/\rho c^2$ where V is the cavity volume. The pressure difference $p_1 - p_2 = p(1 - e^{jkd})$ and we can approximate this at frequencies for which d is only a small fraction of the wavelength by $p_1 - p_2 \approx jpkd\cos\theta = j(pd\omega/c)\cos\theta$. The directional factor provided by the numerator of (9.26) is thus

$$F(\theta) \approx R + j\omega L + \frac{d\cos\theta}{cC} \tag{9.27}$$

It is immediately clear that, to attain optimal directionality, we must make the port a simple opening, so that L is negligibly small, and cover it with fine mesh to increase its resistance to a value $R = d/cC = \rho cd/V$. The angular response then has the simple cardioid form $(1 + \cos\theta)$, which is just what is required.

This example illustrates the way in which two-port systems can be analyzed, and also the way in which such an analysis can be used to design desirable characteristics into acoustic systems.

9.6. Conclusion

This is as far as we can conveniently go in using network analogs to calculate the behavior of acoustic systems. The next step in sophistication is to recognize that all systems are, in fact, three-dimensional, so that we must be concerned with waves propagating in three orthogonal directions and with mode functions of three-dimensional extent. The network approach is not adapted to dealing with more than one dimension because of the geometrically linear nature of circuits, so that such problems must be attacked by using finite-element methods if the geometry is such that exact analytical solutions are not possible. Such methods must be left to specialized textbooks.

Despite this limitation, network methods are of immense value in analyzing all sorts of practical acoustic systems in a quantitative manner, and are much used in practical design. Examples of practical applications are given by Beranek (1954), Olson (1957), and Fletcher (1992). The methods build upon detailed physical knowledge of the behavior of acoustic elements and integrate these conveniently to describe the system as a whole. Specialized computer programs are available to perform the network analysis once the configuration of the system has been decided, but for many practical cases the network is sufficiently simple that a straightforward solution using the algebra of complex variables is possible, and a simple computer program can easily be written to calculate and plot the numerical results for particular cases.

References

Beranek, L.L. (1954). "Acoustics." McGraw-Hill, New York. Reprinted 1986 by American Institute of Physics, New York.

Fletcher, N.H. (1992). "Acoustic Systems in Biology." Oxford University Press, New York.

Fletcher, N.H. and Thwaites, S. (1988). Obliquely truncated simple horns: Idealized models for vertebrate pinnae. *Acustica* **65**, 194–204.

Olson, H.F. (1957). "Acoustical Engineering." Van Nostrand, New York.

CHAPTER 10

Microphones and Other Transducers

In many practical applications it is necessary to convert acoustical signals into electrical signals, or vice versa, and for this purpose a variety of transducers may be used. The most common are microphones and loudspeakers, but other important transducers include accelerometers, which convert vibrational signals to electrical signals, and force transducers, which measure vibrational forces. The purpose of this chapter is to examine the principles underlying the most common examples of each type of transducer and to see how they operate. The technical details, which are extensive and varied, need not concern us here, though they are important in the practical world.

The basic designs underlying most of these transducers date back nearly 100 years, and even their present forms were usually developed nearly 50 years ago. For this reason, many classic texts, such as those by Hunt (1954), Beranek (1954, 1988), and Kinsler et al. (1982), are still highly relevant, though papers in more modern publications such as the *Encyclopedia of Acoustics* (Crocker, 1997) and the specific references cited later should also be consulted. What is new is the associated electronics, which 50 years ago involved vacuum tubes (valves) and now relies upon transistors and integrated circuits. Important though these electronic developments are, consideration of them would take us too far outside the field of this small book. Most of the chapter is devoted to microphones and loudspeakers, but we will also make brief mention of other transducers.

10.1. Microphone Principles

One of the most important acoustic transducers is the microphone, the essence of which is to convert an acoustic pressure signal into a mechanical displacement, and then to convert this displacement into an electrical signal. There are excellent books on this subject such as those edited by Gayford (1994), by Wong and Embleton (1995), and by Busch-Vishniac and Hixson (1997).

Because acoustic pressures are generally small, the first stage of this conversion almost always involves a thin, light foil of some sort. There are, as we shall see, two different varieties of foil diaphragm—one is tautly elastic and responds essentially to a pressure signal, and the other is slack and responds more nearly to velocity. The problem is then to convert the motion of this diaphragm into an electrical signal, and again there are two different approaches—either the displacement is detected electrostatically using a capacitance effect, or else the motion velocity is used to generate a voltage with the aid of a magnetic field. We disregard early telephone microphones in which the diaphragm motion applied pressure to a container of carbon granules and altered their contact resistance, and a discussion of optical microphones, in which optical techniques are used to detect diaphragm motion (Keating, 1994), would lead us too far afield.

The first type of transducer, generally called a "condenser microphone," is now the most common type for precise measurement microphones, flexible studio microphones, and inexpensive electret microphones for general use. Here the conducting diaphragm acts as one plate of a charged capacitor, and the electrical signal resulting from its motion is detected. There are two different types of electromagnetic microphones, the robust "dynamic microphone," in which motion of the diaphragm is communicated to a coil in the strong field of a permanent magnet, and the more delicate "ribbon microphone" that was once largely used as a directional microphone in demanding studio applications. These are considered in turn in the following sections.

The directional response of a microphone is very important in many applications. An "omnidirectional" microphone responds to the total sound pressure, and this is useful in many applications, for example in the measurement of noise levels. For applications in which a signal needs to be isolated from surrounding sound, however, a selective directional response is an advantage. The two basic directional responses that can be achieved are the "figure-eight" pattern defined as $(\cos\theta)^2$ and the "cardioid" or heart-shaped pattern $(1+\cos\theta)^2$, where $\theta = 0$ is the sensitivity axis of the microphone. These patterns are illustrated in Fig. 10.1. In practice, these directional patterns cannot be achieved exactly at all frequencies, and even a nominally omnidirectional microphone generally has a response that is concentrated increasingly along the normal to the diaphragm plane at frequencies high enough that the wavelength is comparable to the diaphragm diameter. Intermediate patterns of the form $(1+\alpha\cos\theta)^2$, with different values of the parameter α, are also possible. If $\alpha > 1$ so that the pattern is intermediate between a cardioid and a figure-eight response, the microphone is called a hypercardioid, while if $\alpha < 1$ it is called a wide cardioid and is like an off-center omnidirectional pattern. Some studio microphones allow selection between all these patterns, as will be discussed later.

There is one important feature of directional microphones that is easily overlooked, and that is the "near-field" or "proximity" effect. Referring back to Section 6.2, we see that the pressure in an outgoing spherical wave

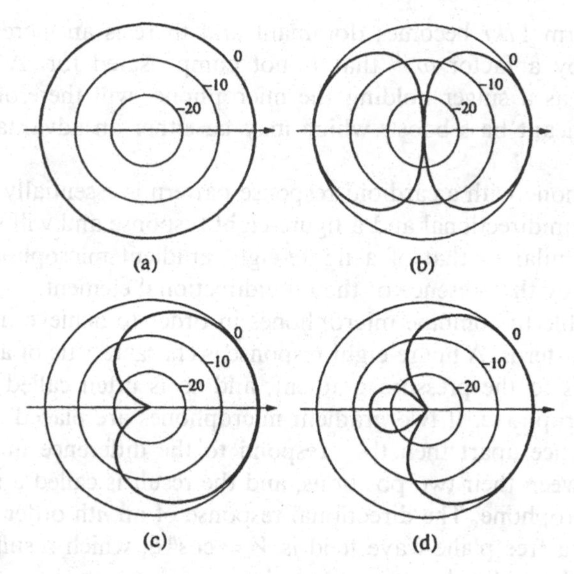

Fig. 10.1. Basic directional response patterns for microphones: (a) omnidirectional, (b) figure-eight, (c) cardioid, (d) hypercardioid with $\alpha = 1.5$. In each case the directional pattern is plotted in decibels relative to the on-axis response. Intermediate patterns are also possible.

from a simple source has the form

$$p(r) = \frac{A}{r} e^{j(\omega t - kr)}, \tag{10.1}$$

and this pressure is the stimulus that is detected by our ears. An omnidirectional pressure-sensitive microphone can convert this pressure signal into an electrical replica.

A directional microphone with a figure-eight pattern, however, is responding to either the acoustic velocity u or to the pressure gradient ∇p. If the microphone is pointed directly at the source, then these signals have forms like

$$\frac{dp}{dr} = -\frac{jkA}{r}\left(1 + \frac{1}{jkr}\right) e^{j(\omega t - kr)} \tag{10.2}$$

$$u(r) = \frac{j}{\rho\omega}\frac{\partial p}{\partial r} = \frac{A}{r\rho c}\left(1 + \frac{1}{jkr}\right) e^{j(\omega t - kr)}. \tag{10.3}$$

In the far field, where the microphone is several wavelengths away from the source, the term $1/jkr$ can be neglected in comparison with unity. The pressure gradient is then simply proportional to ω at all frequencies and the velocity is independent of frequency. A pressure-gradient microphone can be designed to compensate for this smooth frequency variation. At distances less than about one-sixth of a wavelength from the source, however, the additional term $1/kr$ becomes dominant and there is an increase at lower frequencies by a factor ω^{-1} that is not compensated for. A close signal source, such as a singer holding the microphone, will therefore appear to have a significant bass boost, which may be either an advantage or a disadvantage.

A microphone with a cardioid response pattern is essentially a superposition of an omnidirectional and a figure-eight response and will show a near-field effect similar to that of a figure-eight gradient microphone, modified only slightly by the presence of the omnidirectional element.

It is possible to combine microphones in order to achieve more extreme directional patterns. A figure-eight response is characteristic of a microphone that responds to the pressure gradient, and so is often called a first-order gradient microphone. If two gradient microphones are placed back-to-back a small distance apart then they respond to the difference in the pressure gradient between their two positions, and the result is called a second-order gradient microphone. The directional response of an nth order gradient microphone in a free plane-wave field is $V = \cos^n \theta$, which results in a much more narrowly confined pattern if $n > 1$.

10.2. Omnidirectional Condenser Microphones

The simplest form of omnidirectional condenser microphone, as used for measurements and for some studio applications, is shown schematically in Fig. 10.2 and discussed in detail by Nielsen (1994) and by Zuckerwar (1995). A metal diaphragm, generally about 5–10 μm in thickness, is stretched tightly across a ring support with a tension sufficiently high that its resonance frequency is at the top of the desired measurement range. This frequency is typically between 10 and 50 kHz, depending on the diameter of the microphone, which is usually between 3 and 25 mm, the smaller microphones having higher resonance frequencies. Beneath the diaphragm, and supported at a distance of 20–30 μm from it by an insulating disc, is a plane metallic electrode. This electrode is perforated by a number of holes, so that the motion of the diaphragm is not too much impeded by the air film between it and the electrode. The whole assembly is mounted in an airtight metal case, with just a small controlled leak to allow equilibration with atmospheric pressure.

Suppose that the area of the electrode is A and its spacing from the diaphragm d, then the electrical capacitance of the microphone is $C = \varepsilon_0 A/d$, where $\varepsilon_0 = 8.85 \times 10^{-12}$ F/m is the permittivity of the vacuum. If the electrode is connected to a very high-impedance voltage source, typically 100 V at an impedance approaching 1 GΩ, the charge induced on it is $Q = CV$. Now suppose the diaphragm is deflected inwards a distance δ by the acoustic pressure. Then the electrical capacitance increases to $C' = \varepsilon_0 A/(d - \delta)$ and, since the charge cannot change because of the high source impedance, the

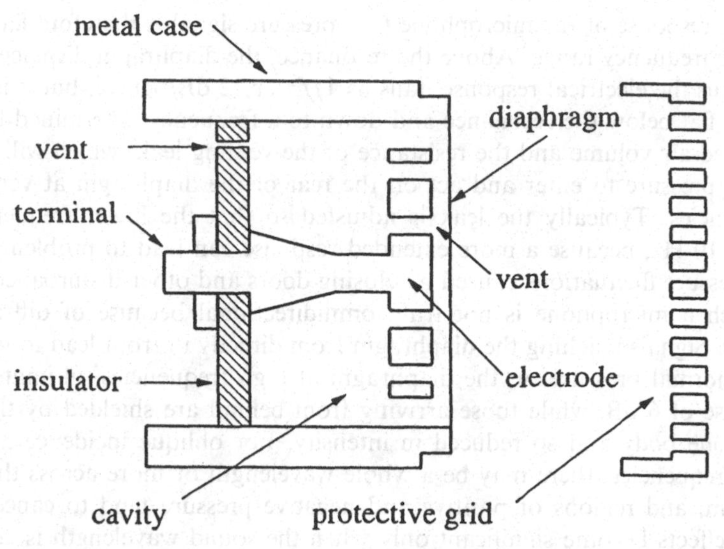

Fig. 10.2. Schematic design of an omnidirectional pressure microphone such as is used for precision sound measurements.

voltage decreases to

$$V' = \frac{Q}{C'} = V\left(1 - \frac{\delta}{d}\right). \tag{10.4}$$

The voltage on the electrode therefore follows the deflection of the diaphragm, which in turn follows the acoustic pressure.

This treatment oversimplifies the response, however, in several ways. Clearly the deflection of the diaphragm at its edge is zero, so that δ must be taken as the diaphragm deflection averaged over its area, but this is no problem. Much more significant is the way in which δ at a given acoustic pressure varies with frequency because of the mechanical properties of the diaphragm. As discussed in Section 3.3, the first resonance frequency of the diaphragm is at a frequency

$$f_0 = \frac{2.405}{2\pi a}\left(\frac{T}{\sigma}\right)^{1/2}, \tag{10.5}$$

where a is the diaphragm radius, T its tension, and σ its mass per unit area. This is a slight underestimate for a measuring microphone with a metal diaphragm, because there is also a small restoring force from the air enclosed within the microphone cavity, but we ignore this for the present. The resonance frequency f_0 is typically in the range 10–30 kHz, depending on the diameter of the microphone. The resonance would be a problem except for the fact that it is strongly damped by the necessity for the diaphragm motion to expel the air between the diaphragm and the electrode, and the holes in the electrode are so designed that this damping almost completely removes the resonance.

The response of the microphone to a pressure signal is therefore flat over a large frequency range. Above the resonance, the diaphragm displacement, and thus the electrical response, falls as $1/f^2$ or 12 dB/octave, but it is then nearly flat below the resonance and down to a frequency determined by the enclosed air volume and the resistance of the venting leak, which will allow sound pressure to enter and act on the rear of the diaphragm at very low frequencies. Typically the leak is adjusted so that the lower frequency is about 10 Hz, because a more extended response can lead to problems with air pressure fluctuations caused by closing doors and other disturbances.

Such a microphone is not truly omnidirectional because of diffraction effects. Signals reaching the diaphragm from directly in front lead to greater than normal pressure on the diaphragm at high frequencies, ultimately an increase of 6 dB, while those arriving from behind are shielded by the microphone body and so reduced in intensity. For oblique incidence at very high frequencies, there may be a whole wavelength or more across the diaphragm, and regions of positive and negative pressure tend to cancel. All these effects become significant only when the sound wavelength is comparable to the microphone dimensions.

Sometimes measuring microphones are further specialized into "free-field" types, which have flattest response for sound incident normally on the diaphragm, and "diffuse field" types, whose parameters have been optimized to give flattest response for a random or diffuse field. The high-frequency response of a typical "half-inch" (12 mm) condenser microphone is shown in Fig. 10.3. Condenser measuring microphones range from 1 inch (25 mm) down to 1/8 inch (3 mm) in diameter, smaller sizes having higher frequency response (up to 60 kHz for a 3 mm microphone) but lower sensitivity.

Omnidirectional studio microphones differ from measuring microphones principally because the diaphragm is made from thin mylar plastic film with an evaporated gold conducting layer. The lower mechanical yield strength of the mylar means that it can only support a much lower tension, so that the free resonance frequency of the diaphragm is only about 1 kHz. The microphone must therefore rely upon the restoring force provided by the air enclosed in the capsule to raise this frequency to 15 kHz or more. Suppose that the enclosed air volume is V and the diaphragm area A, then an average displacement δ of the diaphragm increases the pressure from its normal atmospheric value p_0 and gives a restoring force

$$F = \frac{\gamma A^2 \delta p_0}{V}, \tag{10.6}$$

where $\gamma = 1.4$ is the ratio of the specific heats of air, since the rapid compression is adiabatic. This restoring force raises the resonance frequency of the mounted diaphragm, and an enclosed volume of about 1 cm^3 is generally appropriate. This may either be simply an enclosed volume behind the perforated electrode plate, or sometimes cylindrical cavities drilled into a thick metal electrode, as illustrated later.

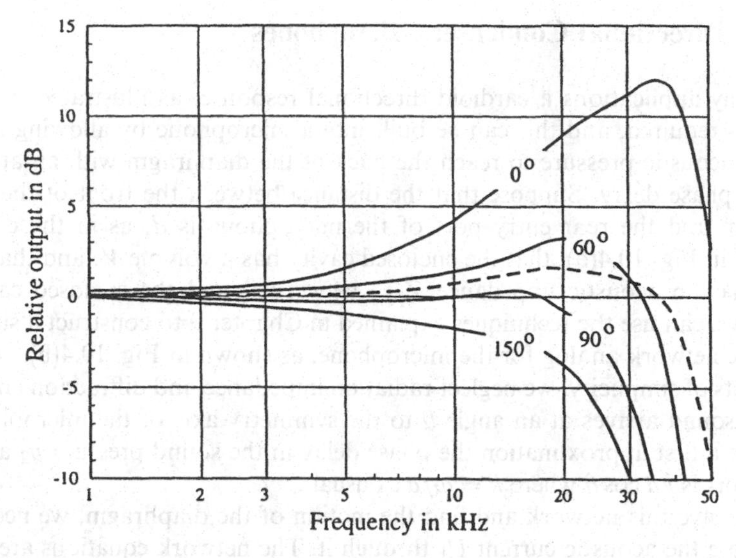

Fig. 10.3. Response of a typical 12 mm ("half-inch") measuring microphone. The diaphragm resonance is at about 30 kHz. The curves show high-frequency response for different sound incidence directions relative to the axis. The broken curve shows the response for a random or diffuse sound field. At low frequencies the response is flat down to about 10 Hz.

To give a feel for some of the dimensions involved, the separation between diaphragm and electrode plate in a measuring microphone is typically about 20 μm and in a studio microphone about 50 μm, the difference being because of the use of plastic film for the studio microphone diaphragm. Such microphones will tolerate sound pressure levels up to perhaps 150 dB, corresponding to nearly 1 kPa, without diaphragm collapse, so that the diaphragm amplitude at normal "loud" levels of say 90 dB is only about 0.02 μm.

Electret microphones (Sessler and West, 1973) differ from standard condenser microphones only by having a thin polarized ferroelectric film on the surface of the electrode, or by having the diaphragm itself made from ferroelectric plastic. The electrical polarization is built into this film at manufacture by applying a strong electric field at high temperature and then quenching to low temperature with the field still applied. This arrangement provides the necessary potential between diaphragm and electrode and obviates the need for an external power supply. Such microphones are now used nearly universally in many applications such as hearing aids, where the saving in space and battery power is of great significance. Other miniature microphones have been developed more recently in which the diaphragm is etched from crystalline silicon using techniques developed for computer chips.

10.3. Directional Condenser Microphones

In many applications a cardioid directional response, as illustrated in Fig. 10.1, is required, and this can be built into a microphone by allowing additional acoustic pressure to reach the back of the diaphragm with an appropriate phase delay. Suppose that the distance between the front of the diaphragm and the rear entry port of the microphone is d, as in the design shown in Fig. 10.4(a), that the enclosed cavity has a volume V, and that the port has an acoustic impedance Z_{port} between it and the enclosed cavity. Then we can use the techniques explained in Chapter 9 to construct a simple electric network analog for the microphone, as shown in Fig. 10.4(b). In the interests of simplicity, we neglect radiation impedance and diffraction effects. If the sound arrives at an angle θ to the symmetry axis of the microphone, then to a first approximation the phase delay in the sound pressure p_2 at the rear port is $kd \cos\theta$ where $k = \omega/c$ as usual.

To solve this network and find the motion of the diaphragm, we need to calculate the acoustic current U_1 through it. The network equations are

$$p_1 = (Z_{dia} + Z_{cav})U_1 - Z_{cav}U_2 \tag{10.7}$$

$$p_2 = Z_{cav}U_1 - (Z_{cav} + Z_{port})U_2, \tag{10.8}$$

where the Z_i are the acoustic impedances of the elements concerned. Solving

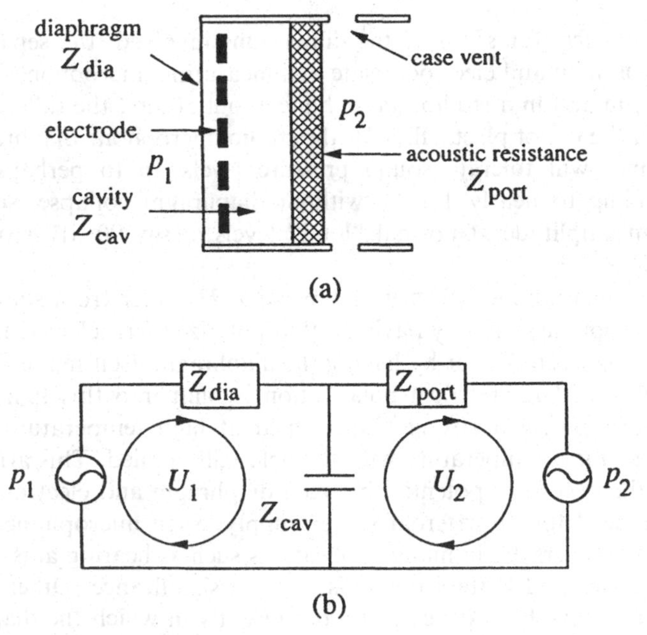

Fig. 10.4. (a) A simple cardioid response "directional" microphone; and (b) the analog network for the microphone in (a).

these equations to find the acoustic volume flow U_1 through the diaphragm, which corresponds to its average motion, gives

$$U_1 = \frac{Z_{\text{port}} p_1 + Z_{\text{cav}}(p_1 - p_2)}{Z_{\text{port}} Z_{\text{dia}} + Z_{\text{cav}} Z_{\text{port}} + Z_{\text{cav}} Z_{\text{dia}}}. \tag{10.9}$$

The denominator of this expression essentially determines the frequency response, and has no directional content—all the directionality is contained in the numerator. We note in passing that, if we close off the port so that $Z_{\text{port}} = \infty$, then this equation simplifies to $U = p_1/(Z_{\text{dia}} + Z_{\text{cav}})$, which is just the non-directional response expected from the microphone of Fig. 10.2.

The impedance of the cavity, which is small compared with the wavelength, is $Z_{\text{cav}} = -j/\omega C$ where $C = V/\rho c^2$ is the compliance of the cavity, ρ is the air density, and V is the cavity volume. The pressure difference $p_1 - p_2 = p(1 - e^{j(\omega d/c)\cos\theta}) \approx j(p\omega d/c)\cos\theta$ provided d is small compared with the sound wavelength. The desired final result can be obtained if the port is designed so as to be nearly purely resistive, for example by making it a simple opening covered by fine mesh, so that $Z_{\text{port}} = R$. The directional factor provided by the numerator of (10.9) is then

$$F(\theta) \approx R + \frac{d\cos\theta}{cC}, \tag{10.10}$$

and this takes on the simple cardioid form $(1 + \cos\theta)$ if $R = d/cC = \rho cd/V$.

A neat way of implementing this arrangement is to use, instead of a separate cavity and acoustic resistance, a block of porous sintered material which has the form of a distributed RC delay line. In some microphone designs, this material is gold-plated on one surface and actually serves as the electrode.

10.4. Studio Condenser Microphones

High quality studio microphones are often designed so as to have a sensitivity pattern that can be changed from omni-directional through figure-eight to cardioid by a simple switching operation (Zuckerwar, 1978). Such microphones, which were originally designed many years ago, have two identical diaphragms and electrodes mounted back-to-back and connected by an acoustic resistance, as shown in Fig. 10.5(a). The cavity behind each diaphragm is usually in the form of blind holes in the thick metal electrode, as shown, but other designs are possible, for example using thick porous electrodes. In the case of solid electrodes with holes, some of these must lead through to the central region to provide an interconnection between the two halves of the microphone. In the design shown, the narrow offset holes in the electrodes and the narrow central cavity form an acoustic resistance path with negligible capacitance.

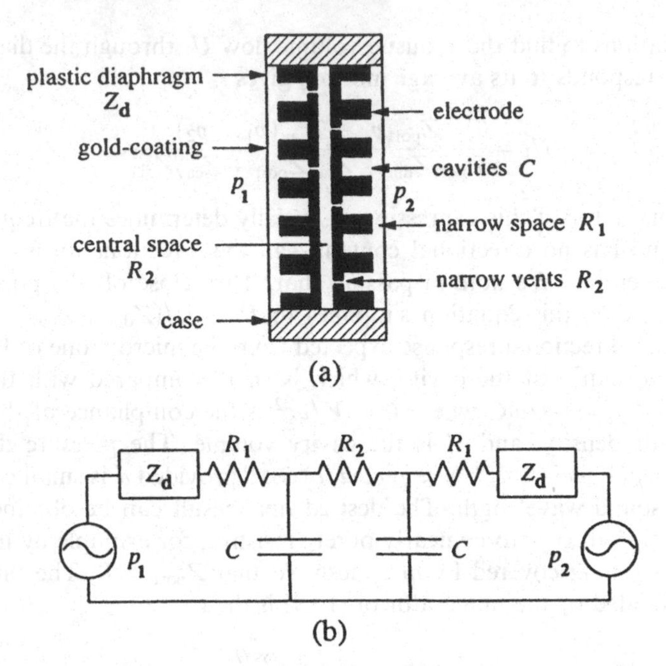

Fig. 10.5. (a) Outline design of a twin-diaphragm studio microphone cartridge; (b) simplified network analog allowing for easy computation of response.

Once again the design can be expressed as an electric network analog, a simplified version of which is shown in Fig. 10.5(b), and this network can be solved to give the motions of the two diaphragms when the acoustic pressures p_1 and p_2 are given. Whatever the values of these pressures, we can decompose them into a superposition of two cases—one in which the two pressures are equal, and one in which they are opposite, the relative weighting of the two cases giving the actual pressures. These two cases lead to two normal modes for the diaphragms, one in which they both move in and out in-phase, and one in which one diaphragm moves out while the other moves in. In the first mode, the diaphragm motion is controlled largely by the compressive stiffness $1/C$ of the air in the electrode cavities, and we have a simple omnidirectional pressure response, which can be detected by making the polarizing voltages on the two diaphragm equal. In the second mode, since the plastic diaphragms are not very stiff, the motion is limited primarily by the resistance $2R_1 + R_2$ impeding air flow from beneath one diaphragm to beneath the other. In this case the velocity of the diaphragms is proportional to the pressure difference $p_1 - p_2$ so that the diaphragm displacement, which generates the electrical signal, varies as $(p_1 - p_2)/\omega$, thus canceling out the frequency weighting ω (or k) imposed by the pressure gradient according to (10.2).

When the design has been optimized by appropriate choice of physical parameters, there are thus three possibilities. If the voltages on the two diaphragms are made equal, then the response is omni-directional; if the voltages are equal and opposite, then the response is figure-8; while if one of the voltages is zero, then with proper design the response is cardioid. This is a very convenient arrangement for studio purposes. The microphone is generally quite large and the condenser cartridge is mounted vertically inside a metal screening case.

10.5. Piezoelectric Microphones

Many crystalline materials are anisotropic and mechanical distortion causes the positive and negative charges in their structure to move in slightly different ways so that an electrical potential is developed between their faces. This effect is called piezoelectric behavior, and two of the best known piezoelectric materials are quartz and ammonium dihydrogen phosphate (ADP). Slices can be cut from these crystals in such orientations as to maximize the effect. Often two such cuts are cemented together, sometimes separated by a thin metal sheet for mechanical stability, and such a structure is called a "bimorph." Recently ceramic materials, such as barium titanate, have been made to show similar effects by polarizing them in a strong electric field at high temperature and then cooling to room temperature.

Piezoelectric microphones are made by connecting the center of a slightly conical diaphragm to the end of a cantilever of piezoelectric material, usually in bimorph form, as shown in Fig. 10.6(a). They have the advantage of simplicity, reasonably high output at moderate impedance, and do not require a voltage supply.

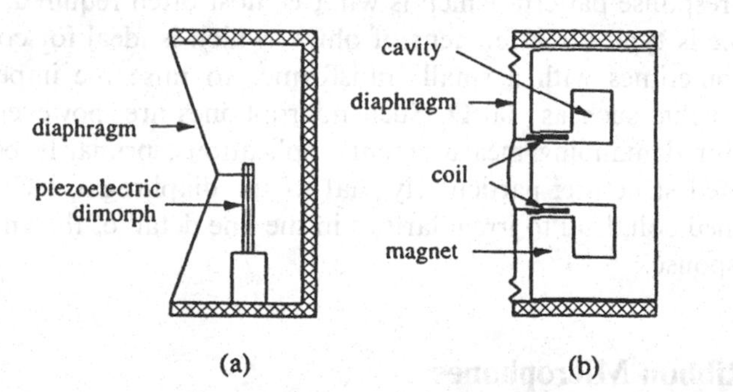

Fig. 10.6. (a) Design of a piezoelectric microphone. The diaphragm is conical, to increase its stiffness, and its apex is connected to a bimorph strip which it excites by bending. (b) Design of a dynamic microphone. Motion of the diaphragm moves a coil, rigidly connected to it, between the poles of a permanent magnet.

10.6. Dynamic Microphones

In a "dynamic" microphone, the transduction principle is that the moving diaphragm is directly connected to a wire coil, symmetrical about the diaphragm axis, and this coil moves in the small circular gap between the poles of a permanent magnet, also designed to be symmetrical about the diaphragm axis, as shown in Fig. 10.6(b) and discussed in more detail by Busch-Vishniac and Hixson (1997). Motion of the coil generates a voltage between its end that is proportional to the velocity of motion, so the microphone design should attempt to make this velocity independent of frequency for a given acoustic pressure.

The diaphragm and its attached coil clearly have appreciable mass, and must be supported on some sort of elastic mounting, generally a corrugated ring around the edge of the diaphragm and connecting it to the microphone frame. The diaphragm motion will then be that of a simple mass–spring resonant system, with a maximum velocity response at the resonance frequency. Clearly such a response is unsatisfactory, and the next addition is an aerodynamic resistance, generally stretched fine silk cloth, through which air is driven by the motion of the diaphragm. This greatly increases the damping and can remove most of the resonance peak, but the response still falls off at both very low frequencies, where the stiffness of the supporting ring becomes important, and at high frequencies, where the moving mass dominates the impedance. Further refinement of the design by the addition of vented chambers and tubes behind the diaphragm can then further extend the frequency response at the high and low-frequency ends.

Dynamic microphones are widely used because they are robust, not overly expensive, and the frequency response is perfectly adequate for quite demanding sound applications. Modifications of the design by providing a sound path to the rear of the diaphragm as previously discussed, can give a cardioid response pattern, which is what is most often required. The output impedance is typically a few tens of ohms, which is ideal for connection to cables, sometimes with a small transformer to raise the impedance to a standard value such as 600 Ω. Such microphones are, however, not really suitable for demanding measurement applications, primarily because their complicated structure, particularly that of the diaphragm, its support, and the attached coil, lead to irregularities in the fine detail of the amplitude and phase response.

10.7. Ribbon Microphones

An early form of high-quality studio microphone with what was called a "velocity response" was the ribbon microphone (Rosen, 1994). A light and generally corrugated strip of metal foil was mounted between the elongated poles of a permanent magnet in such a way that sound had access to both sides of the ribbon, though with a symmetrical spatial displacement caused by the size of the magnet. The tension in the ribbon was so small that its natural resonance was below the normal audio frequency range, so that the motion of the ribbon was limited essentially by just its mass, according to the equation

$$m\frac{d^2x}{dt^2} = A\frac{dp}{dx}\Delta, \tag{10.11}$$

where p is the acoustic pressure, x is the displacement, m is the mass, and is A area of the ribbon, and Δ is the acoustic path length between its two surfaces. Now $dp/dx = \rho\, dv/dt$ for a sound wave, so that single integration of (10.11) leads to the result that

$$\frac{dx}{dt} = \left(\frac{A\,\Delta\rho}{m}\right)v, \tag{10.12}$$

and the velocity of the ribbon follows the particle velocity in the sound wave, at least up to the frequency where Δ becomes more than about one-quarter of the sound wavelength. Since microphones of this type are generally rather large, the frequency response is typically limited to not much above 10 kHz.

The electrical signal is simply the voltage induced in the ribbon by its motion in the magnetic field, and so is proportional to the particle velocity in the sound wave, hence the common name of the microphone. Since the acoustic path length difference $\Delta \approx d\cos\theta$, where d is the physical path length between the two faces of the ribbon and θ is the angle between the sound direction and the normal to the foil surface, the response has a figure-eight pattern. The microphone also shows the same "near-field" effect as other directional microphones, as discussed in equation (10.3).

The electrical impedance of a ribbon microphone is very small, as also is the signal voltage, because the ribbon is only a single conductor rather than a coil. The ribbon output is therefore fed through a step-up transformer to increase the signal voltage and provide a match to normal input impedances of around 600 Ω. Because of their mechanical fragility, ribbon microphones are rarely used today, having been superseded by condenser microphones.

10.8. Electrical Circuitry

Here is not the place to write much about electrical circuitry, but a few remarks are in order. For a condenser microphone 25 mm in diameter and with a diaphragm spacing of 50 μm, the electrical capacitance is only about 10^{-10} F, so that at 1000 Hz the impedance is about 1 MΩ. The microphone preamplifier must therefore have a very high input impedance, and generally consists of a field-effect transistor (FET) mounted inside the microphone case. In fact the whole preamplifier is normally mounted inside the microphone case and is generally arranged so that its output has the standard impedance of 600 Ω. The signal cables connecting the microphone to other circuitry are normally arranged so as to supply the necessary power for the preamplifier and polarization voltage for the microphone cartridge. In a simple electret microphone, no external polarization voltage is required, however, and the preamplifier is normally powered by small batteries included in the microphone case.

10.9. Loudspeakers

All the microphone designs discussed above are, in principle, reversible. This means that if, instead of applying an acoustic pressure signal and observing the electrical output, we apply an electrical signal containing an audio-frequency component, then the transducer will reproduce that sound as an acoustic signal. This applies only to the transducer itself—the associated electronics will not generally operate in reverse. While this reversibility, or "reciprocity," is made use of in some precise microphone calibration methods, it is not otherwise of great practical value, since the sound power produced by a reversed microphone is generally very small. What it means, however, is that there is a close similarity between the basic design of a microphone and that of a loudspeaker, though important practical parameters may be changed.

The most important of these parameters is size. Suppose we have a circular piston of radius a, corresponding to the diaphragm of a loudspeaker, mounted in a large plane baffle, and that it vibrates at frequency ω with velocity amplitude u. Then, as discussed in Section 7.6, the radiated sound pressure at a distance r large compared with both a and the wavelength λ is

$$p(r) = \tfrac{1}{2} j\omega\rho u a^2 \left(\frac{e^{-jkr}}{r}\right)\left[\frac{2J_1(ka\sin\theta)}{ka\sin\theta}\right], \tag{10.13}$$

where $k = \omega/c$ and θ is the direction of the measuring point relative to the normal to the disc surface. The angular factor in square brackets, which is equal to 1 for $\theta = 0$, is nearly independent of θ if $ka \ll 1$ but peaks increasingly around $\theta = 0$ for higher frequencies. The response falls to zero when $ka\sin\theta \approx 3.8$, or equivalently when $\sin\theta \approx \lambda/2a$, where λ is the sound wavelength, and then rises again to subsidiary maxima for larger angles.

This result tells us that, in order to achieve a flat frequency response in the direction in which the speaker is pointing, the velocity amplitude u of the vibrating disc must vary as $1/\omega$, or the vibration amplitude as $1/\omega^2$ for a constant electrical input amplitude. It also tells us that, in order to achieve a high sound pressure and a loud sound, the disc radius a must be large and it must be made to vibrate with a large amplitude. Finally, it tells us that high frequencies in the sound will always tend to be concentrated around the direction of the axis of the diaphragm.

The total acoustic power radiated by the loudspeaker could be obtained by integrating (10.13) over all directions, but there is an easier way to do this, based on the discussion in Section 8.3. A vibrating circular piston of radius a set in a plane baffle feels the acoustic impedance of the air, of which the resistive part R is given by (8.28) and plotted in Fig. 8.7. The more complicated case of a piston set in the end of a pipe is plotted in Fig. 8.8 and is not very different. In each case, R increases as $(ka)^2$, or $(\omega a/c)^2$, for small values of ka, and then saturates for $ka > 2$. If the velocity amplitude of the piston motion is u, then the radiated power is $\frac{1}{2}R(\pi a^2 u)^2$, which rises as ω^2 at low frequencies and saturates at $\frac{1}{2}\rho c \pi a^2 u^2$ when $\omega > 2c/a$ or $\lambda < \pi a$ where λ is the sound wavelength. Translating this into practical values for a loudspeaker 10 cm in diameter, the turnover frequency is about 2 kHz. Clearly the loudspeaker design must compensate for this behavior.

It is interesting to reflect on the acoustic efficiency of loudspeakers—the ratio of the acoustic power output to the electrical power input. This is rarely mentioned by loudspeaker manufacturers for a reason that is immediately apparent—the efficiency is very low indeed! Loudspeakers are advertised instead in terms of simply the electrical power input that they can tolerate before distortion becomes severe. Of course, there is sense in this, for it allows the choice of an appropriate amplifier to drive a chosen loudspeaker.

When acoustic output power is specified, it is usually given in terms of a figure such as "93 dB per watt, on-axis, at 1 meter." When we read this as an intensity level relative to 10^{-12} W/m^2, then the intensity at 1 m from the speaker is 2 mW/m^2 and, if we neglect the fact that the intensity is higher in front of the speaker than behind, and simply multiply by 4π for the area of a sphere of radius 1 m, we get about 25 mW as the total acoustic power output for 1 W of electrical input—an efficiency of 2.5%. A figure like this is typical of most devices that convert mechanical or electrical energy to acoustic energy—devices such as musical instruments, the human voice, and even jet aircraft engines. Hardly any conversion exceeds 10% in efficiency, and most are well below 1%. The reason is primarily the low wave impedance ρc of air. We consider horns as an acoustic matching device below.

There are three basic types of loudspeaker in common use. The first is the dynamic or electromagnetic speaker, which is the most common of all. The second is a small piezoelectric loudspeaker that is often used to supplement dynamic speaker response at very high frequencies. The third is the horn loudspeaker, usually driven by a special electromagnetic driver. Electrostatic speakers in the form of rather large panels have also seen some use, but are not popular today. Because the piezoelectric speaker is very similar in behavior to a small dynamic speaker, we shall not consider it further here, but concentrate on dynamic and horn speakers.

10.10. Dynamic Loudspeakers

The general design of a dynamic loudspeaker is sketched in Fig. 10.7(a) and is discussed in detail by Starobin (1997). The diaphragm is made of light cardboard-like material formed into a cone to give it stiffness and raise the frequencies of its vibrational modes so that it mostly vibrates as a single structure. The cone is supported by flexible corrugated rings around its edges and close to its center. At the center is rigidly affixed a coil of wire on a cylindrical former, which then moves in the small circular gap between the poles of a specially shaped permanent magnet with as high a magnetic field strength as possible. This coil is generally termed the "voice coil."

The application of an oscillating electrical signal to the coil causes it to experience a force in the magnetic field, this force being along the axis of the cone. The force moves the coil, and with it the cone, but this motion in turn induces an electric potential across the coil in such a direction as to oppose the original applied voltage. The complete analysis is very complicated (see the books by Beranek 1954, 1998, Kinsler et al. 1982, or Olson 1957 in the references) and we will not go into it here. By careful design of cone mass, support stiffness, and associated damping, it is possible to get a smooth frequency response over quite a wide frequency range.

Design of the speaker itself is only part of the problem, however. A loudspeaker in open air would radiate nearly equally from the front and rear surfaces of the cone and, since these are in opposite phase, they would tend to cancel at low frequencies. There are several alternative solutions. The first

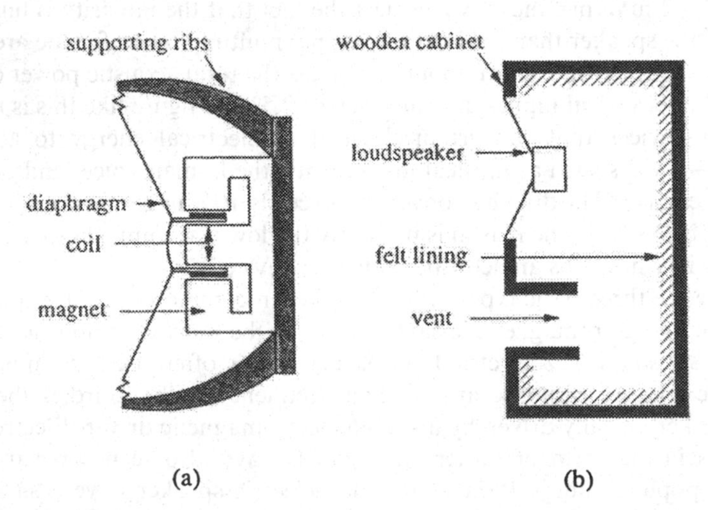

Fig. 10.7. (a) Outline design of a dynamic loudspeaker. The cone is supported by flexible corrugated rings and drives a coil mounted between the poles of a cylindrically symmetric permanent magnet. (b) Bass-reflex cabinet design.

is to fix the speaker in a large baffle, such as the wall of a room, so that rear sound does not interfere. The second is to enclose the rear of the speaker in a closed box lined with absorbent material such as felt, to eliminate the rear-radiated sound. Such an enclosure necessarily raises the resonance frequency of the cone and so may limit the low-frequency response. The third solution is to use a vented rear-enclosure, called a "bass-reflex enclosure," with a resonance designed so as to extend the low-frequency response of the speaker.

This design was discussed briefly in Section 9.1 and is shown in Fig. 10.7(b). If $C = V/\rho c^2$ is the analog capacitance of the box volume V and $L = \rho l/S$ the analog inductance of the vent pipe which has length l and area S, then at frequency ω the outflow U though the vent is related to the inflow U_0 caused by displacement of the speaker cone by

$$U = \frac{U_0}{1 - LC\omega^2}. \tag{10.14}$$

This means that, at frequencies above the box resonance $\omega^* = 1/(LC)^{1/2}$, the duct flow U is 180° out of phase with U_0 and therefore in-phase with the flow driven by the cone into the outside air, so the two sources reinforce. At the box resonance the two flows differ by 90° in phase, so that there is still reinforcement and the resonant flow from the box vent will add to the total sound output. Below the box resonance, however, the net outflow $U - U_0 \rightarrow U_0 LC\omega^2$, so that there is a low-frequency cut of 12 dB/octave even if U_0 does not vary with frequency. The dimensions of the enclosure must therefore be carefully tailored to the properties of the speaker, particularly its cone resonance frequency, in order to provide a smooth extension of the low-frequency response.

In the high-frequency regime, a loudspeaker set against a wall has a directional response similar to that discussed in Section 7.6 for a circular disc set in an infinite plane baffle. Assuming that the speaker design gives a uniform frequency response on-axis, the acoustic intensity I at an angle θ to this axis is

$$I(\theta) = \frac{2J_1(ka \sin\theta)}{ka \sin\theta} I(0), \tag{10.15}$$

where J_1 is a Bessel function of order unity, a is the radius of the speaker cone, and $k = \omega/c$ as usual. Because the speaker radius is typically 5 to 15 cm, this results in a significant weakening of off-axis high frequency sound. The usual solution is to filter off the high-frequency part of the signal and feed it to a separate smaller loudspeaker either mounted independently or sometimes supported in a small enclosure at the center of the larger speaker. Such an arrangement has other advantages, since a speaker can be more easily designed to function efficiently and without distortion over a limited frequency range.

10.11. Horn Loudspeakers

As noted above, the efficiency of most loudspeakers is only a few percent, largely because of the rather low acoustic wave impedance of air. One way to improve efficiency is to use some sort of matching device, and that is what a diaphragm aims to do relative to the voice-coil driver. A horn loudspeaker achieves the same objective by using the properties of a horn to couple a compact and robust electromagnetic driver to a much larger area at the horn mouth (Salmon, 1997). The advantage of a horn loudspeaker over an ordinary diaphragm speaker is that it can be driven much harder, because the complex mechanics of a diaphragm are avoided.

Wave propagation in horns was discussed in Sections 8.6–8.9. Here we limit ourselves to exponential horns, since they are the type most often used in loudspeakers. If the profile of horn radius is defined by

$$a(x) = a_0 \exp(mx), \tag{10.16}$$

where a_0 is the throat radius, x the distance measured from the throat, and m is called the flare constant, then equation (8.47) shows that waves with $k < m$ or $\omega < mc$ are attenuated rather than propagating. Horns are therefore generally deficient in low-frequency response unless they flare rather gently and so are very long. A finite horn has other problems, because waves are reflected to some extent from its open mouth end and cause resonances in its response. Putting these problems aside, the impedance gain in a horn with throat radius a_0 and mouth radius a_1 is $(a_1/a_0)^2$, and the horn behaves like a diaphragm loudspeaker with diaphragm radius a_1. The small compact dynamic driver, however, can be driven to high amplitude without mechanical problems, and the acoustic output can be high. Add to this the fact that the horn mouth can be made large without great difficulty, and we have a speaker that is rather highly directional and therefore suitable for large halls or for outdoor gatherings. In a more modest way, innovative designs have been developed in which the horn is folded to fit in a compact cabinet while still having a mouth diameter of up to 50 cm, and these horns can be used to improve the low-frequency efficiency of domestic audio systems.

10.12. Hydrophones

While in principle hydrophones are similar to microphones or loudspeakers, there is a practical difference caused by the vastly greater wave impedance of water (1.54×10^6 Pa s/m^3) than air (412 Pa s/m^3). This means that, for a given sound intensity in water, the pressure is about 600 times higher and the particle velocity about 600 times lower than for the same sound intensity in air. Underwater transducers, therefore, can be much more robust devices than microphones or loudspeakers, and different designs are appropriate. A general discussion is given by Blue and van Buren (1997).

Two very successful designs make use of either bulk piezoelectric effects or bulk magnetostrictive effects. When an electric field is applied to a piezoelectric crystal or ceramic, its crystal structure distorts and the external dimensions of a macroscopic piece of material change. The same thing occurs when a strong magnetic field is applied to a block of ferroelectric material such as an iron or nickel alloy. The actual dimensional change in a piezoelectric crystal is small, amounting to about 5×10^{-11} per volt/m for barium titanate or, since fields of 10^5 or 10^6 V/m can be applied to small crystal elements, a static dimensional strain of only about 5×10^{-5}. A typical value for the magnetostrictive strain of a ferromagnetic material is 10^{-4} per tesla, and applied magnetic fields will generally be only about 0.1 T so that the maximum strain is less than 10^{-5}.

In many applications, however, underwater transducers are required to operate at only a single frequency, for example to transmit sonar pulses, and mechanical resonance effects can be used to multiply the static response by a factor equal to the Q value of the resonator. Examples of such resonant structures are quarter-wave plates, with one free surface and the other surface bonded to a thin steel plate that in turn communicates the vibrations to the surrounding water. In the case of a piezoelectric transducer, the exciting voltage is easily applied between the metal plate and a thin metallic coating on the other surface of the crystal, while for a magnetostrictive transducer the material may be formed as an array of quarter-wavelength rods, each of which is surrounded by an exciting coil. Other arrangements using different aspects of the piezoelectric or magnetostrictive distortion in structures of different shapes have also been developed.

As in the case of microphones and loudspeakers, the design of the two components is related, except that hydrophones for underwater sound reception as passive listening devices are required to be sensitive over a rather large frequency range, so that resonance enhancement cannot be used in these cases.

10.13. Ultrasonic Transducers

Because the transmission of ultrasound in gases is highly attenuated over quite small distances, most applications of ultrasound are for transmission in liquid or solid media. A particularly important application is medical imaging of the human body, but similar techniques are also used to search for defects in solid structures such as aircraft components (Papadakis, 1997).

In most cases the transducer is required to act over only a very narrow frequency band, so that resonant enhancement techniques can be used, as for underwater sonar transmitters. Piezoelectric transducers made from barium titanate ceramic are often used in this application, the field orientation being chosen so that the transducer is driven in thickness mode, rather than in shear. The frequency in this application is very high, typically 1 MHz or more, so that the wavelength in water is less than 2 mm. This is important, since the wavelength determines the spatial resolution of the image. Since the wavelength of sound in a solid is comparable with that in water, this implies that the active resonant parts of the transducer will also have dimensions in the millimeter or sub-millimeter range. For medical applications, the transducer cannot generally be inserted into the body, but contact is made though a flat plate pressed against the skin, often with an oil film to eliminate air cavities, the presence of which would cause an impedance mismatch.

10.14. Force Transducers and Accelerometers

In vibrational problems, the quantities to be measured are mechanical force and the displacement of a solid object, rather than pressure and acoustic velocity. Some means is also required to impart mechanical vibration to the object under test. In most cases the frequency range of interest is similar to that for ordinary acoustics, but sometimes vibrations of infrasonic frequencies are important. The principles are very much the same as those used in the acoustic case, though details of the devices are different.

To produce a controlled sinusoidal force for the mechanical excitation of an object, the simplest approach is to use an electromagnetic drive nearly identical to that used in loudspeakers and illustrated in Fig. 10.7(a). The difference is that, instead of the drive exciting a conical diaphragm, it is connected directly to the object under test. The connection usually needs to be through an adhesively bonded or bolted pin to prevent "chattering" or loss of contact during the vibration. For examination of small objects, the "shaker" drive is comparable in robustness to that for a normal large loudspeaker, but for larger of heavier structures a much more massive and powerful shaker may be required. The shaker must be made relatively heavy in all cases so that it is the test object that is shaken, rather than just the body of the shaker under the influence of the reaction force.

While the force exerted by the shaker is proportional to the current flowing through its electrical circuit, this current is influenced by the motion of the coil itself and thus by the impedance of the object being examined. It is therefore desirable to have a separate transducer to measure the force actually applied. For this purpose a piezoelectric device is ideal, and all that is required is to sandwich a piece of appropriately polarized piezoelectric material, such as barium titanate ceramic, between two sections of the driving pin so that it bears the full mechanical load, as shown in Fig. 10.8(a). Since a ceramic is brittle, the pin is specially weakened so that it will fail before the transducer is damaged.

The vibration of a solid object is also most easily measured using a piezoelectric transducer. Such a device is generally an accelerometer and con-

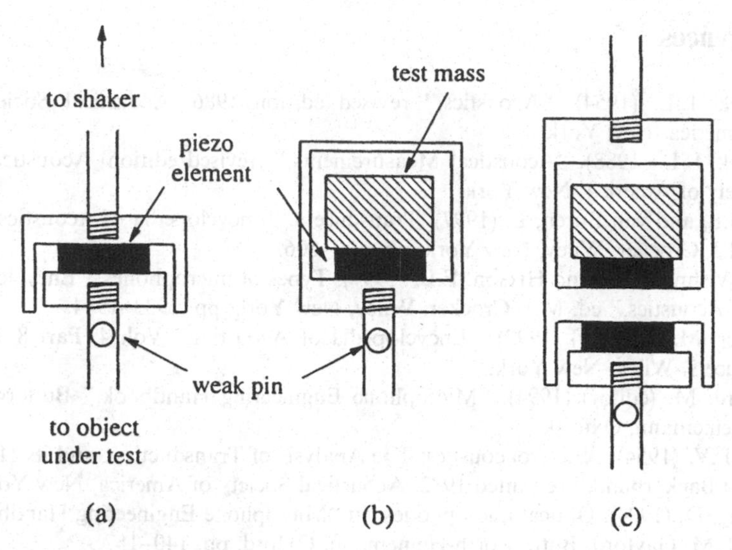

Fig. 10.8. Conceptual designs of mechanical transducers. (a) A force transducer; (b) an accelerometer; (c) an impedance head combining both devices. As shown here, all the piezoelectric elements are excited in compression mode, whereas practical designs sometimes arrange for the excitation to be in shear mode, with the elements disposed in a triangular array around the connecting axis.

sists of a test mass mounted on a piezoelectric support that is itself connected rigidly to the test body, as shown in Fig. 10.8(b). Since the test mass is constrained to move with the vibrating body, the force exerted on the piezoelectric element is proportional to the acceleration of this test mass and the same is then true of the electrical output signal. This signal can then be integrated electronically to give either the velocity or the displacement.

Devices are available that integrate both a force transducer and an accelerometer into the same casing, as in Fig. 10.8(c), so that the two outputs can be combined to give the mechanical impedance (force/velocity) for the object under test. If the signal driving the shaker is wide-band noise, then the two outputs can be subject to further calculation using a fast Fourier transform (FFT) algorithm to give the impedance as a function of frequency, or "impedance spectrum."

In recent times, optical devices such as laser interferometers have been introduced to provide a convenient non-contact means of measuring mechanical vibration. It would take us too far afield to discuss the operation of these methods here, but details are available in books on laser techniques. Apart from the advantage of non-contact measurement, these techniques also allow a significant area of the test body to be viewed simultaneously and measurements made over the whole of this area, thus revealing the shapes of vibrational modes and other details of interest.

References

Beranek, L.L. (1954). "Acoustics," revised edition 1986, Acoustical Society of America, New York.

Beranek, L.L. (1988). "Acoustical Measurements," (revised edition) Acoustical Society of America, New York.

Blue, J.E. and van Buren, L. (1997). Transducers. "Encyclopedia of Acoustics," ed. M.J. Crocker, Wiley, New York. pp. 597–606.

Busch-Vishniac, I.J. and Hixson, E.L. (1997). Types of microphones. "Encyclopedia of Acoustics," ed. M.J. Crocker, Wiley, New York. pp. 1933–1944.

Crocker, M.J. (editor) (1997). "Encyclopedia of Acoustics," Vol. 4, Part 8 Transducers. Wiley, New York.

Gayford, M. (editor) (1994). "Microphone Engineering Handbook," Butterworth-Heinemann, Oxford.

Hunt, F.V. (1954). "Electroacoustics: The Analysis of Transduction, and its Historical Background," reprinted 1982, Acoustical Society of America, New York.

Keating, D. (1994). Optical microphones. In "Microphone Engineering Handbook," ed. M. Gayford, Butterworth-Heinemann, Oxford. pp. 140–157.

Kinsler, L.E., Frey, A.R., Coppens, A.B. and Sanders, J.V. (1982). "Fundamentals of Acoustics," Wiley, New York.

Nielsen, T. (1994). Precision microphones for measurements and sound reproduction. "Microphone Engineering Handbook," ed. M. Gayford, Butterworth-Heinemann, Oxford.

Olson, H.F. (1957). "Acoustical Engineering," Van Nostrand, Princeton NJ.

Papadakis, E.P. (1997). Ultrasonic instruments for nondestructive testing. "Encyclopedia of Acoustics," ed. M.J. Crocker, Wiley, New York. pp. 683–694.

Rosen, G. (1997). Ribbon microphones. "Encyclopedia of Acoustics," ed. M.J. Crocker, Wiley, New York. pp. 267–276.

Salmon, V. (1997). Horns. "Encyclopedia of Acoustics," ed. M.J. Crocker, Wiley, New York. pp. 1925–1932.

Sessler, G.M. and West, J.E. (1973). Electret transducers: A review. *J. Acoust. Soc. Am.* **53**, 1589–1600.

Starobin, B.M. (1997). Loudspeaker design. "Encyclopedia of Acoustics," ed. M.J. Crocker, Wiley, New York. pp. 1903–1924.

Wong, G.S.K. and Embleton, T.E.W. (editors) (1995). "Condenser Microphones," American Institute of Physics, New York.

Zuckerwar, A.J. (1978). Theoretical response of condenser microphones, *J. Acoust. Soc. Am.* **64**, 1278–1285.

Zuckerwar, A.J. (1995). Principles of operation of condenser microphones. "Condenser Microphones," ed. G. Wong and T.E.W. Embleton, American Institute of Physics, New York, pp. 37–69.

CHAPTER 11

Sound in Concert Halls and Studios

When music or speech is heard indoors, most of the sound waves that reach the listeners' ears have been reflected by one or more surfaces of the room or by objects within the room. Typically, sound waves undergo many reflections before they become inaudible. It is not surprising, then, that the acoustical properties of the room play an important role in determining the nature of the sound heard by a listener. Performers in a concert hall, teachers in a classroom, actors in a theater, and speakers in a church or assembly hall all depend upon the acoustics of the room in which they attempt to communicate with their audience. When we listen to recorded music or watch television or home movies in our living rooms, the acoustics of the room also has much to do with the quality of the sound we hear. Recording studios, large and small, have their own special acoustical requirements, and many musicians are creating small studios to make demonstration records.

In this chapter, we will discuss some principles of sound fields in rooms, and see how they might apply to different types of rooms, including large concert halls, small studios, churches, and classrooms. We also consider some ways to control noise in listening rooms.

11.1. Spatial Dependence of the Sound Field

An environment in which the sound pressure is proportional to $1/r$ (where r is the distance from the source) is called a free field. When a sound source is small enough to be considered a point source and is located outdoors away from reflecting objects, a free field results. Sound waves travel away from the source in all directions, the wave fronts having the shape of spheres, as discussed in Section 6.2. The sound pressure p is halved when the distance is doubled. From the definition of pressure level in Section 6.3, we see that the sound pressure level decreases 6 dB each time the distance r is doubled. Figure 11.1 illustrates the way in which sound pressure and sound level decrease with distance in a free field.

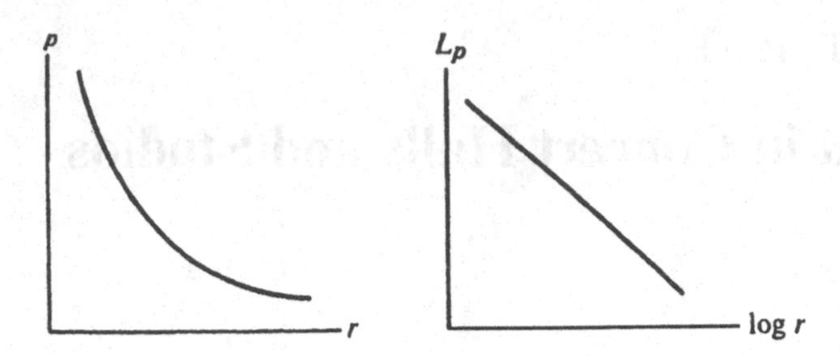

Fig. 11.1. The way in which sound pressure p and sound pressure level L_p decrease with distance r in a free field.

Free-field conditions rarely occur indoors, except in reflection-free anechoic rooms. (Anechoic means "echo-free"; this is generally achieved by covering the walls, ceiling and floor with wedges of sound-absorbing material, as in Fig. 11.2).

In Chapter 6, we learned that the sound intensity I for free progressive waves is proportional to the square of the sound pressure (see Eq. 6.32). At a distance r from a sound source that radiates W watts of sound power in an anechoic field, the sound intensity will be given by

$$I = \frac{W}{4\pi r^2}, \tag{11.1}$$

Fig. 11.2. An anechoic room.

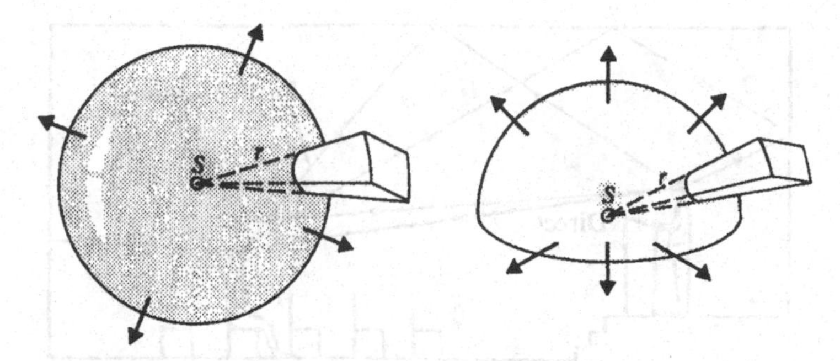

Fig. 11.3. (a) Spherical sound waves in a free field. The power from source S is distributed over a spherical surface $4\pi r^2$ in area. (b) Hemispherical sound waves from a source S on a hard reflecting surface. The power is distributed over a surface $2\pi r^2$ in area.

since the power is distributed throughout the surface of a sphere of radius r. If the sound source is placed on a hard reflecting surface, the sound waves will be more nearly hemispherical, and the intensity will be given by $I = W/2\pi r^2$.

The sound intensity level (see Eq. 6.31), in both a free (spherical) field and a hemispherical field will decrease by 6 dB for each doubling of the distance r, although it starts 3 dB higher in the hemispherical case. Spherical and hemispherical sound fields are illustrated in Fig. 11.3.

11.2. Time Dependence of the Sound Field

Sound reflections do much to determine the acoustical characteristics of a room. We generally do not sense the individual reflections, and it is convenient to characterize the time dependence of the sound we hear in terms of direct, early, and reverberant sound. In a large room, the sound waves that travel directly to the listener (at about 343 m/s) may reach the listener's ears after a time t_0 of anywhere from 20 to 200 ms, depending upon the distance from the source to the listener. A short time later, the same sound will reach the listener from various reflecting surfaces, mainly the walls and ceiling. In Fig. 11.4 these reflections are shown arriving with various time delays t_1, t_2, t_3, etc. The first group of reflections, reaching the listener with about 50–80 ms of the direct sound, is often called the early sound.

After the first group of reflections, the reflected sounds arrive thick and fast from all directions. These reflections become smaller and closer together, merging after a time into what is called reverberant sound. If the source emits a continuous sound, the reverberant sound builds up until it reaches an equilibrium level. In the case of an impulsive sound, the decay begins immediately, and there is no equilibrium level.

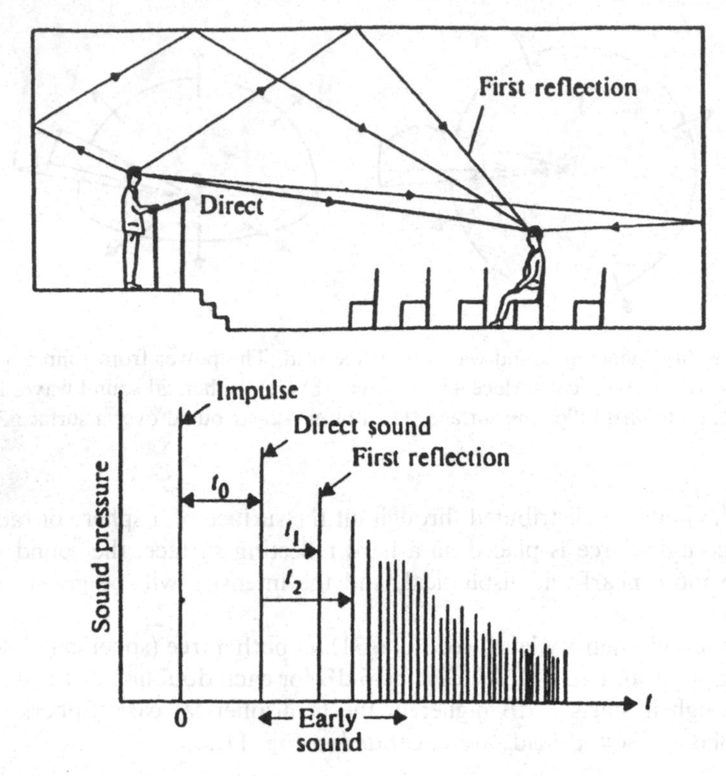

Fig. 11.4. Paths of direct and reflected sound from source to listener with corresponding time delays for a sound impulse.

11.2.1. Direct and Early Sound: The Precedence Effect

Our auditory system has an uncanny ability to determine the direction of a sound source, even in the presence of many distracting sounds. For sounds of low frequency, localization depends mainly on the observation of a very slight difference in the time of arrival (or the phase of steady sounds) at our two ears. For sounds of high frequency (above about 1000 Hz), the difference in sound level at our two ears, due to the shadow cast by our head, provides the main clue.

Imagine that the sound source emits an impulsive sound. Our ears will receive not only the direct sound but several reflections that closely follow the direct sound. The spectrum and time envelope of these reflected sounds will be more or less identical to those of the direct sound, and if they arrive within about 50 to 80 ms of the direct sound, the ear does not hear them as separate sounds. Rather, they tend to reinforce the direct sound, a fact that is especially important to listeners located quite a distance from the source. (For rapidly varying sound, such as speech, the limit is probably around 50 ms, but for more slowly varying music, the limit is more like 80 ms).

Quite remarkably, however, the auditory processor deduces the direction of the source from the first sound reaching the ears, which it interprets as following the direct path, and ignores successive sounds arriving a few milliseconds later from reflecting surfaces. This remarkable ability of our auditory system is called the precedence effect (other names are the "law of the first wavefront" and the "Haas effect"). The source is perceived to be in the direction from which the first sound arrives provided that (1) successive sounds arrive within about 35 ms; (2) the successive sounds have spectra and time envelopes reasonably similar to the first sound; and (3) the successive sounds are not too much louder than the first.

As a result of studying 76 of the world's leading concert and opera halls, Beranek (1996) concluded that a concert hall can be considered "intimate" if the delay time between direct and first reflected sound is less than 20 ms. If the auditorium has the traditional rectangular ("shoebox") shape, this first reflection for most listeners will come from the nearest side wall, although listeners located near the center may receive their first reflection from the ceiling. In some concert halls, a portion of the audience will be too far removed from both ceiling and side walls to receive early reflections within the desirable time interval; in those case reflecting surfaces of some type are often suspended from the ceiling. Studies have shown that early reflections from side walls are not equivalent to early reflections from the ceiling or an overhead reflector, however. One study showed a high preference for concert halls with ceilings sufficiently high that the first lateral reflection reaches the listener before the first overhead reflection (West, 1996). If the total energy from lateral reflections is greater than the energy from overhead reflections, the hall takes on a desirable "spatial responsiveness." Apparently our two ears prefer to receive slightly different sounds ("binaural dissimilarity") rather than identical sounds, as they will when the first reflection comes from directly overhead.

11.2.2. Reverberant Sound

Instead of the impulsive source illustrated in Fig. 11.4, let us switch on a steady source for a time T, after which it is switched off. The growth and decay of the sound levels at the source and listener are shown in Fig. 11.5. The listener's sound level increases in small steps as the direct sound D, and then the reflections 1, 2, etc. arrive. After the first few reflections, the individual steps are difficult to observe, and the sound builds up to its reverberant level. The reverberant level is reached when the rate at which sound energy is supplied by the source (source power) is equal to the rate at which sound is absorbed.

When a steady sound is switched off, the sound pressure at a typical listener location dies away approximately exponentially, which leads to a nearly linear decay of the sound pressure level. The time it takes the level to decrease 60 dB is generally called the reverberation time of the room.

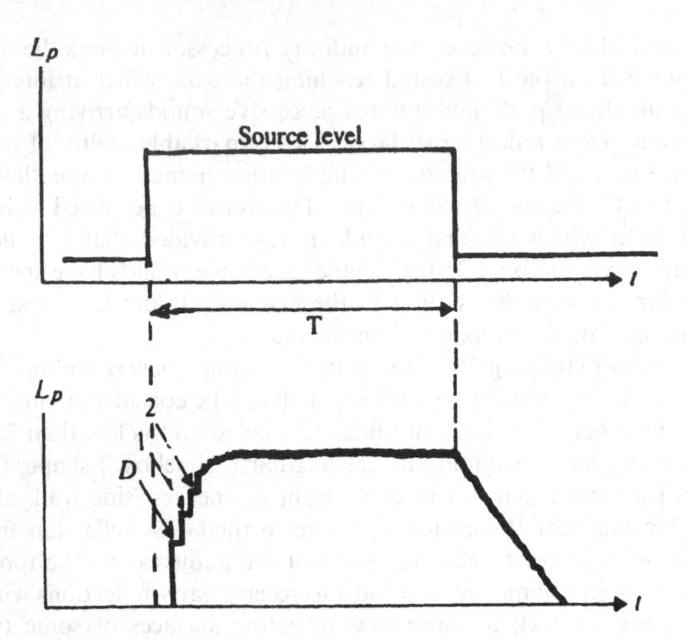

Fig. 11.5. Growth and decay of reverberant sound in a room: D represents direct sound; 1, 2, 3, etc. are early reflections.

In principle, it is easy to determine the theoretical reverberation time of a room, in which the sound is uniformly distributed. The sound energy stored in the room depends on the power of the source and the volume of the room; the rate at which that energy is absorbed depends on the area of all surfaces and objects in the room and their absorption coefficients. In a bare room, where all surfaces absorb the same fraction of the sound that reaches them, the reverberation time is proportional to the ratio of the volume to the area of sound-absorbing surface (Sabine, 1922).

$$T_{60} = \frac{K \times \text{volume}}{\text{area}}. \tag{11.2}$$

In SI units $K = 0.161$, so $T_{60} = 0.161V/A$ in a hypothetical room with A square meters of perfect absorber (such as an open window). (When V and A are expressed in ft^3 and ft^2, then $T_{60} = 0.049V/A$.) In a real room, we can compute the Sabine reverberation time by comparing the absorbing power of each surface to that of the window in our hypothetical room. The window is assumed to absorb all the sound incident on it, so its absorption coefficient α is assumed to be 1. A surface having an area S and an absorption coefficient α has a total absorption given by $A = S\alpha$. The total absorption in a room is found by adding up the contributions from each surface exposed to the reverberant sound:

$$A = S_1\alpha_1 + S_2\alpha_2 + S_3\alpha_3 + \cdots \tag{11.3}$$

Absorption coefficients for various materials at six different frequencies are given in Table 11.1. Note that some surfaces (e.g., carpet on concrete)

Table 11.1. Absorption coefficients for various materials

Material	Frequency (Hz) 125	250	500	1000	2000	4000
Concrete block, unpainted	0.36	0.44	0.31	0.29	0.39	0.25
Concrete block, painted	0.10	0.05	0.06	0.07	0.09	0.08
Glass sindow	0.35	0.25	0.18	0.12	0.07	0.04
Plaster on lath	0.14	0.10	0.06	0.05	0.04	0.03
Plywood paneling	0.28	0.22	0.17	0.09	0.10	0.11
Drapery, lightweight	0.03	0.04	0.11	0.17	0.24	0.35
Drapery, heavyweight	0.14	0.35	0.55	0.72	0.70	0.65
Terazzo floor	0.01	0.01	0.02	0.02	0.02	0.02
Wood floor	0.15	0.11	0.10	0.07	0.06	0.07
Carpet, on concrete	0.02	0.06	0.14	0.37	0.60	0.65
Carpet, on pad	0.08	0.24	0.57	0.69	0.72	0.73
Acoustic tile, suspended	0.76	0.93	0.83	0.99	0.99	0.94
Acoustic tile, on concrete	0.14	0.20	0.76	0.79	0.58	0.37
Gypsum board, 12 mm	0.29	0.10	0.05	0.04	0.07	0.09

absorb sounds of high frequencies well but have little absorption at low frequency (since the carpet is nearly transparent to sounds with long wavelength), while others (e.g., window glass) absorb low frequencies much better than high frequencies (glass sheets flex at low frequency). This affords an architect the opportunity to design an auditorium to have the desired reverberation time over a wide range of frequency.

11.2.3. Absorption by Air, by People, and by Seats

In a large auditorium, the air can contribute a substantial amount to the absorption of sound at high frequencies. The absorption of air depends upon the temperature and relative humidity (see Section 6.5), and an additional term mV, proportional to the volume, is generally added to the total absorption A. The constant m is given in the last two lines of Table 11.2. Also

Table 11.2. Sound absorption by people and seats, and air absorption

Material	Frequency 125	250	500	1000	2000	4000	8000	Unit
Wood or metal seats, unoccupied	0.014	0.018	0.020	0.036	0.035	0.028		m^2
Upholstered seats, unoccupied	0.13	0.26	0.39	0.46	0.43	0.41		m^2
Audience in upholstered seats	0.27	0.40	0.56	0.65	0.64	0.56		m^2
Air absorption per m^3								
20°C, 30% RH	—	—	—	—	0.012	0.038	0.136	m^{-1}
20°C, 50% RH	—	—	—	—	0.010	0.024	0.086	m^{-1}

given in Table 11.2 are values of absorption (in m^2) that should be added to the total absorption of the walls, ceiling, and floor to take into account the absorption by seats and people. Note that an occupied seat absorb pretty much the same amount of sound as an unoccupied upholstered seat, since the occupant covers up most of the upholstered surface. (*Note*: Values of sound absorption are given in m^2; to convert to ft^2, multiply by 10.8. Values of air absorption are given in m^{-1}; to convert to ft^{-1}, divide by 3.3.)

11.3. Sound Fields in Real Rooms

The rather straightforward calculation of reverberation time described in the preceding section is based on the idealization of Sabine (1922) that sound "fills" a reverberant room in such a way that the sound energy is uniformly distributed throughout the room. This approximation applies quite well to rooms whose dimensions are larger than the sound wavelength, whose absorbing surfaces are well distributed throughout the room, and whose total absorption is not too great. Within such a room, the sound field can be regarded as a superposition of freely propagating plane waves, no two of which are traveling in the same direction. Such a field is known as a diffuse sound field.

When sound waves fall on a surface, their energy is partially reflected and partly absorbed (see Chapter 6). For most surfaces, the absorption coefficient (giving the portion that is absorbed) is dependent on the angle of incidence. In a reverberant room with a diffuse sound field, the sound waves are randomly incident upon absorbing surfaces, and it is possible to define a statistical absorption coefficient α. Most published values of sound absorption coefficients (including those in Table 11.1) are determined by measuring the sound decay rate in a highly reverberant room with and without a sample of the material. Absorption coefficients thus determined generally exceed the true statistical (energy) absorption coefficient. Absorption coefficients determined from decay rates are sometimes called Sabine absorption coefficients.

Diffraction of the sound field incident on an absorbing patch of finite size in the test chamber probably accounts for most of the difference between the statistical absorption coefficient and the Sabine absorption coefficient. In order to minimize the error, the absorbing sample should be as large as possible, and moving reflectors should be used to help ensure diffuseness of the sound field (Embleton, 1971). Another complicating factor is the decrease in sound energy density near a highly absorptive surface, so that a diffuse field does not exist. In the extreme case where the absorption coefficient is unity, there would be no waves from the direction of this surface, so the energy density and sound pressure level near the surface would fall by 3 dB compared to some average point.

Other equations have been proposed as alternatives to the Sabine equation for calculating reverberation times. One is that proposed by Eyring (1930) which is based on the mean free path between reflections. The mean distance traveled between successive reflections from the walls of a rectangular enclosure is $L = 4V/S$, so the number of reflections per second is $N = cS/(4V)$. With each reflection, the sound is reduced in energy by a factor $(1 - \bar{\alpha})$, where $\bar{\alpha}$ is the average random-incidence absorption coefficient of all surfaces. The total attenuation of the energy over a time interval equal to the reverberation time T_{60} is $(1 - \bar{\alpha})^{NT_{60}}$, which results in a reduction of the sound level by 60 dB, so $10 \log[(1 - \bar{\alpha})^{NT_{60}}] = -60$. This leads to the so-called Norris-Eyring reverberation equation (Norris was responsible for the derivation of it):

$$T_{60} = -\frac{4V}{cS \ln(1 - \bar{\alpha})}. \tag{11.4}$$

For live rooms with $\bar{\alpha} \ll 1$, the Norris-Eyring equation gives the same result as the Sabine equation (as can be seen by expanding the natural logarithm). For rooms with one or more very absorbing surfaces, the Norris-Eyring equation usually gives a value of T_{60} closer to the observed value, and the extra effort is justified. Further discussion of refinements in the reverberation formula can be found in Cremer, Müller, and Schultz (1982), in Pierce (1981), and in a number of other books on architectural acoustics.

11.3.1. Coupled Rooms

Some large auditoria, churches, and concert halls, although they comprise a single air volume, are divided architecturally into several subspaces, so the sound energy will not be uniformly distributed, especially during the buildup and decay of the sound field. In the example shown in Fig. 11.6, the source room with a volume V_1 and the second room with a volume V_2 are connected by an opening with area S. If A_{10} denotes the equivalent absorption of area of room 1 (except for the opening S) and A_{20} denotes the equivalent absorption of room 2 (without the opening), then the respective amounts of

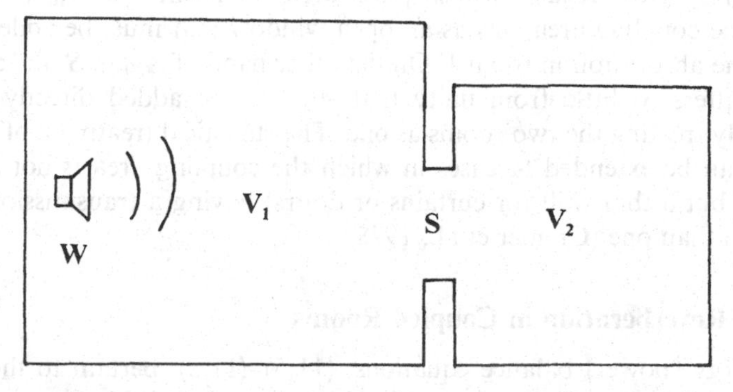

Fig. 11.6. Two rooms coupled by an opening with area S.

sound power actually absorbed in the two rooms (assuming diffuse sound fields) are $A_{10}E_1c/4$ and $A_{20}E_2c/4$, respectively, where E_1 and E_2 are the energy densities in the two rooms. The power transferred from room 1 to room 2 is $SE_1c/4$ and that transferred from room 2 to room 1 is $SE_2c/4$. If the source power is W, we can write power balances for the two rooms as:

$$W - (c/4)A_{10}E_1 - (c/4)SE_1 + (c/4)SE_2 = 0 \tag{11.5}$$

$$(c/4)SE_1 - (c/4)A_{20}E_2 - (c/4)SE_2 = 0. \tag{11.6}$$

Letting $A_{11} = A_{10} + S$ and $A_{22} = A_{20} + S$ (which means including, as part of the absorption in each room, the coupling area S with absorption coefficient of unity), we can write (11.5) and (11.6) in the form:

$$4W/c = A_{11}E_1 - SE_2 \tag{11.7}$$

$$0 = -SE_1 + A_{22}E_2. \tag{11.8}$$

Solving these equations gives the energy density in room 1 containing the source:

$$E_1 = \frac{4P_1/c}{A_{11} - S^2/A_{22}}. \tag{11.9}$$

If these two rooms had been treated as a single space, we would have an energy density

$$E = \frac{4P_1/c}{A_{10} + A_{20}}. \tag{11.10}$$

The denominator of equation (11.9) can be written $(A_{10} + A_{20}S/A_{22})$. Comparing this with the denominator of (11.10), we see that the equivalent absorption area A_{20} of room 2 does not enter into the energy balance in its full amount but is diminished by the factor S/A_{22}, which can be called the "coupling factor" from room 2 to room 1. This factor, which characterizes the difference between the "single room" and "coupled room" analyses, depends not only on the ratio of the coupling area to the total area of the room but also on the absorption coefficients of all the surfaces in that room.

If $A_{20} \gg S$, the resultant absorption area for room 1 is $A_{10} + S$, which means the coupling area acts as an open window that must be added to the rest of the absorption in room 1. On the other hand, if $A_{20} \ll S$, the coupling factor differs so little from unity that A_{20} can be added directly to A_{10}, effectively treating the two rooms as one. The statistical treatment of coupled rooms can be extended to cases in which the coupling area is not an open window but a thin wall (or curtains or doors) having a transmission coefficient less than one (Cremer et al., 1978).

11.3.2. Reverberation in Coupled Rooms

The energy (power) balance equations, (11.5)–(11.8), pertain to the steady state. If we set $W = 0$, the energy densities E_1 and E_2 in the two rooms de-

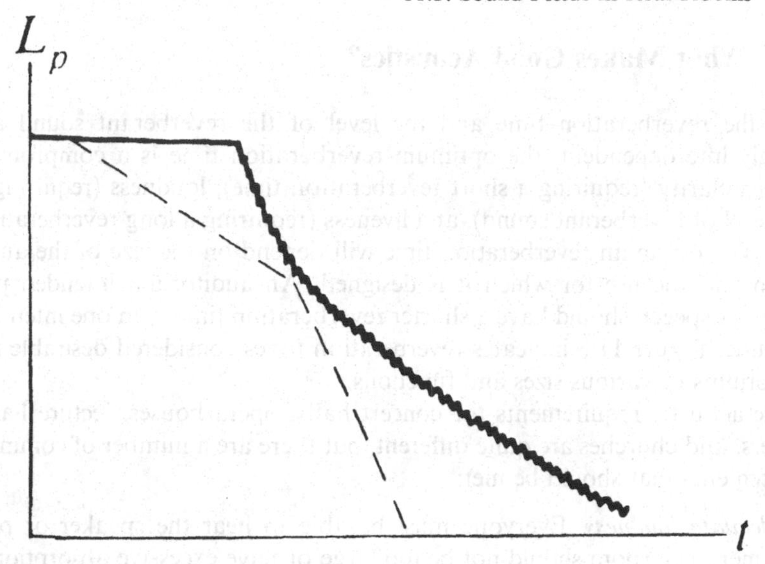

Fig. 11.7. Decay of reverberant sound in a room with different reverberation times in two coupled subspaces.

crease with time and we can write

$$\frac{dE_1}{dt} = -\frac{c}{4V_1}(A_{11}E_1 - \tau SE_2) \tag{11.11}$$

$$\frac{dE_2}{dt} = -\frac{c}{4V_2}(-\tau S_{12}E_1 + A_{22}E_2), \tag{11.12}$$

where τ is the transmission coefficient I_t/I_0 for the common wall, as defined by equation (6.40). It can be shown, as discussed by Cremer et al. (1978), that the solutions to these equations can lead to compound reverberation decay curves that have two different slopes due to different reverberation times in the two coupled rooms. Such a decay curve is shown in Fig. 11.7. A listener might tend to characterize the hall as "dry" on the basis of the more rapid initial decay rate (when the sound is loudest), even though the full 60-dB decay time is fairly long. For this reason, the early decay time is an important parameter in concert halls and auditoria. Furthermore, concert halls are usually provided with irregular surfaces (statues and other decorations served well in classic concert halls) to help promote diffusion of sound throughout the hall.

In a study of 22 European concert halls, Schroeder et al. (1974) found that the greater the early decay time, the greater the audience preference for the hall, up to a reverberation time (determined from extrapolating the early decay rate) of 2 s. Above 2 s, the preference for the hall decreased with increasing reverberation time.

11.4. What Makes Good Acoustics?

Since the reverberation time and the level of the reverberant sound are strongly interdependent, the optimum reverberation time is a compromise between clarity (requiring a short reverberation time), loudness (requiring a high level of reverberant sound), and liveness (requiring a long reverberation time). The optimum reverberation time will depend on the size of the auditorium and the use for which it is designed. An auditorium intended primarily for speech should have a shorter reverberation time than one intended for music. Figure 11.8 indicates reverberation times considered desirable for auditoriums of various sizes and functions.

The acoustic requirements for concert halls, opera houses, lecture halls, theaters, and churches are quite different, but there are a number of common requirements that should be met:

1. *Adequate loudness.* Everyone must be able to hear the speaker or performer. The room should not be too large or have excessive absorption.
2. *Uniformity.* Listeners in all parts of the room should hear as nearly the same sound as possible. There must be a sufficient number of sound diffusing surfaces to avoid "dead spots."
3. *Clarity.* There must be sufficient absorbing surfaces that the reverberant sound does not mask following sounds.
4. *Reverberance or liveness.* The listener should feel bathed in sound from all sides, but at the same time be able to localize the sound source.

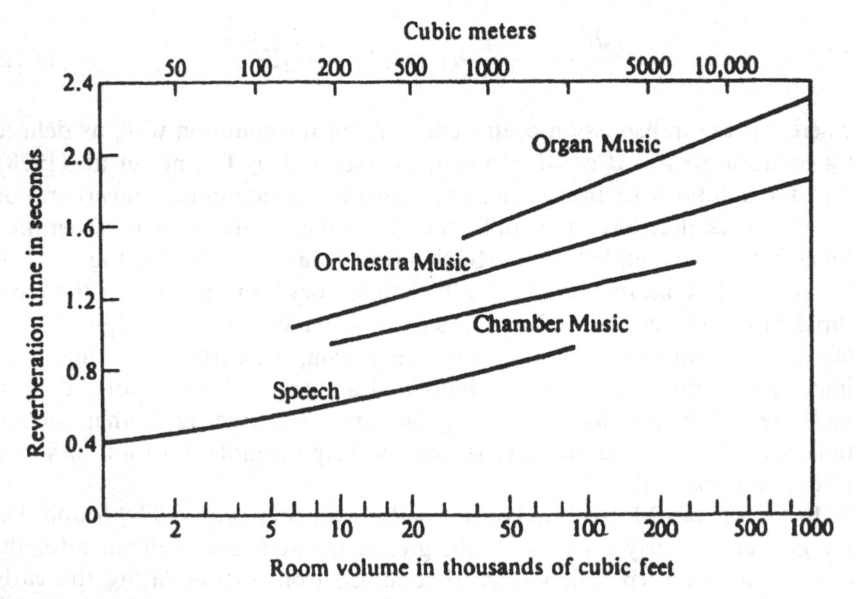

Fig. 11.8. Desirable reverberation times for auditoriums of various sizes and for various functions.

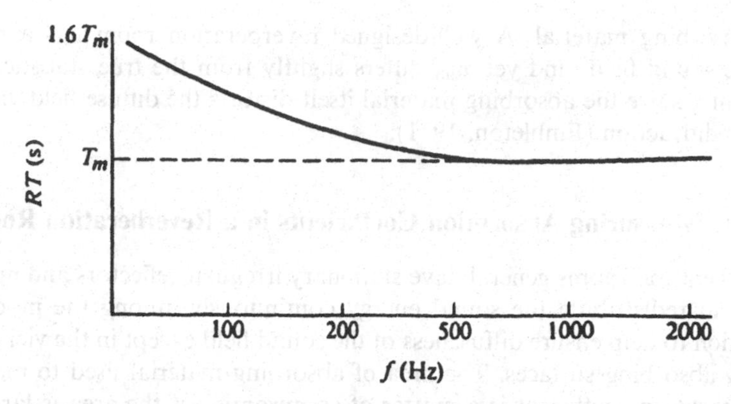

Fig. 11.9. Variation of reverberation time with frequency in a good concert hall.

5. *Freedom from echoes.* Reflected sound should arrive early enough to reinforce the direct sound but not be perceived as a separate echo.
6. *Low level of background noise.* The noise from heating and ventilating systems and from external sources should be kept very low.

A feeling of liveness or reverberance is especially important at low frequency to give support to bass notes. Fortunately, many building materials have lower absorption coefficients at low frequencies. Figure 11.9 shows how the reverberation time may vary with frequency in a good concert hall.

Several studies have stressed the importance of having sufficient reflected sound arriving from the sides. Such lateral reflections arriving with time delays of from 25 to 80 ms add to the feeling of spaciousness, whereas overhead reflections during the same period add mainly to the early sound (Barron, 1971; Reichardt et al., 1975).

11.5. Measuring Sound Absorption Coefficients

When sound waves fall on a surface or object, their energy is partially reflected and partially absorbed, as discussed in Chapter 6. The sound-absorbing efficiency of the surface is given in terms of an absorption coefficient α. A more detailed analysis, however, shows that there are several different absorption coefficients.

The *sound absorption coefficient for a given angle of incidence* α_θ is the ratio of the sound energy absorbed by a surface to the sound energy incident upon that surface at a given angle θ.

A *statistical sound-absorption coefficient* α is defined as the ratio of sound energy absorbed to the sound energy incident in a perfectly diffuse field, which implies a random distribution of incidence angles.

The *Sabine sound-absorption coefficient* α_{sab} is the ratio of absorbed energy to incident energy in a reverberation room with and without a patch of the absorbing material. A well-designed reverberation room has a nearly diffuse sound field, and yet α_{sab} differs slightly from the true statistical coefficient α since the absorbing material itself distorts the diffuse field, mainly due to diffraction (Embleton, 1971).

11.5.1. Measuring Absorption Coefficients in a Reverberation Room

Reverberation rooms general have stationary irregular reflectors and moving vanes to redistribute the sound energy continuously among the modes of vibration to help ensure diffuseness of the sound field except in the vicinity of highly absorbing surfaces. The area of absorbing material used to measure its absorption coefficient is a matter of compromise: if the area is large, its absorption becomes too great, and the incident sound field is not sufficiently diffuse. If the area is too small, then the correction of the measured values to account for diffraction becomes large. In general, small test samples give larger absorption coefficients than large samples. These considerations argue for large reverberation rooms with large patches of test material (typically of about 7–12 m^2).

11.5.2. Measuring Absorption Coefficients in an Impedance Tube

The normal impedance of a material can be measured by placing a sample in a heavy-walled tube and noting its effect on the sound field. Generally a sound source is placed at one end of the tube and the sample at the other. A movable microphone is used to determine the sound pressure as a function of position. Such an instrument is called an impedance tube. Measurements are restricted to the frequency range over which plane-wave propagation is assured (see Section 8.1). For a cylindrical tube, this is given by $f = 0.586c/d$, where d is the diameter.

Two examples of impedance tubes are shown in Fig. 11.10. In both tubes, a small loudspeaker is at one end and the sample to be measured is mounted at the other end. In the first one, a small probe tube is used to sample the sound pressure within the main tube and transmit this pressure to a microphone. The microphone assembly is moved along a scale to determine the position. In the second tube, a sound level meter is placed in a sliding section of a tube to determine the sound pressure level at various positions along the impedance tube.

In both tubes the positions of pressure maxima and minima are determined, along with sound pressure levels. The ratio of maximum to minimum sound pressure is the standing wave ratio n. The normal incidence sound absorption α_{90} is given by:

$$\alpha_{90} = 1 - \left(\frac{n-1}{n+1}\right)^2 = 1 - \left(\frac{10^{L/20} - 1}{10^{L/20} + 1}\right)^2. \tag{11.13}$$

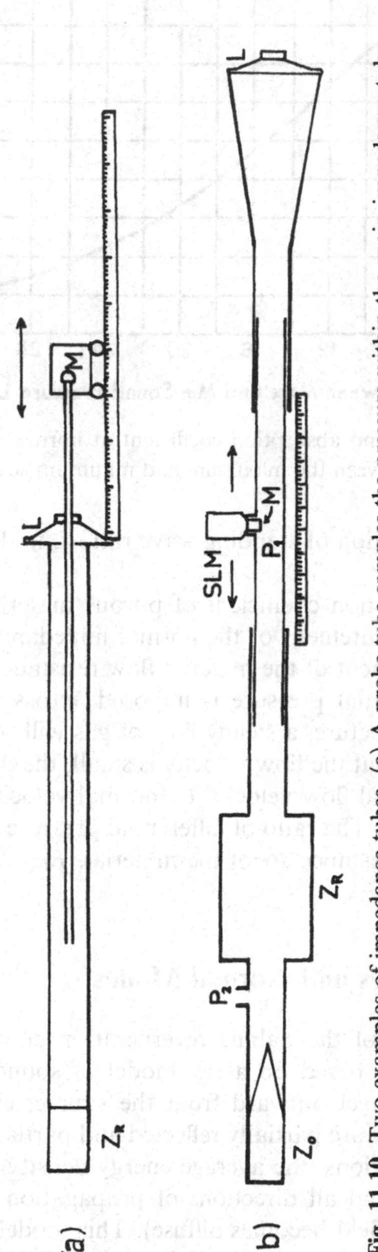

Fig. 11.10. Two examples of impedance tubes. (a) A probe tube samples the pressure within the main impedance tube and transmits this to a microphone on a calibrated scale. (b) A sound level meter is mounted in a sliding section of the impedance tube to determine the sound pressure level. (From Rossing, 1982. Reprinted with permission.)

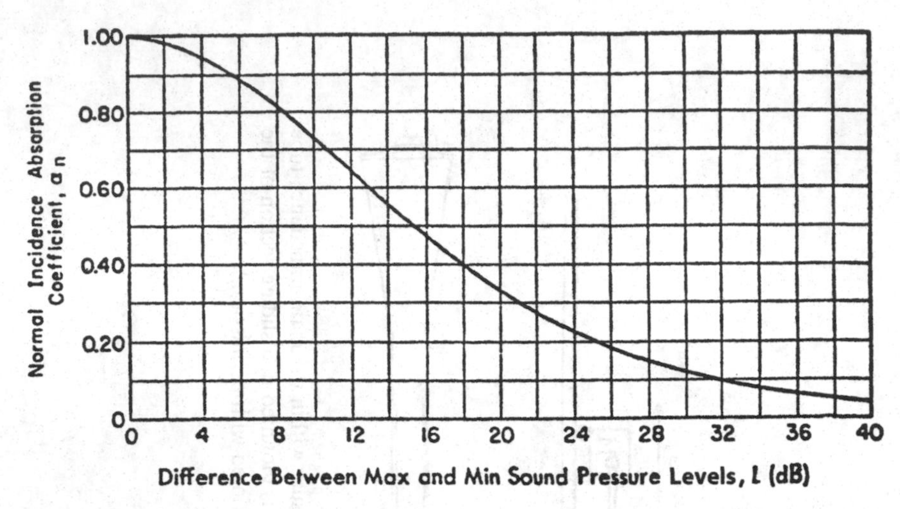

Fig. 11.11. Relation of sound absorption coefficient at normal incidence α_{90} to the difference L in decibels between the maximum and minimum sound levels.

A graph of α_{90} as a function of standing wave ratio L in decibels ($20 \log_{10} n$) is given in Fig. 11.11.

The statistical absorption coefficient of porous materials may be calculated either from a measurement of the normal impedance using an impedance tube or a measurement of the material flow resistance.

If a constant differential pressure is imposed across a layer of porous material of open cell structure, a steady flow of gas will be induced through the material. Provided that the flow velocity is small, the differential pressure p and the induced normal flow velocity U (normal velocity per unit surface area) are linearly related. The ratio of differential pressure to normal velocity is known as the flow resistance R_f of the material. $R_f = pA/U$, where A is the sample area.

11.6. Standing Waves and Normal Modes

Theoretical derivations of the Sabine reverberation equation (and similar equations) are generally based on a ray model of sound, in which sound rays are assumed to travel outward from the source; each time they encounter a boundary, they are partially reflected and partially absorbed. After a large number of reflections, the average energy density becomes the same throughout the room, and all directions of propagation are equally probable (that is, the sound field becomes diffuse). This model oversimplifies the behavior of sound in a room, particularly at low frequencies, because it neglects the existence of normal modes, the distribution of absorbing materials, and the shape of the room. In some rooms, particularly in small ones, these factors may become important.

The wave equation can be readily solved for simple enclosures, such as rectangular rooms, and real rooms can often be approximated by a simple enclosure. Solutions to the Helmholtz equation (see section 6.2) are interpreted as room modes. In a large room, the frequencies of these modes are packed closely together, but in a small room they can sometimes be observed individually.

The sound pressure in a rectangular enclosure with sides of lengths a, b, and c is given by equation (6.55), and the normal mode frequencies by equation (6.56). A room mode is designated by (l, m, n). Modes in which two of the integers l, m, and n of that equation are zero are called axial modes, because they consist of standing waves propagating back and forth parallel to pairs of room boundaries and being reflected by another pair. Modes in which only one of the integers l, m, and n is zero are called tangential, and they consist of waves reflecting off two pairs of surfaces (like balls on a billiard table). Modes in which none of the integers l, m, and n is zero are called oblique modes. In general, axial modes store more acoustic energy than the other types because the sound waves travel farther between reflections.

Contours of equal sound pressure for the $(2, 0, 0)$ axial mode and the $(3, 2, 0)$ tangential mode are shown in Fig. 11.12. Note the pressure maxima along the boundaries and the pressure maxima that occur in the corners. Pressure maxima occur in the corners for all room modes, which suggests corners as good locations for sound absorbers. Placing a loudspeaker or microphone near a corner maximizes the response, but only at such low frequencies that the distance to the corner is smaller than a quarter wavelength or so.

11.7. Small Rooms and Studios

From (6.56) it is easy to see that the frequency distribution of the modes of a room is determined by its dimensions. Distributions of mode frequencies for a cubic and a rectangular room are compared in Fig. 6.3. The cube has a very "peaky" response with many coincident modes, whereas a rectangular room with dimensions in the ratio 1:2:3 has a more even spread of resonances. Even better mode distribution can be obtained by avoiding integer multiples entirely. The golden ratio 1.618:1:0.618 gives a very smooth response. Rooms with oblique walls do also, but rectangular rooms are generally preferred for other reasons. As frequency increases, the number of modes greatly increases. The number $N(f)$ of modes with frequencies within the range 0 to an upper limit f is given approximately by the expression

$$N(f) \approx \frac{4\pi}{3} V \left(\frac{f}{c}\right)^3 + \frac{\pi}{4} S \left(\frac{f}{c}\right)^2 + \frac{L'}{8} \frac{f}{c}, \tag{11.14}$$

where V is the volume of the room, S is its area, $L' = 4(L + W + H)$ is the

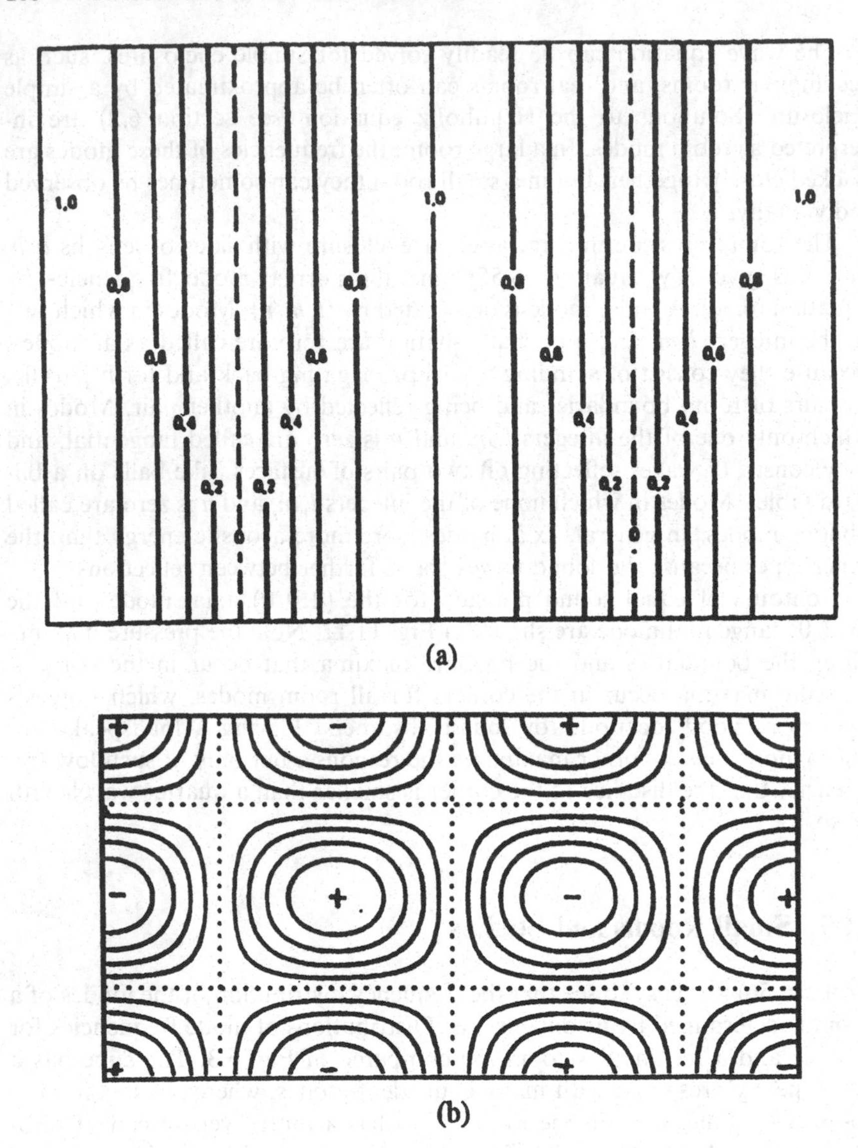

Fig. 11.12. Contours of equal sound pressure in a rectangular room. (a) (2, 0, 0) axial mode; (b) (3, 2, 0) tangential mode. Dashed lines represent nodes.

sum of the lengths of all the edges of the room, and c is the speed of sound (Kuttruff, 1997).

When the resonance peaks are closer together than the bandwidth of each peak, the resonances are less evident. Since the average spacing decreases with increasing frequency, there is a frequency above which the resonance peaks can be regarded as a smoothed-out continuum (Schroeder, 1954). This frequency, called the Schroeder cutoff frequency, is given by

$$f_{sc} = \left(\frac{c^3 T_{60}}{4V \ln 10} \right)^{1/2}. \tag{11.15}$$

Above the Schroeder cutoff frequency, a sum over mode indices can be approximated by an integral (Pierce, 1981).

In a small room, such as a home listening room, walls and ceiling are generally so close to the listener that many reflections arrive within a few milliseconds after the direct sound. It is, therefore, not so important to distinguish between early reflected sound and reverberant sound. Achieving "intimacy" (which depends upon a short time delay between direct and first reflected sound) is no problem at all in a small room. In fact, the inherent intimacy in a small room makes it difficult to simulate the acoustics of a larger space.

We can walk into a strange room blindfolded and rather quickly and accurately estimate its size by listening to the sound in the room. Presumably, when reflected sounds follow the direct sound with little delay, an auditory impression of smallness is created. Conversely, when the reflections arrive with a distribution in time and space that is characteristic of a large concert hall, a feeling of spaciousness is created.

In addition to home listening rooms, the design of home theaters, small sound-recording studios and their control rooms requires the application of the acoustics principles we have discussed. In sound-recording studios careful control of reverberation time and diffusion are important as is noise isolation. In order to lengthen the initial time delay in the control room, so that the sound mixer can hear the sound from the studio more clearly, early reflections from the front end of the control are sometimes minimized so that the first reflection comes from the rear of the room. Such an arrangement is often called a live-end dead-end (LEDE) control room in which the front half of the room is very absorptive and the rear half as diffusive as possible.

11.8. Sound Diffusers and Absorbers

The sound we hear in a listening room is a combination of the direct sound and reflections from many surfaces in the room. In a large auditorium, reverberant sound comes from a complex mixture of reflections from many surfaces, and there is adequate delay between the direct and reflected sound. In small rooms, the sound is often absorbed before there is a uniform mixture, and sound diffusors are especially important.

Geometric shapes attached to room surfaces help to scatter and diffuse the sound. Rather than reflecting sound in a single direction, as a flat surface does, triangular shapes, rectangular protrusions, and semi-cylindrical surfaces scatter sound in many directions, resulting in a diffuse sound field, even

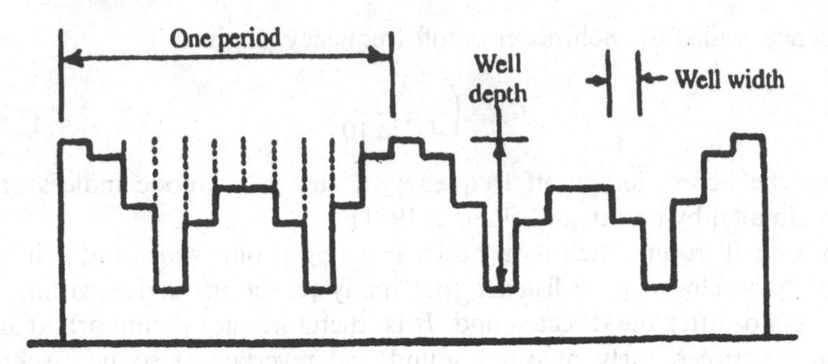

Fig. 11.13. Cross section of a grating (quadratic residue) sound diffuser.

in a small room. Using the finite difference time domain method, Yokota et al. (2002) have calculated how various types of diffusing reflectors increase the diffusivity of a rectangular room. A grating of slits can also effectively diffuse sound due to diffraction.

In 1975, Schroeder suggested the use of phase-grating sound diffusers. The theory of phase-grating sound diffusers is based on number theory; Schroeder used a scheme called maximum-length sequences, a stream of fixed-length digital 1 s and 0 s with some interesting statistical properties. Based on this theory, *quadratic-residue diffusers* and *primitive-root diffusers* have been developed. A cross section of a phase-grating sound diffuser consisting of a structure with a sequence of wells to scatter sound within a certain frequency band is shown in Fig. 11.13. The sequence is repeated along the diffusers. The maximum depth of the wells determines the effective low-frequency limit of the diffusers. The well depth should be 12 times the wavelength at the lowest frequency. The highest frequency is determined by the well width, which is half a wavelength at the highest frequency. The actual sequence of wells is determined by number theory (Schroeder, 1986).

Sound absorption in a room depends mainly upon the surface area of the walls, ceiling, and floor, and the nature of the surfaces. Porous materials, such as drapery, carpets, glass fiber, and acoustical tile convert acoustic energy to heat as the vibrating air particles interact with the tiny fibers in the absorber. Porous materials absorb very well at high frequency (Table 11.1). Panels of wood, glass, gypsum board, and even plaster on lath, on the other hand, flex in response to sound waves of long wavelength and thus absorb rather well at low frequency but have very little absorption at high frequency.

Another type of absorber, which depends upon the principle of the Helmholtz resonator (section 9.1), can provide absorption over a selected frequency band. A Helmholtz resonator absorbs sound energy near its resonance frequency. One way to achieve this is to cover cavities with a perforated panel. The resonance frequency is given approximately by

$f = 200tp^{1/2}/d$, where p is the perforation percentage (hole area/panel area $\times$ 100), t is the effective hole length with the correction factor (panel thickness + 0.8 $\times$ hole diameter), and d is the depth of the air space.

11.9. Rooms for Worship

Churches, temples, and synagogues share many of the same requirements for good acoustics discussed in section 11.4. Figure 11.7 illustrated how different the optimum reverberation time is for speech and organ music. Obviously some compromise is called for in church design. Most of the old cathedrals in Europe have long reverberation times, as the spoken word was generally not as important as it is in most contemporary worship. In many old churches (especially in northern Europe), the pulpit is strategically placed in the center of the congregation and has a large canopy to reflect the preacher's voice.

It is important that the background-noise level in churches be kept as low as possible in concert halls. Heating, ventilating, and air-conditioning (HVAC) systems must be designed with great care. Electronic reinforcement of sound is often necessary in larger churches, although it too frequently is overdone. Smaller churches should be able to function without electronic reinforcement except, perhaps, sound reinforcement for the hard of hearing.

11.10. Classrooms

The need for good acoustics in classrooms is simple: students must be able to understand the teacher and each other. Providing adequate speech intelligibility is a matter of controlling three types of classroom sound: reverberation; heating ventilation and air-conditioning (HVAC) noise; and noise from outside the classroom.

Studies have shown that reverberation times in a quiet classroom should be 2 s or less in order to avoid speech interference. Special consideration should be given to the speech intelligibility range, 500 to 4000 Hz. Deciding on the maximum acceptable noise level of classroom HVAC systems is critical. According to the American National Standards Institute (ANSI), the noise rating (NC) for lecture halls and classrooms should be NC-25 to NC-30. A teacher using a normal voice will produce a sound level of about 46 dB at the ears of a student 30 ft away. NC-30 corresponds roughly to a background noise level of about 36 dB, which should produce the 10-dB difference in speech and noise level required for speech intelligibility. Setting criteria for exterior noise is somewhat more complex. Traffic noise tends to be more or less constant, and the criteria for HVAC noise are appropriate.

Requirements for intermittent noise, such as aircraft flyovers, can be slightly less stringent. Speech interference levels (SIL) of SIL-50 are generally acceptable; these are roughly equivalent to NC-50. Lower levels are clearly desirable if learning is to be maximized. The Department of Education, the Acoustical Society of America, and other professional organizations have organized several conferences on the important topic of classroom acoustics, and increased attention will be paid to it in the future (ASA 2000, 2003). We urge readers of this book to become activists in their own communities. Students deserve the opportunity to learn in quiet classrooms!

11.11. Walls and Noise Barriers

When an airborne sound wave strikes a solid wall, the largest part is reflected, whereas smaller portions are absorbed and transmitted through the wall. The coefficients of reflection, absorption, and transmission are determined by the physical properties of the wall and by the frequency of the sound and its angle of incidence to the wall (see section 6.4).

The transmission coefficient τ is defined as the ratio of transmitted to incident intensity

$$\tau = I_T/I_0 \tag{11.16}$$

and the transmission loss TL in decibels as

$$\mathrm{TL} = -10\log_{10}\tau. \tag{11.17}$$

Sound waves striking a wall can bend it, shake it, or both. (These motions can be described as flexural or compressional waves in the wall). At low frequency, the sound transmission loss in a solid wall follows a mass law; it increases with increasing frequency and mass density M of the wall. For waves that approach a wall of large dimensions with normal incidence, the transmission loss is

$$\mathrm{TL}(0^\circ) = 10\log_{10}\left(1 + \frac{\pi M f}{400}\right), \tag{11.18}$$

where M is the wall mass density (in kg/m^2) and f is the frequency (in hertz). In a room, it is a good approximation to assume the sound waves of low frequency to be randomly distributed over all angles from 0 to 80°. This decreases the transmission loss by about 5 dB, so that

$$\mathrm{TL} = 10\log_{10}\left(1 + \frac{\pi M f}{400}\right) - 5. \tag{11.19}$$

From these formulas it is clear that the common wall between adjacent rooms should be as heavy as possible, and that low-frequency sounds are the

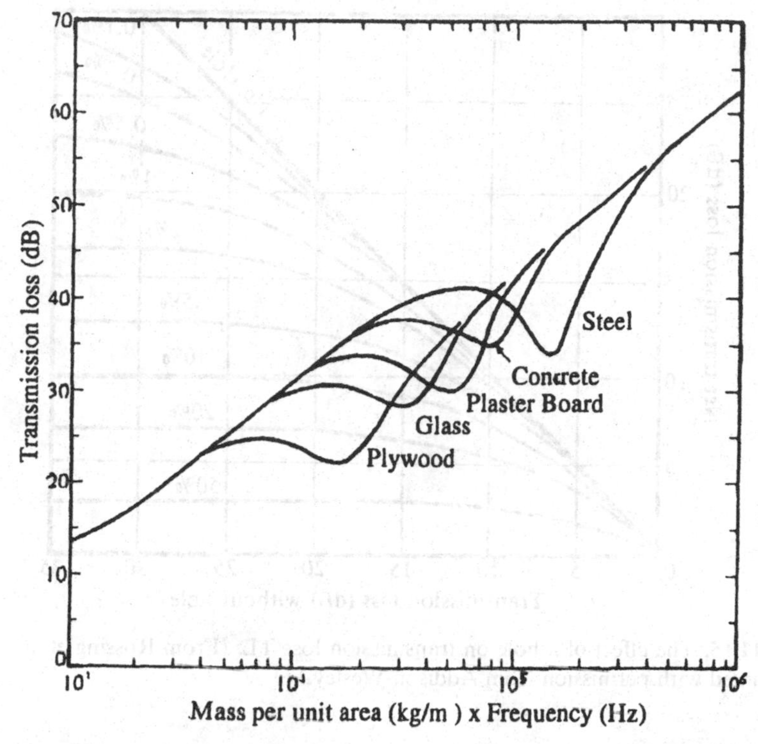

Fig. 11.14. Transmission loss TL of a wall as a function of mass and frequency. Note the drop in TL near the critical frequencies. (From Rossing et al., 2002. Reprinted with permission from Addison-Wesley.)

most difficult to block (no surprise if you have heard sound from a neighbors' stereo through a common wall).

Transmission loss for a wall may fall considerably below that predicted by the mass law, due to any of the following effects:

1. Wall resonances that occur at certain frequencies;
2. Excitation of bending waves at the *critical frequency*, where they travel at the same speed as certain sound waves in air;
3. Leakage of sound through holes and cracks.

The transmission losses for walls of several materials are shown in Fig. 11.14. Note the dip in TL at the critical frequency, which is different for each material.

Leakage of sound through small holes or cracks in walls tends to be underestimated all too often in building construction. Openings around pipes and ducts and cracks at the ceiling and floor edges of walls allow the leakage of airborne sound. Common causes of leakage in party walls separating apartments may include back-to-back electrical outlets or medicine cabinets.

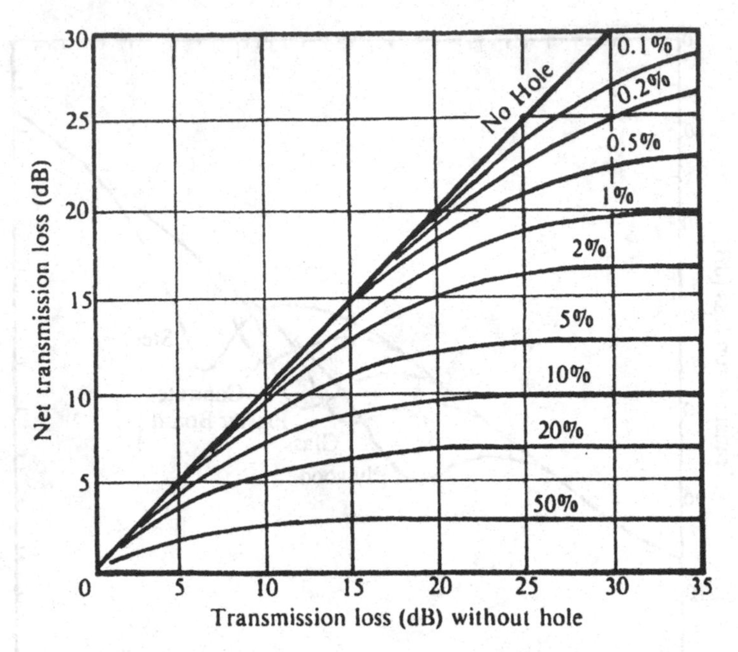

Fig. 11.15. The effect of a hole on transmission loss TL. (From Rossing et al., 2002. Reprinted with permission from Addison-Wesley.)

Cracks under doors are especially bad. Figure 11.15 illustrates the effect of holes of various sizes on the transmission loss of walls.

Sound barriers, which block the direct sound path from source to receiver, can result in appreciable noise reduction, both indoors and outdoors. Sound transmission through a barrier is generally less important than sound transmission around a barrier. A typical situation in an indoor office is shown in Fig. 11.16. There are three types of transmission paths to be considered: transmission through the barrier (path SCR), diffraction around the barrier (path SBR) and reflection from the ceiling (paths SAR, SDER, etc.).

Transmission through the barrier, which is similar to that through a full wall of the same construction, will generally be much less than will transmission by diffraction and reflection. The diffraction of sound of a given frequency around a barrier depends on the Fresnel number N, which is expressed as $N = 2/\lambda(A + B - d)$, where λ is the wavelength, and $A, B,$ and D are shown in Fig. 11.16(b). Transmission by reflection (Fig. 11.16(c)) depends on the acoustic properties of the ceiling, the size of the opening above the barier, and the nature of the walls on the source and receiver sides of the barrier. A highly absorbent ceiling is essential in an open-plan office or school.

The use of barriers to attenuate noise outdoors will be discussed in Chapter 12.

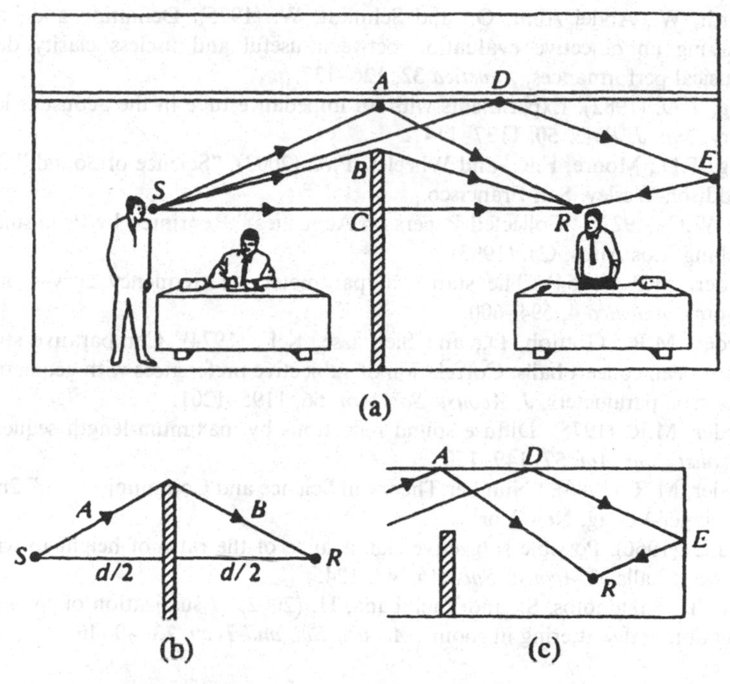

Fig. 11.16. (a) Transmission paths through and around a barrier. (b) Diffraction around the barrier. (c) Reflection paths around a barrier.

References and Suggested Reading

ASA (2000, 2003). "Classroom Acoustics I" and "Classroom Acoustics II," *Acoust. Soc. Am.*, Melville, N.Y.

Barron, M. (1971). The subjective effects of first reflections in concert halls—The need for lateral reflections, *J. Sound Vib.* **15**, 475–494.

Beranek, L.L. (1996). "Concert and Opera Halls: How They Sound." Acoustical Society of America, Melville, N.Y.

Bies, D.A. and Hansen, C.H. (1997). Sound Absorption in Enclosures. "Encyclopedia of Acoustics," vol. 3, ed. M.J. Crocker, Wiley, New York, Chap. 92.

Cremer, L., Müller, H.A., and Schultz, T.J. (1982). "Principles and Applications of Room Acoustics," vol. 1. Applied Science Publisher, London.

Embleton, T.F.W. (1971). Sound in Large Rooms. "Noise and Vibration Control," ed. L.L. Beranek, McGraw-Hill, New York, Chap. 9.

Eyring, C.F. (1930). Reverberation time in dead rooms, *J. Acoust. Soc. Am.* **1**, 217–241.

Kuttruff, H. (1973). "Room Acoustics." Applied Science, London.

Kuttruff, H. (1997). Sound in Enclosures, in "Encyclopedia of Acoustics," vol. 3, ed. M.J. Crocker, Wiley, New York.

Pierce, A.D. (1981). "Acoustics." McGraw-Hill, New York.

Reichardt, W., Abdel Alim, O., and Schmidt, W. (1975). Definition and basis of making an objective evaluation between useful and useless clarity defining musical performances, *Acustica* **32**, 126–137.

Rossing, T.D. (1982). Experiments with an impedance tube in the acoustics laboratory. *Am. J. Phys.* **50**, 1137–1141.

Rossing, T.D., Moore, F.R., and Wheeler, P.A. (2002). "Science of Sound," 3rd ed. Addison-Wesley, San Francisco.

Sabine, W.C. (1922). "Collected Papers on Acoustics." Reprinted by Peninsula Publishing, Los Altos, Ca. (1993).

Schroeder, M.R. (1954). The statistical parameters of frequency curves of large rooms, *Acustica* **4**, 594–600.

Schroeder, M.R., Gottlob, D., and Siebrasse, K.F. (1974). Comparative study of European concert halls: Correlation of subjective preferences with geometric and acoustic parameters, *J. Acoust. Soc. Am.* **56**, 1195–1201.

Schroeder, M.R. (1975). Diffuse sound reflections by maximum-length sequence, *J. Acoust. Soc. Am.* **57**, 149–150.

Schroeder, M.R. (1986). "Number Theory in Science and Communication," 2nd edn. Springer-Verlag, New York.

West, J.E. (1966). Possible subjective significance of the ratio of height to width of concert halls, *J. Acoust. Soc. Am.* **40**, 1245.

Yokota, T., Sakamotos, S., and Tachibana, H. (2002). Visualization of sound propagation and scattering in rooms. *Acoust. Sci. and Tech.* **23**, 40–46.

CHAPTER 12

Sound and Noise Outdoors

In a free field, the sound pressure at a distance r from the source is proportional to $1/r$ and the sound intensity to $1/r^2$, as we have discussed in Chapters 6 and 11. Ideally, an outdoor environment away from sound-reflecting buildings and other hard surfaces could be considered a free field. However, effects such as atmospheric absorption, turbulence, refraction, diffraction, and the proximity of the ground make the description of sound propagation outdoors somewhat more complicated and interesting. In this chapter we will consider outdoor sound propagation with particular attention to the propagation of noise outdoors, which has become such an important environmental consideration. Some consideration will be given to the control of noise outdoors.

12.1. Sound Propagation in the Atmosphere

When sound propagates in an ideal uniform atmosphere, its intensity deceases with distance, both because of inverse-square-law spreading and also because of losses due to thermal conduction, viscosity, and molecular absorption. The intensity $I(r)$ at a distance r from a point source is therefore

$$I(r) = \frac{A}{r^2} e^{-\alpha_a r}, \tag{12.1}$$

where α_a is the attenuation coefficient for sound in air, as discussed in Section 6.5. The attenuation increases with increasing frequency ω at a rate that varies somewhat depending upon humidity (Bass et al., 1995) but is, over the range 10 Hz–100 kHz, roughly proportional to ω^n with $n \approx 1.5$. The absolute values are around 0.3 dB/km at 100 Hz, rising to 300 dB/km at 10 kHz. Over the more limited range 100 Hz–10 kHz, $n \approx 1$, or the range can be split as in equations (6.48). The effect of this frequency variation of α_a is clearly heard in the sound of thunder. A nearby thunderclap makes a sizzling snap, with acoustic energy spread over the entire frequency spectrum, as is to be expected from a sharp pulse of energy; a distant thunderclap, on the other hand, is a dull rumbling with only very low frequencies still audible.

Such a semi-infinite uniform atmosphere does not, however, occur in practice, and many other effects must be considered (Sutherland and Daigle, 1997). In a stable atmosphere, the pressure decreases exponentially with altitude h with a scale height H of about 6 km, so that $p(h) = p(0)\exp(-h/H)$. The temperature also decreases with height at about 0.6°C per 100 m, the so-called "adiabatic lapse rate." At altitudes above about 10,000 m, above the tropopause and into the stratosphere, the lapse rate reverses and the atmospheric temperature begins to rise again. The real atmosphere is, however, not as simple as this. Weather conditions such as frosts or fogs can lead to formation of a temperature inversion, with cold air close to the ground and warmer air above, and of course the atmosphere is not stationary, but disturbed by winds, the strength and direction of which may change with altitude.

A change in atmospheric pressure does not itself change the propagation speed of sound, but the associated change in temperature does. Equation (6.13) shows that c is proportional to $T^{1/2}$, where T is the absolute temperature, so that c decreases by as much as 10% between ground level and the typical flight altitude of jet aircraft. Because we are familiar with optical ray propagation, and because optical refractive index is the ratio of the speed of light in a material to its speed in a vacuum, it is sometimes helpful to think of the atmosphere as a medium with a refractive index that increases with height above ground level. This allows us to draw the generalized ray trajectories shown in Fig. 12.1(a) and (b) for the case of a stable standard atmosphere. Sound from a source on the ground tends to be deflected upwards, and the same is true of a source high in the air. The range at which such sources might be heard at ground level is therefore reduced from what would be expected from (12.1).

In the contrary case of a temperature inversion, there is cold air close to the ground and warmer air above, as on frosty or foggy mornings, the temperature gradient in the lower air being inverted from its normal behavior in the lower region. Sound rays then behave as in Fig. 12.1(c), and those propagating at low angles to the horizontal are bent back towards the ground, making distant sounds louder at ground level. This can often be noticed in the case of noise from a distant freeway on foggy or frosty mornings, though increased attenuation in the fog may also have an influence in the opposite direction.

Sound at ordinary frequencies is attenuated so rapidly by atmospheric absorption that little penetrates into the stratosphere, where again the temperature gradient is reversed. This is not true of infrasound below about 10 Hz, however, at which frequencies atmospheric attenuation is very small. Infrasound can therefore be refracted back in the stratosphere towards the surface of the Earth and detected at very large distances (Gabrielson, 1997).

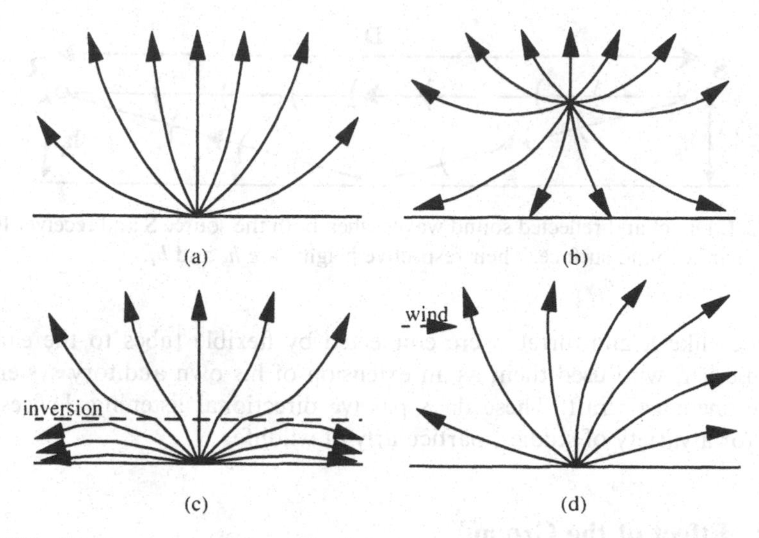

Fig. 12.1. (a) Ray propagation paths for a ground-level sound source in a standard atmosphere with temperature decreasing steadily with increasing altitude. (b) Ray paths for a high-level source in a standard atmosphere. (c) Ray paths for propagation in an atmosphere with a low-altitude temperature inversion. (d) Ray paths for propagation in a standard atmosphere with a wind gradient, wind velocity increasing with altitude.

Finally, consider the case in which there is a wind that increases steadily with altitude—quite a normal situation. The sound propagation is effectively tied to the local motion of the air through which it is propagating, so that it is convected downstream as shown in Fig. 12.1(d).

Since much of our common experience with sound is limited to behavior near the ground, it is also relevant to examine the effect of different ground covers. As might be expected, smooth hard surfaces introduce little additional attenuation, and even increase the radiated intensity by 6 dB over that from a point source in an infinite medium by ground reflection, soft or porous surfaces, such as grass, have a similar effect at frequencies below about 100 Hz, but introduce increasing attenuation at higher frequencies, reaching about 40 dB/km at 1 kHz. Trees and foliage add rather little to the attenuation at low frequencies, but contribute to intensity reduction by scattering when the sound wavelength becomes comparable to the dimensions of the tree trunks or leaves, typically above about 1 kHz.

Sonar techniques can be used in the atmosphere for a variety of purposes. Turbulence, for example, as well as generating low-frequency sound, can also scatter high-frequency sound, so that radar-like sonar systems can be used near airports to detect hazardous atmospheric conditions. Passive listening devices with microphones arranged in arrays to give directional properties can also be used to detect the sources of particular noises. In early devices of this type, two large horns, separated by several meters and mounted on

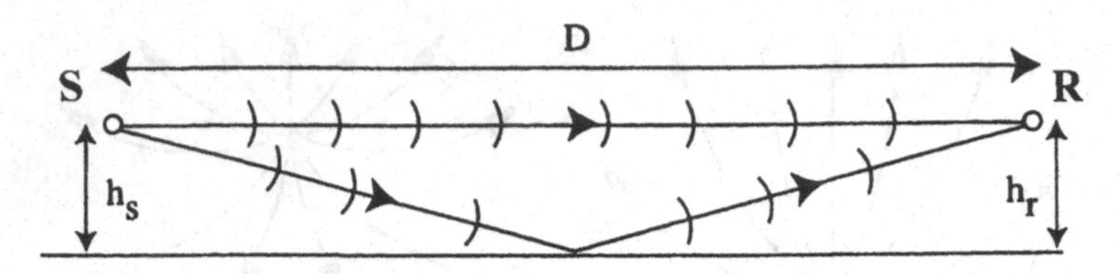

Fig. 12.2. Direct and reflected sound waves when both the source S and receiver R are near a hard ground surface. Their respective heights are h_s and h_r.

a device like a gun-turret, were connected by flexible tubes to the ears of an observer, who used them as an extension of his own auditory system to locate enemy aircraft. These days passive directional listening devices are used for a variety of studies, particularly of wildlife.

12.2. Effect of the Ground

12.2.1. Hard Surface

The most basic effect of the ground on the sound field is that of interference between the direct and reflected sound wave as shown in Fig. 12.2. At a hard surface, such as asphalt or concrete, there is a large impedance mismatch ($z_1 \ll z_2$) and the reflection coefficient (Eq. 6.37) is essentially 1, so little or no phase change occurs on reflection. Thus the observed effects are due to the difference in path lengths between the direct and reflected waves. Interference effects lead to a sound spectrum of the type shown in Fig. 12.3, which is often described as "comb filtering." The greater the source height h_s, the closer together in frequency the interference minima will lie.

12.2.2. Soft Ground

Measurements of sound propagation over a grass-covered surface show another interesting effect. If the source is placed on the ground, the direct and reflected paths are equal, and the path length interference effects illustrated in Fig. 12.2 are avoided, as in the $h_s = 0$ case at the top. For high frequencies, however, another type of attenuation is noted, as seen in Fig. 12.4. With the receiver very close to the ground, the sound pressure level remains essentially constant until about 800 Hz, and then decreases very rapidly with frequency.

At an acoustically soft surface, such as grass-covered ground, $z_2 \ll z_1$, so the reflection coefficient becomes approximately -1 (Eq. 6.37), and so the incident and reflective waves are opposite in phase and destructively interfere. This is essentially what happens above 800 Hz, as shown in Fig. 12.4. But why is the received signal nearly full strength below 800 Hz? This is because of the so-called acoustical "ground wave" (which is similar, in some

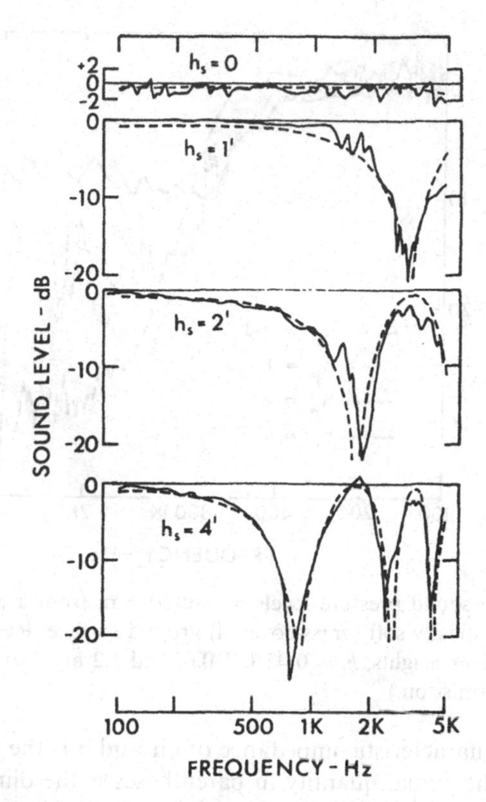

Fig. 12.3. Relative sound pressure levels measured 15.2 m from a point source and 1.2 m above an acoustically hard ground (asphalt). Results are shown for four different source heights, $h_s = 0, 0.3, 0.6$, and 1.2 m, respectively. (From Embleton et al., 1976. Reprinted with permission.)

respects, to the ground wave in electromagnetic wave propagation). The ground wave is that part of the reflected sound field that is not accounted for by the plane–wave reflection coefficient, and it occurs whenever the incident waves are not plane waves. A reasonably simple description of the ground wave can be obtained by considering the frequency dependence of the ground impedance. The specific acoustic impedance of a ground surface for normal incidence is the complex ratio of the acoustic pressure at the surface and the resulting normal component of particle velocity into the ground. For a semi-infinite medium, this specific acoustic impedance is the same as the characteristic impedance throughout the medium.

Delany and Bazley (1970) showed that the normalized characteristic impedance $Z_c/\rho c$ of a wide range of absorbent porous materials can be described by the expression

$$\frac{Z_c}{\rho c} = \left[1 + 0.0511\left(\frac{f}{\sigma}\right)^{-0.75}\right] - j\left[0.0768\left(\frac{f}{\sigma}\right)^{-0.73}\right] \tag{12.2}$$

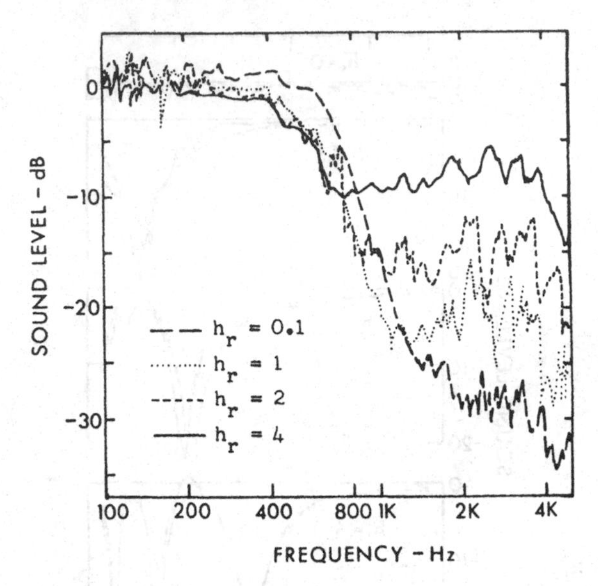

Fig. 12.4. Relative sound pressure levels measured 5 m from a point source at the surface of an acoustically soft (grass-covered) ground surface. Results are shown for four different receiver heights, $h_r = 0.02, 0.3, 0.6$, and 1.2 m. (From Embleton, 1996. Reprinted with permission.)

where ρc is the characteristic impedance of air and σ is the flow resistivity at frequency f. (The actual quantity in parentheses is the dimensionless $f\rho/\sigma$, but the numerical value of ρ has been substituted.) For a grass lawn the flow resistivity σ is about 1.25×10^5 to 3×10^5 Pa s/m^2. A fairly detailed summary of this is given by Sutherland and Daigle (1997).

A sort of cutoff frequency, the frequency at which the relative sound pressure level in Fig. 12.4 has fallen 3 dB, is shown in Fig. 12.5. The slope of this graph implies that the acoustic impedance of the grass surfaces is inversely proportional to the square root of the frequency. Also the position of this line depends on the magnitude of the frequency-dependent ground impedance (Embleton, 1996).

12.2.3. Acoustic Surface Waves

For distances greater than about 150 m, another component of the sound field, called a "trapped surface wave" must be included. This is a low-frequency component of the sound field, which at 300 Hz and at a large distance from the source, is reduced by 30 dB from its level at very low frequencies (Embleton, 1996). A surface wave only exists when the reactive component of the acoustic impedance exceeds its real part. Furthermore, the reactance must be a compliance, as it is for porous ground surfaces.

The trapped surface wave has a maximum amplitude at the ground that decreases exponentially with height. The rate of decrease with height depends on the softness (acoustic compliance) of the surface, and on frequency. The

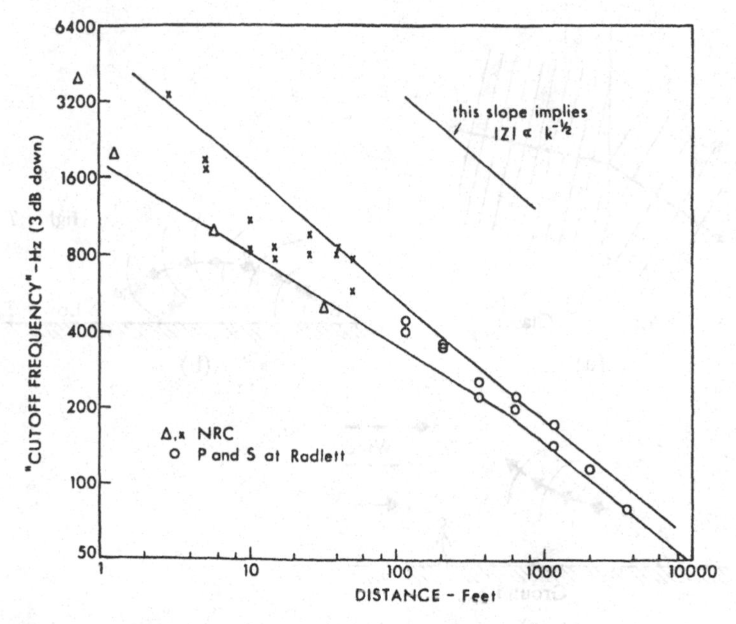

Fig. 12.5. Cutoff frequency (3 dB-down) versus distance for propagation over grass. Different symbols are from different locations. (From Embleton et al., 1996. Reprinted with permission.)

surface wave extends to greater height for softer ground and lower frequency. The wave spreads only horizontally during propagation and so decreases by 3 dB per doubling of distance (as compared to 6 dB per doubling of distance for spherical components of the wave). Hence at some distance the surface wave becomes dominant (Embleton, 2001).

12.2.4. Foliage and Trees

The attenuation of sound through a dense forest may be as great as 20 dB per 100 m. The main effect at low frequencies is to enhance ground attenuation, the roots making the ground more porous (Aylor, 1972). At high frequencies, where the dimensions of leaves become comparable with the wavelength, there is also a significant attenuation caused by scattering (Embleton, 1963). In a forest, the vertical gradients of wind and temperature are reduced at elevations up to approximately the height of the trees, thus reducing the effects of refraction from such gradients (Sutherland and Daigle, 1997).

12.3. Effect of Refraction

When the speed of a wave changes, refraction may result in a change in the direction of propagation or a bending of the waves, as discussed in Section

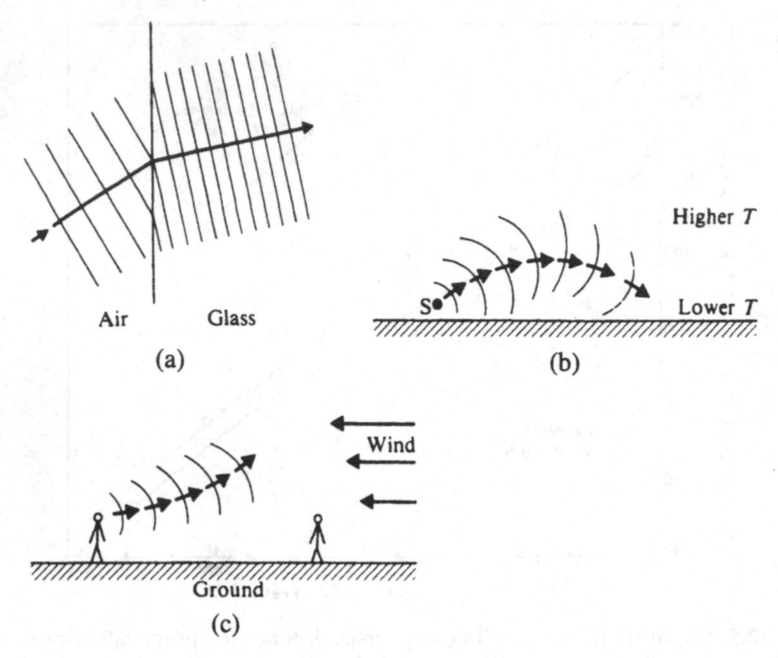

Fig. 12.6. Refraction of sound: (a) passing from helium to air; (b) in the atmosphere when temperature varies with height; (c) sound traveling against the wind (Rossing et al., 2002).

12.1. The change of speed may occur abruptly as the wave passes from one medium to another, or it may change gradually if the medium changes gradually. Examples of these are shown in Fig. 12.6.

The situation illustrated in Fig. 12.6(b), which sometimes occurs during the cool evening hours, causes sounds to be heard over great distances. Because the speed of sound increases with temperature (see Chapter 6), the sound travels faster some distance above the ground where the temperature is greater. This results in a bending of sound downward as shown. Sound that would ordinarily be lost to the upper atmosphere is refracted aback toward the ground.

Figure 12.6(c) shows why it is difficult to be heard when yelling against the wind. (It is not because the wind blows the sound waves back; even a strong wind has a speed much less than that of sound). Refraction results because the wind speed is less near the ground than it is some distance above it. Because the speed of sound with respect to the air (in this case the moving air) remains the same, the ground speed of the sound changes with altitude. The resulting refraction causes some of the sound to miss its target. Normally, temperature decreases with altitude: thus thee is an upward refraction, since sound travels faster in the warm air near the surface of the earth. Two examples of temperature inversion that will cause downward refraction are

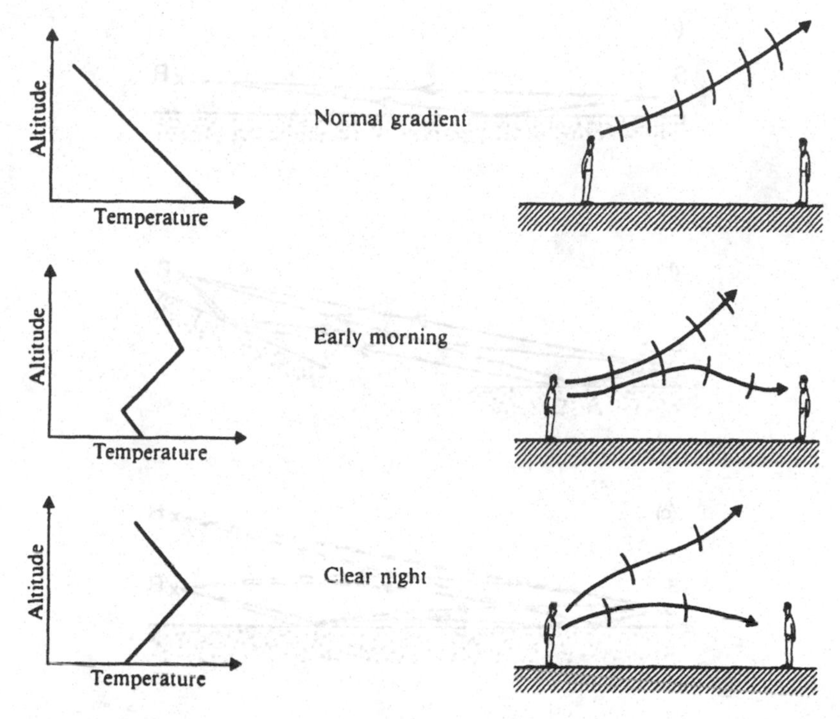

Fig. 12.7. Refraction of sound under different conditions (Rossing et al., 2002).

illustrated in Fig. 12.7. Mining companies, for example, carefully monitor the weather conditions to decide on the best time for blasting operations in order to minimize noise levels in surrounding communities.

12.3.1. Analogy with Nonflat Ground

There is an interesting analogy between a refracting atmosphere and ground surfaces that curve either downwards or upwards. Fig. 12.8(a) illustrates the basic idea of a direct field and a reflected field. When the atmosphere is nonrefracting and the ground surface is flat, both ray paths are straight and there is one point of specular reflection.

Figure 12.8(b) and (c) illustrates the analogy between propagation in a nonrefracting atmosphere over upwardly curving ground and in a temperature inversion above flat ground. In both cases there is interference between one direct wave and three that are reflected from the ground. The ground reflections in both cases are now located near the source and near the receiver. In both cases, the grazing angles at the ground increase compared with Fig. 12.8(a).

There is also a good analogy between propagation in a nonrefracting atmosphere having straight ray paths over downwardly curving ground such

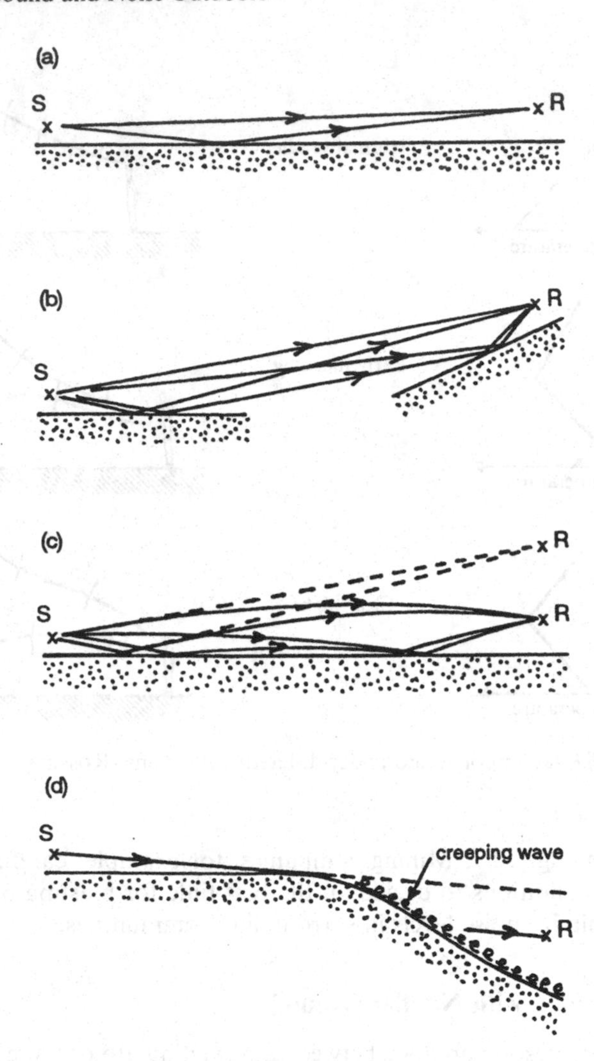

Fig. 12.8. Analogy between a refracting atmosphere and ground surfaces that are not flat. (a) propagation in a nonrefracting atmosphere above flat ground (compare with Fig. 12.2); (b) propagation in a nonrefracting atmosphere above upwardly concave ground; (c) propagation in a temperature inversion or downwind above a flat ground; (d) propagation in a nonrefracting atmosphere above upwardly convex ground. (From Embleton, 1996. Reprinted with permission.)

as a hill (Fig. 12.8(c)) and in a temperature lapse over flat ground. There is a shadow region into which no sound can penetrate directly, according to ray theory. In both cases, sound energy penetrates into the shadow by propagating near the ground as a creeping wave, and some of this energy is shed upwards at an appropriate point to reach a receiver. Progressive shedding of sound energy provides a sound field throughout the shadow region (Embleton, 1996).

12.3.2. Outdoor Sound Propagation in the U.S. Civil War

An interesting example of the dramatic effects of outdoor sound propagation is related in a book *Trial by Fire: Science, Technology and the Civil War* (Ross, 1999). Before electrical and wireless communications became available, the sound of battle was often the quickest and most efficient method by which a commander could judge the course of a battle. Troop dispositions were often made based on the relative intensity of the sounds from different locations on the battlefield. Acoustic 'shadows' due to atmospheric absorption, wind shear, temperature inversions, ground effects, or all of these apparently influenced several key battles in the U.S. Civil War.

At the battle of Gettysburg, for example, Confederate General Ewell was apparently unable to hear the artillery of General Longstreet, and hence did not move his troops. As a result, Union General Meade was able to shift his troops in the nick of time to defeat Longstreet's attack. On the previous day, Meade had been unable to hear the Gettysburg fighting from his position at Taneytown (12 miles away), yet the battle was clearly audible in Pittsburgh, 150 miles from Gettysburg.

12.4. Diffraction and Sound Barriers

Sound barriers are widely used to reduce highway noise and airport noise. If a large solid body blocks a sound field, the ray theory of sound propagation predicts a shadow region behind the body with sharply define boundaries. In practice, however, sound "leaks" across this sharp boundary due to diffraction.

The diffraction of sound of a given frequency around a barrier depends on the Fresnel number N, which is given by

$$N = \frac{2}{\lambda}(d_1 + d_2 - d) \tag{12.3}$$

where λ is the wavelength of sound and lengths d_1, d_2, and d are shown in Fig. 12.9. The attenuation by diffraction A_d due to diffraction is then given as a function of the Fresnel number:

$$A_\mathrm{d} = 10 \log_{10}(20N) \tag{12.4}$$

This equation is also plotted in Fig. 12.9, assuming a thin barrier and no ground and then empirically allowing for the presence of the ground by reducing the sound level loss by 2 dB (Sutherland and Daigle, 1997). The prediction curve is not exact, because the empirical correction does not ac-

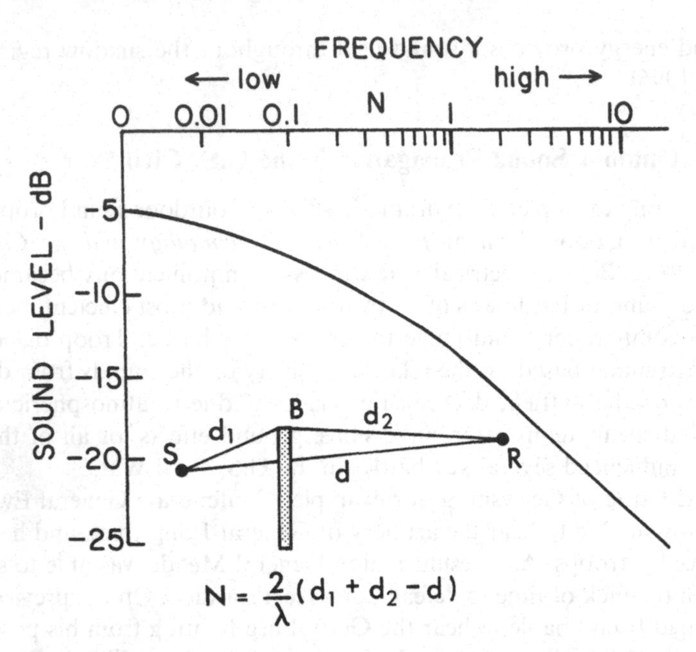

Fig. 12.9. Fresnel number in terms of the path difference and wavelength λ. The curve is the barrier attenuation as a function of Fresnel number N. (From Sutherland and Daigle, 1997. Reprinted with permission.)

count for the frequency dependence of the ground reflection interference in a specific configuration.

In order to obtain a more exact prediction of the sound field behind the barrier, the complex interference spectrum resulting from the sum of four paths shown in Fig. 12.10 must be calculated. Nevertheless, the curve in Fig. 12.9 is correct to about ± 5 dB in most cases (Isei et al., 1980).

Studies of traffic noise reduction by barriers generally show average insertion loss of 5–8 dB and rarely exceed 10 dB. When barriers are used to attenuate sound, it is good practice to locate them as closely as possible to either the source of the receiver. A barrier of a given height then results in the greatest value of path difference $(d_1 + d_2 - d)$. The insertion loss of a barrier is limited by the effects of atmospheric turbulence to about 15–25 dB (Sutherland and Daigle, 1997).

12.5. Atmospheric Turbulence

Turbulence is always present near the ground, even on a calm day. Temperature fluctuations of 5°C are not uncommon. Similarly the wind fluctuations can be one third of the average velocity. When sound waves propagate through the atmosphere, these fluctuations scatter the sound energy. The

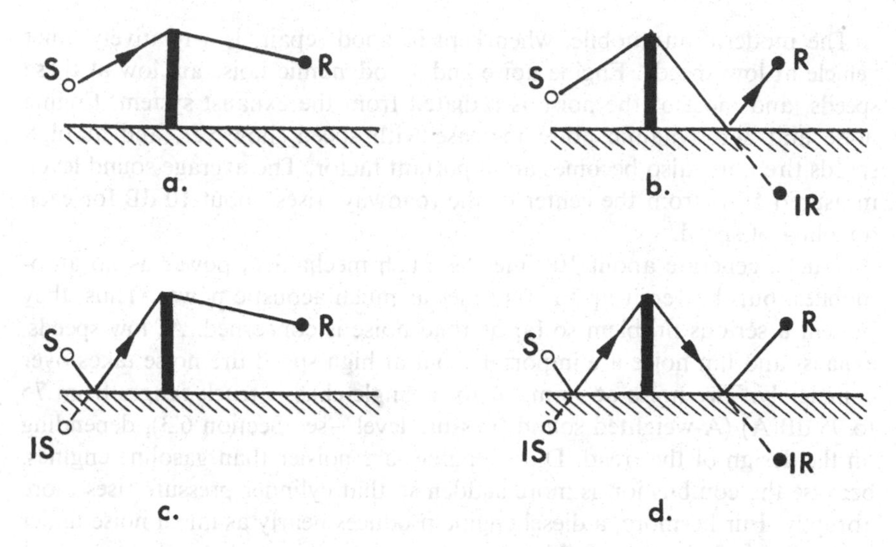

Fig. 12.10. Four paths contributing to the sound field behind a barrier above ground. (From Sutherland and Daigle, 1997. Reprinted with permission.)

total sound field is the sum of these scattered waves, resulting in random fluctuations in amplitude and phase. Turbulence can be visualized as a series of eddies from millimeters up to meters in diameter. As the eddy size becomes smaller, most of the energy is dissipated as heat. In addition to natural turbulence, common artificial causes of turbulence include the wake of moving vehicles, wind blowing around obstacles, jet engines or other exhaust pipes.

The effects of turbulence on sound fields can be large and result from two processes: one is the scattering of sound energy by each eddy; the other is the destruction of the constant phase relations that would otherwise determine sound level. The effects of turbulence on a sound field increase with distance (approximately as the 11/12 power of the distance up to a value of about 6 dB) and with frequency (approximately as the 7/12 power of frequency) (Embleton, 1996).

12.6. Motor Vehicle Noise

With more than 100 million passenger cars in the United States traveling over 10^{12} miles annually, automobiles generate megawatts of acoustic power. Fortunately, much of this power is radiated to areas with low population density, but an appreciable portion is generated in urban areas. Sources of noise in an automobile or truck include the engine, the cooling fan, the drive train, the tires, aerodynamic turbulence, body vibrations, and the intake and exhaust systems.

The modern automobile, when kept in good repair, is a relatively quiet vehicle at low speeds. Engine noise and aerodynamic noise are low at these speeds, and most of the noise is radiated from the exhaust system. Engine noise and aerodynamic noise increase with speed, however, and at high speeds tire noise also becomes an important factor. The average sound level, measured 50 ft from the center of the roadway, rises about 10 dB for each doubling of speed.

Trucks generate about 10 times as much mechanical power as do automobiles, but they emit up to 100 times as much acoustic power. Thus, they present a serious problem so far as road noise is concerned. At low speeds, exhaust and fan noise are important, but at high speed tire noise takes over completely. Tire noise at 55 mi/hr for a single-chassis truck ranges from 75 to 95 dB(A) (A-weighted sound pressure level—see Section 6.3), depending on the design of the tread. Diesel engines are noisier than gasoline engines, because the combustion is more sudden so that cylinder pressure rises more abruptly. Furthermore, a diesel engine produces nearly as much noise under no load as it does under full load.

Although regulation of motor vehicle noise is generally left to state and local governments, the Federal government does regulate the noise of new trucks, motor homes, motorcycles and mopeds. Several states and cities set limits for passenger cars. Pass-by noise measurements generally follow procedures defined by the Society of Automotive Engineers in the appropriate standards (SAE J986). For example, noise from passenger cars is measured 15 m from the roadway center and 1.2 m above the ground while the vehicle is undergoing maximum acceleration, and the car is to attain maximum rated engine speed not more than 45 m past the microphone. Most states set the "pass by" limit at 80 dB(A). Noise limits in Europe are somewhat lower (Hickling, 1997).

12.7. Railroad Noise

Noise from railroad operations is not as widespread as noise from highways. Nevertheless, for persons living near a railroad, it can be a great annoyance. Major sources of noise are locomotives, rail-wheel interaction, whistles and horns, yard retarders, refrigerator cars, maintenance operations, and loading equipment.

According to EPA standards in the United States, locomotives manufactured after 1980 should not emit more than 90 dB(A) of noise, measured 100 ft from the track. Railcars (or combination of them) should not emit more than 88 dB(A) at speeds up to 45 mi/hr, or 90 dB(A) at speeds above 45 mi/hr. Some of the possible ways in which railroad noise can be reduced include equipping diesel locomotives with mufflers, use of welded rails, careful maintenance of rolling stock, and barrier walls around retarders in rail yards.

12.8. Aircraft Noise

Control of aircraft noise is one of the most challenging of urban environmental issues. Airlines transport approximately 80% of all intercity passenger traffic traveling by common carrier in the United States. In spite of efforts to develop high-speed ground transport, the number of aircraft takeoffs and landings near major cities continues to grow. Furthermore, as land prices rise, residential dwellings encroach on noise buffer zones near airports in increasing numbers.

The modern jet aircraft is actually a rather inefficient noise source, radiating less than 0.02% of its total power as sound. Nevertheless, this may exceed 1 kW of sound power because of the prodigious amount of mechanical power generated by the engine. The development of the turbofan engine around 1960 led to greater efficiency and somewhat less total noise power, but it added a new source of annoyance: a siren like whine from the fan.

A turbofan jet engine produces two main types of noise. The first is due to the turbulence created when the high-velocity jet of gas reacts with the quiescent atmosphere. This noise, which has a considerable low-frequency component, dominates during takeoff and climb. The second type of noise is the high-pitched whine of the fan, which becomes dominant during a landing approach with reduced power. Two major engineering developments have spearheaded the attack on jet engine noise: the development of acoustic linings for engine nacelles and the high-bypass-ratio engines, now used in most wide body aircraft.

Almost all major airports in the United States have active noise abatement programs (Eldred, 1997). The principal actions that can be taken to abate aircraft noise and its perceived problems are (1) realign runways or take-off and landing paths to avoid residential centers; (2) require operational procedures, particularly throttle setting after takeoff, to minimize noise; (3) restrict hours of airport operation; (4) install sound insulation in affected buildings; and (5) continue to monitor airport noise and its effects.

Although the United States terminated its program to develop a supersonic transport plane (SST) in the 1960s, the Russian TU144 and the Anglo-French Concorde were put into service. SSTs present unique noise problems around those airports that they serve. Both the TU144 and the Concorde used afterburners for increased thrust during takeoff. Afterburners increase the noise emission, especially at low frequency where atmospheric absorption is very low. These low-frequency sounds are also apt to excite building vibration and rattle. The afterburners can be cut out shortly after takeoff, however to reduce noise by sacrificing rate of climb.

A sonic boom is a pressure transient of short duration that occurs during the flyover of an aircraft at a speed that exceeds the speed of sound (approximately 770 mi/hr or 343 m/s at low altitude where the temperature is about 20°C). A sonic shock is analogous to the bow wave produced by a boat moving through water at greater than the speed of water waves. At

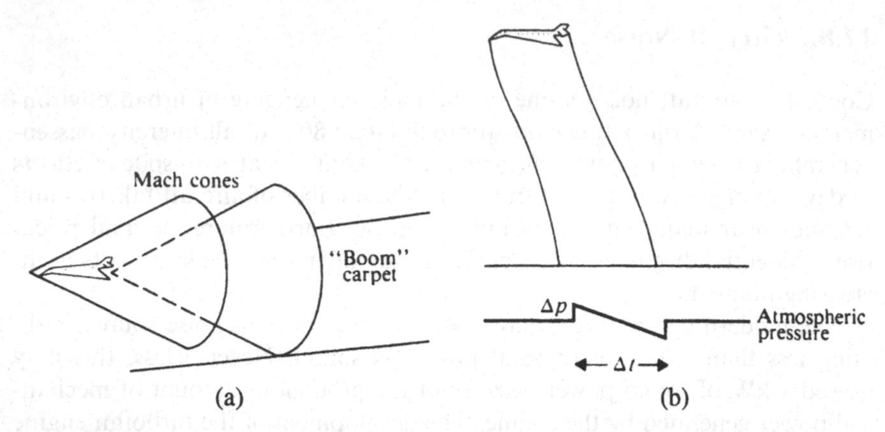

Fig. 12.11. Structure of the sonic boom generated by an aircraft flying at supersonic speed.

ground level, a momentary overpressure of 10 to 100 N/m^2 occurs, followed a moment later by a similar underpressure as the pressure fronts (Mach cones) pass by (see Fig. 12.11). For the Concorde, flying at an altitude of 40,000 ft, the overpressure on the ground is about 20 N/m^2 (about 10^{-4} atm), and the time between overpressure and underpressure is approximately 0.2 to 0.3 s.

12.9. Summary of Factors at Various Distances

Embleton (1996) has summarized the influence of factors at various distances from the source.

Short range, up to 15 m. Standard test procedures for the measurement of motor vehicle noise require that the vehicle be driven along a hard level surface. Measurements are made at a height of 1.22 m. In North America, a 15.2 m distance from the roadway is preferred, but in Europe the distance is 7.6 m. Although most major sound sources such as tail pipe, engine block, and oil pan are small enough to be considered point sources at these distances, interference between the direct sound field and that reflected from ground surface can affect the measured values. Interference effects are especially strong in vehicles with elevated noise sources, such as the exhaust stack of a tractor-trailer.

Intermediate range, up to about 50 m. The insertion loss of roadside noise barriers depends strongly on the type of ground in their vicinity. One reason is that grazing angles of reflection for sound fields diffracted over the top of the barrier are greater than the very small grazing angles without the barrier. This is a major reason why some barriers provide less noise reduction than expected.

Distances of the order of kilometers. In predicting noise around airports or from other strong noise sources, topography and meteorology are more important than at shorter distances. Near one airport, for example, the sound level at the top of a hill 5500 m away was found to be as much as 20 dB greater than at 4100 m at a lower elevation (Embleton, 1996).

References

Aylor, D. (1972). Noise reduction by vegetation and ground. *J. Acoust. Soc. Am.* **51**, 201–209.

Bass, H.E., Sutherland, L.C., Zuckerwar, A.J., Blackstock, D.T., and Hester, D.M. (1995). Atmospheric absorption of sound: Further developments. *J. Acoust. Soc. Am.* **97**, 680–683.

Delany, M.E. and Bazley, E.N. (1970). Acoustical properties of fibrous absorbent materials. *Appl. Acoust.* **3**, 105–116.

Embleton, T.F.W. (1963). Sound propagation in homogeneous deciduous and evergreen woods. *J. Acoust. Soc. Am.* **35**, 1119–1125.

Eldred, K.N. (1997). Airport Noise. "Encyclopedia of acoustics," ed. M.J. Crocker, Wiley, New York, Vol. 2, 1059–1072.

Embleton, T.F.W. (1996). Tutorial on sound propagation outdoors. *J. Acoust. Soc. Am.* **100**, 31–48.

Embleton, T.F.W. (2001). Noise Propagation and Prediction Outdoors. Tutorial at the 142nd meeting, Acoustical Society of America, Ft. Lauderdale.

Embleton, T.F.W., Piercy, J.E., and Olson, N. (1976). Outdoor sound propagation over ground of finite impedance. *J. Acoust. Soc. Am.* **59**, 267–277.

Gabrielson, T.B. (1997). Infrasound. "Encyclopedia of Acoustics," ed. M.J. Crocker, Wiley, New York, Vol. 1, pp. 367–372.

Hickling, R. (1997). Surface Transportation Noise. "Encyclopedia of Acoustics," ed. M.J. Crocker, Wiley, New York, Vol. 2, 1073–1081.

Isei, T., Embleton, T.F.W., and Piercy, J.E. (1980). Noise reduction by barriers on finite impedance ground. *J. Acoust. Soc. Am.* **67**, 46–58.

Ross, C.D. (1999). "Trial by Fire: Science, Technology and the Civil War," White Mane, Shippensburg, Pa.

Rossing, T.D., Moore, F.R., and Wheeler, P.A. (2002). "Science of Sound," 3rd ed. Addison Wesley, San Francisco.

Sutherland, L.C. and Daigle, G.A. (1997). Atmospheric sound propagation. "Encyclopedia of acoustics," ed. M.J. Crocker, Wiley, New York, Vol. 1, 341–345.

CHAPTER 13

Underwater Sound

Although oceans cover over 70% of the earth's surface, only in recent years has oceanography become a major science. Sound waves are widely used to explore the oceans, because they travel much better in sea water than do light waves. Likewise, sound waves are used, by humans and dolphins alike, to communicate under water, because they travel much better than do radio waves.

In this chapter, we will present a brief introduction to some principles of underwater acoustics, including a discussion of how sound propagates in the sea and how sound waves are used in devices such as sonar. Acoustical oceanography has many military, as well as commercial applications. Much of our understanding of underwater sound has been a result of research conducted by the Navy during and following World War II.

13.1. Underwater Sound Propagation

Sound propagates in water very much as in air, but there are a few important differences. The sound velocity is $(K/\rho_W)^{1/2}$ where K is the reciprocal of the compressibility and ρ_W is the liquid density. This is straightforward. The sound speed in sea water, which is about 1500 m/s, increases with increasing static pressure by about 1 part per million per kilopascal, or about 1% per 1000 m of depth, assuming temperature remains constant. The variation with temperature is an increase of about 2% per degree C temperature rise. In temperate parts of the ocean, the sound speed is around 1520 m/s in the warm water at the surface and stays fairly constant in the mixed surface layer of depth about 500 m. Below that depth the temperature falls and with it the sound speed, typically reaching a minimum of around 1490 m/s at a depth of about 1000 m. Below this, the ocean is nearly isothermal, the effects of increasing pressure dominate, and the sound speed rises steadily to reach 1520 m/s again at a depth of about 4000 m. In the polar regions, however, the warm surface layer is absent, the sound speed at the surface is about 1440 m/s, and this speed simply increases steadily with depth to reach about 1520 m/s at 4000 m, as in warm oceans.

Sound absorption coefficient α_w increases with increasing frequency as ω^2 for pure water, but for sea water there are several dispersion steps, giving an overall variation of α_w about as $\omega^{1.5}$. This behavior is very similar to the variation with frequency of α_a for normally humid air, but the absolute value of α_w is only about 0.001 dB/km at 100 Hz and about 1 dB/km at 10 kHz. This attenuation coefficient is thus less than that for air by a factor of about 300.

Sound propagation in the oceans is different from that in the atmosphere for another important reason. The lower part of the atmosphere terminates abruptly at the Earth's surface, but its upper regions fade off exponentially into space. The oceans, in contrast, terminate abruptly both at the sea bed and at the surface, so that sound transmission takes place in a confined duct of variable thickness. In addition to this, the variation of salinity and temperature with depth may create ducts of smaller vertical extent, rather like the inversion layers in the lower part of the atmosphere. Sound can propagate even more efficiently in such ducts because there is no absorption at the irregular ocean bottom to interfere. This means that the intensity $I(r)$ of sound in oceans varies with distance r about as

$$I(r) = \frac{A}{r} e^{-\alpha_w r}, \tag{13.1}$$

provided the distance r is very much greater than the ocean depth. For distances smaller than the ocean depth, of course, the spreading behavior is as $1/r^2$ rather than $1/r$.

One of the problems with duct propagation when the width of the duct (in this case the depth) is much greater than the wavelength of sound, is that higher modes can propagate in addition to the plane-wave mode. Such higher modes were discussed for the case of a cylindrical duct in Section 8.1, but here we need to consider the behavior of an extended planar duct. If the wave equation for propagation in the x direction in such a two-dimensional planar duct is written

$$\frac{\partial^2 p}{\partial t^2} = c^2 \left(\frac{\partial^2 p}{\partial x^2} + \frac{\partial^2 p}{\partial z^2} \right), \tag{13.2}$$

and we impose the boundary conditions that $p = 0$ at the surface $z = 0$ and $\partial p / \partial z = 0$ at the ocean floor at $z = h$, then the solution looks like

$$p(x, z) = A \sin \left[\frac{(2n-1)\pi z}{2h} \right] e^{-j\omega(t - x/v)}, \tag{13.3}$$

where v is the phase velocity of the wave in the x direction. Substituting this expression into (13.2) gives

$$\left[\frac{(2n-1)\pi}{2h} \right]^2 + \left[\frac{\omega}{v} \right]^2 = \left[\frac{\omega}{c} \right]^2, \tag{13.4}$$

for $n = 1, 2, 3, \ldots$. The phase velocity v is thus greater than c by an amount that increases with the mode number n. This is because waves are propagating obliquely and being reflected at the seafloor and the surface as shown in Fig. 13.1—an oblique cut across such a wave by a horizontal plane makes it appear to have a longer wavelength and thus a faster phase-propagation velocity. The group velocity, which is the velocity with which a pulse of several cycles of the wave will propagate down the duct, is however less than c, as would be expected for such an oblique wave path. Calculation of the group velocity is algebraically a little complicated, and it differs for each mode. If the wavelength is much less than the ocean depth, or the duct depth if ocean layering is responsible for confining the sound, then we can simply use ray propagation and count up the total path covered by obliquely propagating rays. For short wavelengths there are many path options, so that a pulse will be spread out into a long signal at a distant point.

Although the seafloor does reflect acoustic waves without a great deal of attenuation, its wave impedance ρc, where ρ is its density and c is the acoustic wave velocity within it, is within a factor 3 or so of the wave impedance for sound in water. This contrasts with the situation in the atmosphere, where the wave impedance in solid earth is around 10,000 times that in air. This means that acoustic waves launched in the water propagate some distance into ocean floor sediments until they are reflected out by rocks or other discontinuities. Sound waves can therefore be used to examine both the surface shape of the ocean floor and also, to some extent, hidden geological features. If the frequency is in the high kilohertz range so that the wavelength is short, then sonar techniques can be used to map the ocean floor at quite high resolution.

Sonar also takes the place of radar under water, for electromagnetic waves are strongly attenuated because of the significant electrical conductivity due to dissolved salts. Apart from the different nature of the waves carrying the information, sonar techniques are very much like those developed for radar. Multiple transmitting transducers arranged in an array are used to produce a narrow beam, and a similar technique is used for the receiving transducers. The reflection strength of a target, such as a submarine, is roughly proportional to its surface area, or rather its cross-section normal to the beam, provided the target is much larger than the sound wavelength. The returned echo strength is also proportional to the inverse fourth of the target distance, because of the inverse square-law spreading experienced in both propagation directions.

13.2. Underwater Waveguides

It is possible to transmit sound for long distances under water due to "waveguides" which occur at various ocean depths due to reflection and refraction. During World War II, a "deep channel" was discovered in which sound waves could travel distances in excess of 3000 km. All rays originating near the axis of the deep sound channel and making small angles with the hori-

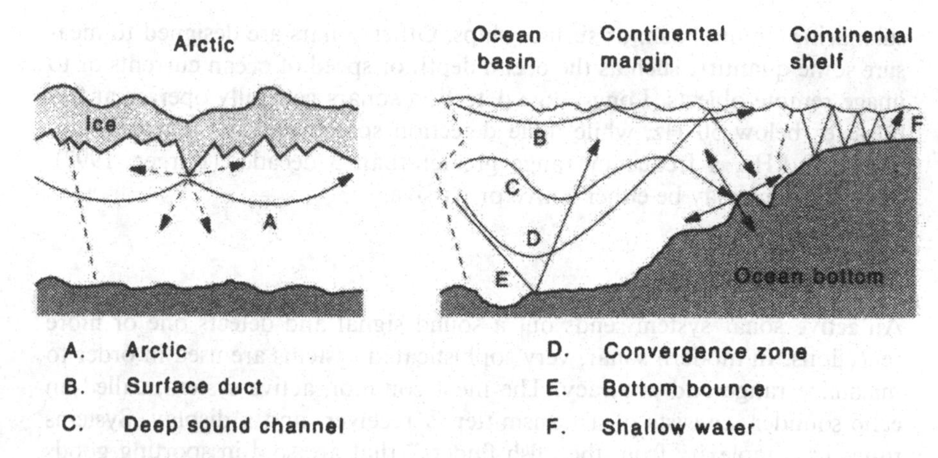

Fig. 13.1. Schematic representation of various types of sound propagation in the ocean. (From Kuperman, 1997. Reprinted with permission.)

zontal return to the axis without reaching either the surface or the bottom. This phenomenon gives rise to the deep channel or SOFAR (for "SOund Fixing And Ranging") channel. SOFAR has been used to locate, by acoustic means, airmen downed at sea.

Figure 13.1 shows the basic types of propagation in the ocean resulting from sound speed profiles. According to Snell's law, sound waves bend toward the regions of lower sound speed, so waves are trapped in regions of low sound speed. Paths A and B correspond to surface duct propagation where the minimum sound speed is at the ocean surface, while path C propagates in the deep sound channel whose axis is at the sound speed minimum. For mid-latitudes, far away from the Arctic where the local minimum tends to become shallow, sound in the deep channel can propagate long distances without encountering lossy boundaries. Path D, which has a steeper angle than those associated with path C, is termed convergence zone propagation, which results from the upward refracting nature of the deep sound speed profile (Kuperman, 1997).

The depth at which the sound speed is the same as it is at the surface is called the critical depth, and defines the lower limit of the deep sound channel. The bottom bounce path E, in which sound reflects from the ocean bottom, is also periodic. At the right-hand side of the figure are paths in a shallow region such as a continental shelf (Kuperman, 1997).

13.3. Sonar

One of the most important applications of underwater acoustics is Sonar (SOund Navigation And Ranging), which dates back almost 100 years. The purpose of most sonar systems is to detect and localize a target, such as submarines, mines, fish, or surface ships. Other sonars are designed to measure some quantity, such as the ocean depth or speed of ocean currents or to image remote objects. Long-range detection sonars generally operate at frequencies below 50 Hz, while mine detection sonars operate at frequencies above 50 kHz, a frequency range greater than 3 decades (Barger, 1997). Sonar systems may be either active or passive.

13.3.1. Active Sonars

An active sonar system sends out a sound signal and detects one or more reflections. In modern sonar, very sophisticated systems are used in order to maximize range and accuracy. The most common active system, called an echo sounder, consists of a transmitter, a receiver, and a display. Systems range in complexity from the "fish finders" that are sold in sporting goods stores to the multibeam systems that are used by commercial fishermen and navies.

A typical echo sounder has a time (or depth) varying gain to compensate for the $1/r^2$ dependence of the echo amplitude on range. Echoes also come from the multiple reflection path from the ocean bottom to the surface and back to the receiver.

A side-scanning sonar is an echo sounder that is pointed sideways. The sending transducer produces a fan-shaped beam, and the receiver has a time-variable gain to compensate for range. Side-scanning sonars are used in geological studies to give images of rough features on the sea floor and also to locate objects such as sunken ships (Medwin and Clay, 1998). A very high resolution side-scanning sonar, using a 1.5 MHz beam, has been used for mine detection. The forward motion of the towed system produces the second dimension for a plot of the searched area (McKinney, 2002).

13.3.2. Multibeam Sonar

Mapping with sonar is quite different from mapping with radar. Radar pulse travel time, for a range of 30 km, is only 2×10^{-4} s so a complete 360° image can be made in less than 0.1 s, during which an airplane (flying at about the speed of sound) has traveled about 30 m. On the other hand, the time required for sonar to range to 30 km is about 4 s. In a sequential data acquisition system that takes one echo at a time, several hours would be required for one 360° image at comparably high resolution. Thus sonar data must be acquired simultaneously from many different beams.

A multiple sonar system for sea-floor mapping is shown schematically in Fig. 13.2. Since these systems are usually mounted on the hull of a ship, the receiving array points in different directions as the ship pitches and rolls. The data-reduction system must compensate for the ship motions and the direction in which the receiving array is pointing when the echoes arrive (Medwin and Clay, 1997).

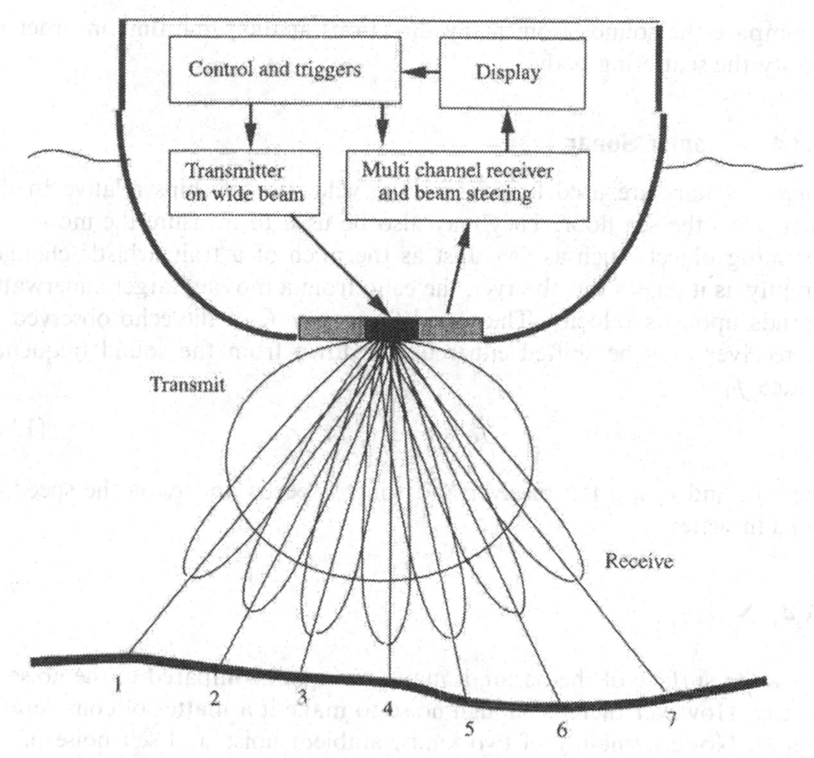

Fig. 13.2. Schematic of a multichannel sonar system. The transmitter sends a broad beam. By adjusting time delays of the receiving elements, the multi element receiving array is preformed to a set of narrow beams that scan from port to starboard and measure the depths to various positions. (Reprinted from Medwin and Clay, 1998 with permission from Elsevier.)

13.3.3. Passive Sonars

Passive sonars listen to and interpret the sounds that are in the ocean. The sound receiver may be a single hydrophone or it may be an array of hydrophones. The receiver is generally located in a quiet place, such as the ocean bottom, or towed behind a quiet ship. Signals may follow a direct path from the source, reflect off the ocean bottom, or reflect from some other surface.

Receiver arrays, beam formers, and signal conditioners do not differ significantly from those used in active systems. The weaker signal usually requires some type of filter plus a rather long time averaging. The phrase "acoustic daylight" has been applied to natural and man-made sounds in the sea that scatter off objects in the ocean. Since the water is acoustically transparent, images can be constructed from the scattered sound, just as scattered light allows us to see objects. Since natural sounds are variable in frequency spectrum, amplitude, and phase, the task is more difficult than viewing an image in daylight. Generally signal processing equipment is used to compare the sounds from many directions at the same time in order to identify the scattering body.

13.3.4. Doppler Sonar

Doppler sonars are used to measure the velocities of ships relative to the water or to the sea floor. They may also be used to measure the motion of swimming objects such as fish. Just as the pitch of a train whistle changes abruptly as it passes the observer, the echo from a moving target underwater depends upon its velocity. The sound frequency f_r of the echo observed at the receiver may be shifted either up or down from the sound frequency emitted f_0.

$$f_r = \frac{c_w \pm v_r}{c_w \mp v_s} f_0 \tag{13.5}$$

where v_r and v_s are the receiver and source speeds and c_w is the speed of sound in water.

13.4. Noise

Below the surface of the ocean, it may seem quiet compared to the noise of the city. However there is enough noise to make it a matter of considerable interest. Noise is mainly of two kinds: ambient noise and self noise of the receiver and its supporting system or carrier.

13.4.1. Ambient Noise

Ambient noise sources include wind noise, noise of ocean mammals, noise of ships, and turbulence. Fig. 13.3 shows survey of noise in the open ocean by Wenze (1962). Although it was done in 1962, it is still considered the most definitive survey. Wenze distinguishes between prevailing and intermittent (or local) noise mechanisms. The prevailing noises are generally considered the more important.

At the lowest frequencies, turbulence can induce pressure fluctuations at a hydrophone; they depend upon the hydrophone size and shape as well as the nature of the turbulence. Within the operating range of most sonars, however, these are generally overshadowed by other noise. Noise radiated by ships is distributed broadly in frequency, but at long distances its spectrum is shaped by the low-pass filtering of the ocean, so that it peaks in the 10–100 Hz range. In the oceans around the United States and Europe, distant shipping is the dominant noise in this frequency range, while in the southern oceans around Australia noise from wind, waves and biological sources is much more important (Cato and McCauley, 2002).

In the range of about 100 to 10,000 Hz, noise from the formation and oscillation of air bubbles created by breaking waves dominate the spectrum.

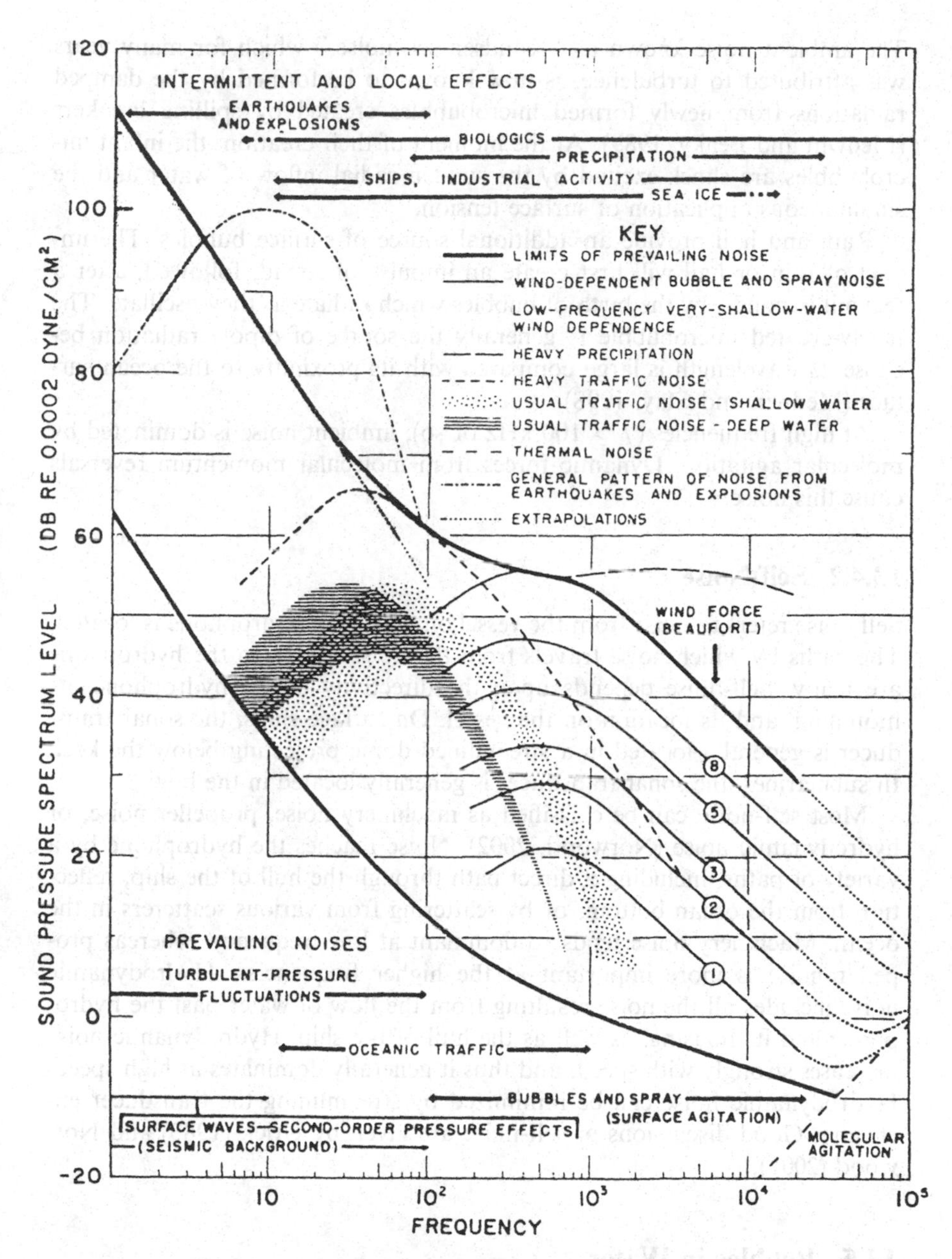

Fig. 13.3. Composite of ambient-noise spectra, summarizing results and conclusions about spectrum shape, level, and probable sources of ambient noise between 1 Hz and 100 kHz (Wenze, 1962). Note that 0.0002 dyne/cm^2 = 20 μ Pa.

The ambient noise known as "Kundsen sea noise," which for many years was attributed to turbulence, is now known to be formed by the damped radiations from newly formed microbubbles created by spilling breakers (Medwin and Beaky, 1989). At the moment of their creation, the infant microbubbles are shock-excited by the sudden radial inflow of water and the simultaneous application of surface tension.

Rain and hail provide an additional source of surface bubbles. The impact of rain or hail will first create an impulse of sound, followed, after a few milliseconds, by the birth of bubbles which radiate as they oscillate. The newly-created microbubble is generally the source of dipole radiation because its wavelength is large compared with its proximity to the ocean surface (Medwin and Clay, 1998).

At high frequencies ($f > 100$ kHz or so), ambient noise is dominated by molecular agitation. Dynamic forces from molecular momentum reversals cause this noise.

13.4.2. Self-Noise

Self-noise refers to noise from the vessel on which the hydrophone is located. The paths by which noise travels from the noise source to the hydrophone are many. Self-noise depends upon the directivity of the hydrophone, its mounting, and its location on the vessel. On surface ships, the sonar transducer is generally located in a streamlined dome projecting below the keel. In submarines, the sonar transducer is generally located in the bow.

Most self-noise can be classified as machinery noise, propeller noise, or hydrodynamic noise (Norwood, 2002). Noise reaches the hydrophone by a variety of paths, including a direct path through the hull of the ship, reflection from the ocean bottom, or by scattering from various scatterers in the ocean. Machinery noise tends to dominant at low frequency, whereas propeller noise is more important at the higher frequencies. Hydrodynamic noise includes all the noise resulting from the flow of water past the hydrophone and its housing, as well as the hull of the ship. Hydrodynamic noise increases strongly with speed, and thus it generally dominates at high speed. Hydrodynamic noise can be minimized by streamlining the transducer enclosure. Good discussions of self-noise are given by Urick (1983) and Norwood (2002).

13.5. Bubbles in Water

We have already mentioned ocean bubbles as a source of ambient noise (Section 13.4.1). Bubbles have several other acoustical properties as well. They scatter sound, and they affect the speed of sound and the attenuation as well. Medwin and Clay (1998) include a chapter on bubbles, and this brief discussion will borrow heavily from that.

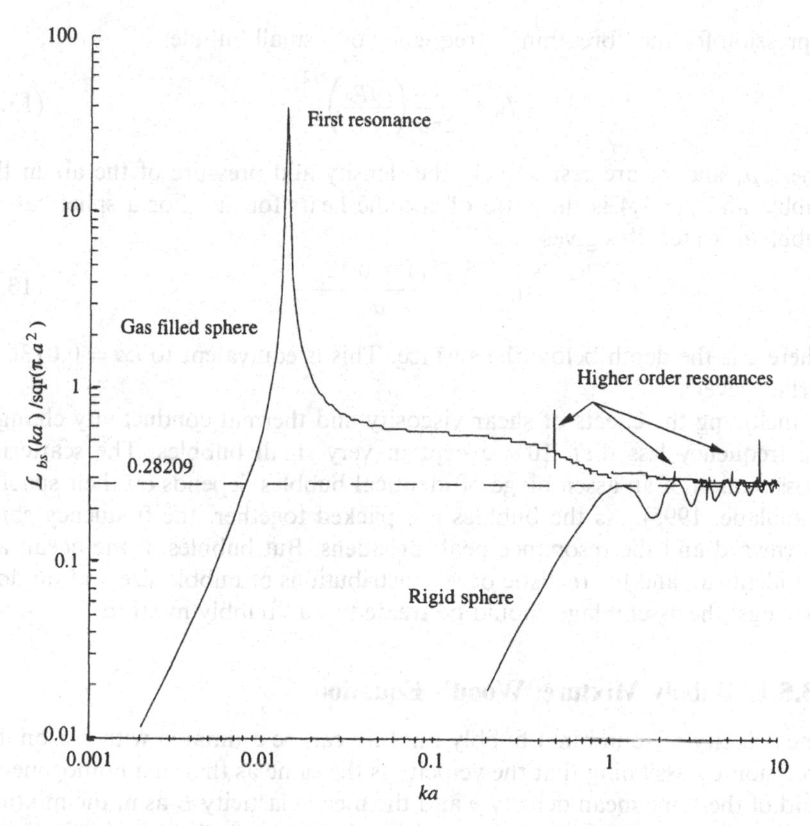

Fig. 13.4. Relative back scattering length of a spherical air bubble at sea level compared with that of a rigid sphere. (Reprinted from Medwin and Clay, 1998 with permission from Elsevier.)

While investigating the military implications of sound scatter from rough seas in the 1950s, underwater acousticians discovered that much of the scatter came from below the surface, presumably from bubbles created by the breaking waves. The scattering of sound by a fluid-filled sphere is quite different from that of a rigid sphere, as shown in Fig. 13.4. The large resonance peak at $ka = 0.0136$ was noted by Minnaert (1933) in research on the musical sound of running water. For $ka \ll 1$, a spherical bubble can be treated as having an equivalent mass, stiffness, and mechanical resistance. The equivalent mass is due to the inertia of the adjacent layer of water that has essentially the same radial displacement as the bubble surface. The compressibility of the bubble volume (plus the surface tension for very small bubbles) determines the spring stiffness (Medwin and Clay, 1998).

It can be shown (Medwin and Clay, 1998) that the effective mass m is given by

$$m = 4\pi a^3 \rho_w, \tag{13.6}$$

where a is the radius and ρ_w is the density of the water. This gives an expression for the "breathing" frequency of a small bubble:

$$f_b = \frac{1}{2\pi a}\left(\frac{3\gamma p_a}{\rho_A}\right)^{1/2}, \tag{13.7}$$

where ρ_a and p_a are respectively the density and pressure of the air in the bubble and $\gamma = 1.4$ is the ratio of specific heats for air. For a spherical air bubble in water, this gives:

$$f_b = \frac{3.25(1 + 0.1z)^{1/2}}{a}, \tag{13.8}$$

where z is the depth below the surface. This is equivalent to $ka = 0.0136$ at surface level.

Including the effects of shear viscosity and thermal conductivity changes the frequency less than 10% except in very small bubbles. The scattering cross section of an assemblage of identical bubbles depends on their spacing (Feuillade, 1995). As the bubbles are packed together, the frequency shifts downward and the resonance peak broadens. But bubbles in the ocean are not identical, and for realistic ocean distributions of bubble size and random spacings, the assemblage should be treated as a "bubbly mixture."

13.5.1. Bubbly Mixture: Wood's Equation

The velocity of sound in a bubbly mixture can be estimated with reasonable precision by assuming that the velocity is the same as that in a homogeneous fluid of the same mean density ρ and the mean elasticity E as in the mixture. Letting ρ_1, E_1 and ρ_2, E_2 represent the density and elasticity of the constituents 1 and 2 and x the proportion of the first constituent by volume, Wood (1955) shows that the mean velocity will be:

$$c = \left(\frac{E}{\rho}\right)^{1/2} = \left\{\frac{E_1 E_2}{[xE_2 + (1 - x)E_1][x\rho_1 + (1 - x)\rho_2]}\right\}^{1/2}. \tag{13.9}$$

The resulting velocity of sound, from pure water to pure air, is shown in Fig. 13.5. Note that the velocity shows a minimum around 10% air, which is a fairly high bubble concentration. The minimum velocity is only about 1/15 of the velocity of sound in air. Under such conditions the mixture may be regarded as "froth," the water serving to load the air-bubbles (the springs) and consequently to lower the velocity of sound.

Wood (1955) also shows that the simple mixture model predicts a large reduction in sound intensify in a bubbly mixture. This is observed in the sea. The noise of a ship's propeller is reduced by the bubbly water in the wake. The loss increases rapidly as the proportion of air to water increases.

13.5.2. Instant Coffee: A Bubbly Mixture

The dependence of the velocity and absorption of sound on bubble density can be easily demonstrated by adding some instant coffee to hot water in a

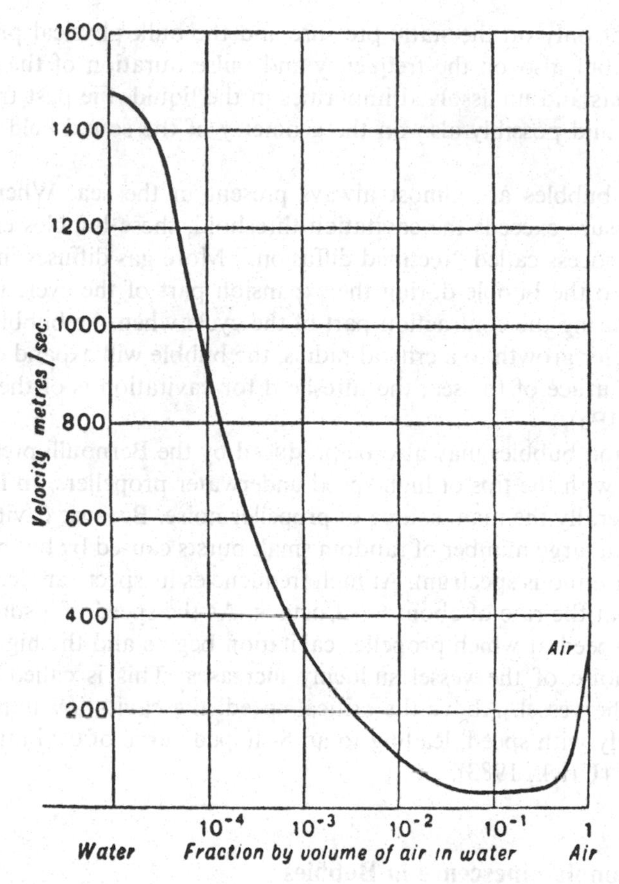

Fig. 13.5. Velocity of sound in an air-water mixture (Wood, 1955).

coffee cup. Tapping the bottom of the cup with a spoon before the powder is added gives a clear tone with an easily identified musical pitch. As the instant coffee foams up, the pitch drops (by two octaves or more), and the clear ring gives way to a rather dull "thud." As the bubbles density decreases, the pitch slowly returns to the original water-only pitch (Morrison and Rossing, 2002). A similar but smaller effect is also observed in beer (Bragg, 1968) and hot chocolate (Crawford, 1982).

13.5.3. Cavitation

An interesting phenomenon called cavitation occurs when sound waves of high intensity propagate through water. When the rarefaction tension phase of the sound wave is great enough, the medium ruptures and cavitation bubbles appear. It might be expected that sound would induce cavitation when the peak sound pressure exceeds the difference between the static and vapor pressures, but that is only approximately correct. The cavitation onset depends not only on the static pressure and the bulk physical properties of the liquid, but also on the frequency and pulse duration of the sound, the dissolved gas and undissolved impurities in the liquid, the past treatment of the liquid, and possibly also on the geometry of the sound field (Strasberg, 1959).

Minute bubbles are almost always present in the sea. When the peak sound pressure exceeds the cavitation threshold, these bubbles expand rapidly by a process called "rectified diffusion." More gas diffuses inward from the liquid to the bubble during the expansion part of the cycle than moves outward during the contraction part of the cycle when the bubble surface is smaller. After growth to a critical radius, the bubble will expand explosively. Near the surface of the sea, the threshold for cavitation is of the order of 1 atm (0.1 MPa).

Cavitation bubbles may also be produced by the Bernoulli pressure drops associated with the tips of high-speed underwater propellers. In fact, cavitation is generally the main source of propeller noise. Because cavitation noise consists of a large number of random small bursts caused by bubble collapse, it has a continuous spectrum. At high frequencies its spectrum decreases with frequency at the rate of about 6 dB/octave. As the speed of a ship increases, there is a speed at which propeller cavitation begins and the high-frequency radiated noise of the vessel suddenly increases. This is called the critical speed of the vessel. Above the critical speed, the cavitation noise increases more slowly with speed, leading to an S-shaped curve of cavitation noise vs ship speed (Urick, 1983).

13.5.4. Sonoluminescence in Bubbles

Sonoluminescence is a process in which gas bubbles in a liquid are nucleated, caused to grow, and then collapse by ultrasound. As the bubbles collapse, they give off flashes of light, which indicate that internal temperatures have reached thousands of degrees. The extremely short flashes of light occur at regular intervals, indicating that the bubbles grow and collapse repeatedly.

Although sonoluminescence has been known since the 1930s, it was not until Gaitan and colleagues developed a technique for acoustically levitating a single bubble and driving it at its resonance frequency that our understanding of sonoluminescence began. At a certain amplitude, the bubble begins to give off flashes of light (Gaitan et al., 1992). In single-bubble sonoluminescence, a small gas bubble that has been acoustically levitated and driven into large amplitude volume oscillations by the sound field radiates visible light each acoustic cycle (Crum, 1994). The collapsing bubble apparently generates an imploding shock wave that compresses and heats the gas in the bubble. The light emissions are extremely short in duration (<50 ps) and their spectrum suggests that the gas temperature exceeds that of the surface of the sun.

References

Bragg, W. (1968). "The World of Sound." Dover, New York.

Barger, J.E. (1997). Sonar systems. "Encyclopedia of Acoustics," ed. M.J. Crocker. Wiley, New York. Chap. 49.

Cato, D.H. and McCauley, R.D. (2002). Australian research in ambient sea noise. *Acoustics Australia* **30**, 13–20.

Crawford, F.S. (1982). The hot chocolate effect. *Am. J. Phys.* **50**, 389–404.

Crum, L.A. (1994). Sonoluminescence. *Phys. Today* **47**(9), 22–29.

Dyer, I. (1997). Ocean ambient noise. "Encyclopedia of Acoustics," ed. M.J. Crocker. Wiley, New York. Chap. 48.

Feuillade, C. (1995). Scattering from collective modes of air bubbles in water and the physical mechanism of superresonances. *J. Acoust. Soc. Am.* **98**, 1178–1190.

Gaitan, D.F., Crum, L.A., Roy, R.A. and Church, C.C. (1992). Sonoluminescence and bubble dynamics for a single, stable cavitation bubble, *J. Acoust. Soc. Am.* **91**, 3166–3188.

Kuperman, W.A. (1997). Propagation of Sound in the Ocean. "Encyclopedia of Acoustics," ed. M.J. Crocker. Wiley, New York. Chap. 36.

McKinney, C.M. (2002). The Early History of High Frequency, Short Range, High Resolution, Active Sonar. *ECHOES* **12**(2), 4–7 (also presented at the 142nd ASA meeting, Dec. 2001).

Medwin, H. and Beaky, M.M. (1989). Bubble sources of the Knudsen sea noise spectra. *J. Acoust. Soc. Am.* **86**, 1124–1130.

Medwin, H. and Clay, C.S. (1998). "Fundamentals of Acoustical Oceanography." Academic Press, Boston.

Minnaert, M. (1933). On musical air bubbles and the sounds of running water. *Phil. Mag.* **16**, 235–248.

Morrison, A. and Rossing, T.D. (2002). Sound of a cup with and without instant coffee. Paper 3aMU3, 143rd meeting, Acoust. Soc. Am.

Norwood, C. (2002). An introduction to ship radiated noise. *Acoustics Australia* **30**, 21–25.

Strasberg, M. (1959). Onset of ultrasonic cavitation in tap water. *J. Acoust. Soc. Am.* **31**, 163–176.

Urick, R.J. (1983). "Principles of underwater sound," 3rd ed. McGraw-Hill, New York. Chap. 11.

Wenze, G.M. (1962). Ambient noise in the ocean: spectra and sources. *J. Acoust. Soc. Am.* **34**, 1936–1956.

Wood, A.B. (1955). "A textbook of sound." Macmillan, New York.

Select Bibliography

This select bibliography refers the interested reader to sources for further reading in acoustics topics not covered in any detail in the present book.

General Acoustics

Fundamentals of Acoustics 3rd ed. by L.E. Kinsler, A.R. Frey, A.B. Coppens and J.V. Sanders (Wiley, New York 1982)

The Science of Sound 2nd ed. by T.D. Rossing (Addison Wesley, Reading, MA 1990)

Theoretical Acoustics

Acoustics by A.D. Pierce (McGraw-Hill, New York 1981; reprinted Acoustical Society of America 1989)

Theoretical Acoustics by P.M. Morse and K.U. Ingard (McGraw-Hill, New York 1968)

Vibration and Sound 2nd ed. by P.M. Morse (McGraw-Hill, New York 1948; reprinted Acoustical Society of America 1981)

Applied Acoustics

Acoustics by L.L. Beranek (McGraw-Hill, New York 1954; reprinted Acoustical Society of America 1986)

Physical and Applied Acoustics by E. Meyer and E-G Neuman (Academic Press, New York 1972)

Vibrations

Simple and Complex Vibratory Systems by E. Skudrzyk (Pennsylvania State University Press, University Park 1968)

Sound, Structures and Their Interaction 2nd ed. by M.C. Junger and D. Feit (MIT Press, Cambridge, MA 1986)
Structure-Borne Sound by L. Cremer, M. Heckl and E. Ungar (Springer-Verlag, Berlin 1988)

Musical Acoustics

Acoustics and the Performance of Music by J. Meyer (Verlag Das Musikinstrument, Frankfurt 1978)
Fundamentals of Musical Acoustics by A.H. Benade (Oxford University Press, New York 1976; reprinted Dover, New York 1992)
The Physics of Musical Instruments by N.H. Fletcher and T.D. Rossing (Springer-Verlag, New York 1991)

Biological Acoustics

Acoustic Systems in Biology by N.H. Fletcher (Oxford University Press, New York 1992)
Hearing by W.L. Gulick, G.A. Gescheider and R.D. Frisina (Oxford University Press, New York 1989)

Architectural Acoustics

Music, Acoustics and Architecture by L.L. Beranek (Wiley, New York 1962)
Principles and Applications of Room Acoustics (Vols 1, 2) by L. Cremer and H.A. Müller (trans. T.J. Schultz) (Applied Science, London 1982)
Room Acoustics by H. Kuttruff (Applied Science, London 1973)

Problems

Chapter 1

1.1. A vibrating system is described by the expression: $x = 2\cos(12.57t + 1.05)$ cm. What are the displacement, velocity, and acceleration at $t = 0$ and $t = 2$ s? What are the maximum values of displacement, velocity, and acceleration?

1.2. For a body executing simple harmonic motion around $x = 0$ at a frequency of 3 Hz, the displacement and velocity at $t = 0$ are 0.25 m and 2 m/s.
   (a) Find the amplitude, the maximum velocity, the maximum acceleration, and the phase angle.
   (b) Write the equation of motion and an expression for x.

1.3. Two particles are oscillating along parallel tracks with the same frequency and amplitude. What is their relative phase angle if they pass each other moving in opposite directions
   (a) at their equilibrium position;
   (b) where their displacements are half their amplitude?

1.4. Given that $\tilde{x} = a + jb$ and $\tilde{y} = c + jd$,
   (a) Find $\mathrm{Re}\,\tilde{x} \cdot \mathrm{Re}\,\tilde{y}$; $\mathrm{Re}(\tilde{x}\tilde{y})$; magnitude $(\tilde{x}\tilde{y})$; magnitude $\tilde{x} \cdot$ magnitude $\tilde{y}$.
   (b) Express $\tilde{x}\tilde{y}$ in the form $Ae^{i\theta}$.

1.5. Derive Eqs. (1.24) and (1.25) from Eq. (1.22).

1.6. A vibrating system consisting of a 2-kg mass and a spring with a spring constant of 19.74 N/m is set into oscillation with an amplitude of 1 m. Find the potential energy, the kinetic energy, and the total energy at $t = 0$, $t = 0.25$ s, $t = 0.5$ s, and $t = 1$ s.

1.7. An oscillator with negligible damping has a natural frequency of 10 Hz. What is the natural frequency of this same oscillator with damping characterized by $\alpha = 3\ \mathrm{s}^{-1}$? If the phase angle is $\pi/2$, write an expression for the displacement of the damped oscillator.

1.8. Show that two springs with spring constants K_1 and K_2 have an effective spring constant $K_1K_2/(K_1 + K_2)$ when connected in series and $K_1 + K_2$ when connected in parallel.

1.9. A mass of 0.25 kg oscillates at 3 Hz when supported by a spring of length 0.3 m. If this spring is to be replaced by an air spring of the same length, what diameter should its piston have?

1.10. A uniform spring having a length L and a spring constant K is cut into two pieces, one piece being n times as long as the other, where n is an integer. Write the spring constants of the two segments in terms of L, n, and K. What are the values of K_1 and K_2 for $n = 3$?

1.11. At what frequency does x in Eq. (1.60) have its maximum magnitude?

1.12. Draw equivalent electrical circuits for the mechanical systems shown, and derive expressions for their resonance frequencies:

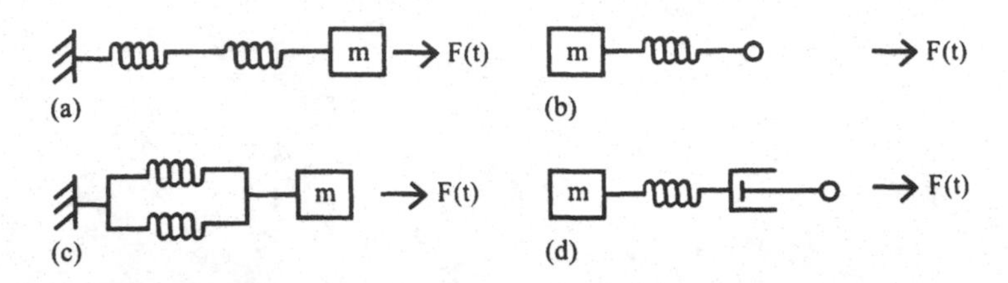

1.13. A force $f(t) = F \sin 2\pi ft$ is abruptly applied to an initially stationary oscillator having a natural frequency f_0 and $Q = 10$. Compare the beat rates during the transient for $f/f_0 = 0.2, 0.8, 1.0, 1.2, 2.0$, and 4.0 to f and f_0.
How many cycles of oscillation will occur before the transient behavior has decreased to 37% ($1/e$) of its initial value? (Compare your solution to Fig. 1.15).

1.14. Make a graph of A vs. ω from Eq. (1.80) using the following values: $m = 0.5$ kg, $F_0 = 2$N, $K = 50$ N/m, $b = 0.2$ N/m³.

Chapter 2

2.1. What is the speed of transverse waves on a 2-mm diameter steel string ($\rho = 7700$ kg/m³, $E = 19.5 \times 10^{10}$ Pa) having a tension of 500 N?

2.2. (a) A nylon guitar string has a mass of 8.3×10^{-4} kg/m and a tension of 56 N. Find the speed of transverse waves on the string.
(b) If the string is 63 cm long, what is its fundamental frequency of vibration?

2.3. By carrying out the integrals in Eq. (2.20), show that the amplitudes of the first six harmonics of a string plucked with amplitude h at one-fifth its length are $0.744h$, $0.301h$, $0.1338h$, $0.0465h$, 0, and $0.0207h$.

2.4. Applying the coefficients in Prob. 2.3 to the string in Prob. 2.2, find the energy of each mode when the string is plucked with an amplitude of 5 mm at one-fifth its length.

2.5. Find the characteristic impedance of the steel string in Prob. 2.1 and the guitar string in Prob. 2.2.

2.6. Calculate the input impedance of the guitar string in Prob. 2.2 driven sinusoidally at the center at its fundamental frequency. [Hint: consider it to be two strings of length $L/2$ driven in parallel.]

2.7. Suppose that one end of the guitar string in Prob. 2.2 is fastened to a bridge that can be characterized as having an effective mass of 2 kg at the fundamental frequency of the string. Find the fundamental frequency. How does it compare to the fundamental frequency calculated in Prob. 2.2 (which assumed a completely fixed end)?

2.8. Find the frequencies of the first three longitudinal modes of vibration of the guitar string in Prob. 2.2 (assume $E = 5 \times 10^9$ Pa, $\rho = 1140$ kg/m³). Compare these to the frequencies of the first three transverse modes.

2.9. (a) Find the frequencies of the first three transverse modes of vibration in an aluminum bar 356 × 38 × 9.5 mm ($E = 7.1 \times 10^{10}$ Pa, $\rho = 2700\,\text{kg/m}^3$) with free ends.
(b) Find the frequencies of the first three longitudinal modes of vibration.
(c) Find the frequencies of the first three torsional modes of vibration.

2.10. Sketch the approximate configuration of the nodal lines for each of the nine modes in Prob. 2.9.

2.11. Calculate the inharmonicity constant β and the frequency f_2 of the second partial of the following strings:
(a) A_0 string on a grand piano with a steel core diameter of 1.4 mm, $L = 1.35$ m, $\rho = 7700\,\text{kg/m}^3$, $E = 1.95 \times 10^{11}\,\text{N/m}^2$, $T = 941$ N, $f_1 = 27.5$ Hz [data from Podelsak and Lee, JASA *83*, 305–317 (1988)].
(b) Solid steel A_4 string on an upright piano with a diameter of 1.04 mm, $L = 53.5$ cm, $f_1 = 440$ Hz [data from Fletcher, JASA *36*, 203–209 (1964)].

2.12. By solving Eq. (2.62), show that the bending mode frequencies of a bar with free ends are given by Eq. (2.63).

2.13. What tension would a steel wire 1 mm in diameter require in order that the transverse and longitudinal wave speeds are equal. Is this possible in practice?

Chapter 3

3.1. Find the first four vibrational frequencies of a timpani drumhead 66 cm in diameter, having a surface density of $0.26\,\text{kg/m}^2$ and a tension of 4000 N/m. Are they harmonic?

3.2. Find the first four vibrational mode frequencies of a square membrane having the same area, surface density, and tension as the circular membrane in Prob. 3.1.

3.3. A cylindrical 1-kg mass with a diameter of 2 cm is placed on the drumhead in Prob. 3.1. How large is the deflection due to its weight? How large would the deflection be if the same mass had a diameter of 1 cm?

3.4. Why is the (2, 0) mode the lowest mode in a circular plate with a free edge? [Why are there no (0, 0) and (1, 0) modes?]

3.5. From Eq. (3.20), derive an expression for Poisson's ratio ν in terms of f_+ and f_-. Is measuring f_+ and f_- in a square aluminum plate a practical way to measure ν in aluminum?

3.6. Compare the frequencies of the lowest vibrational modes in a square plate with free edges, with simply-supported (hinged) edges, and with clamped edges.

3.7. Using the values for E, h, a, and ρ following Eq. (3.24), show that the (1, 1) mode is raised about 0.004% by the bending stiffness of the membrane. How much would the frequency change due to bending stiffness be if the membrane were twice as thick?

3.8. Verify that an elliptical plate with $a/b = 2$ has a fundamental frequency 37% greater than a circular plate of the same material having the same area and thickness.

Chapter 4

4.1. Two pendulums consist of 500-g masses suspended by cords 2 m long. They are connected by a 30-cm spring having a spring constant $K = 0.2$ N/m.

(a) Find the frequencies of the two normal modes of vibration.
(b) If the system is started with pendulum A at rest and pendulum B swinging, how long will it be before pendulum B comes to rest?
(c) If pendulum B is displaced 10 cm and released, describe the resulting motion (hint: what are the amplitudes of the two normal modes?)
(d) Would you describe this system as having strong coupling, weak coupling, or neither one?

4.2. In the vibrating system in Fig. 4.6, let $K = 0.2$ N/m, $K_c = 0.1$ N/m, and $m_A = m_B = 500$ g. Let a force $F = 0.5 \cos \omega t$ newtons be applied to mass A.
(a) Find the frequencies of all the resonances and antiresonances.
(b) Find the amplitudes of mass A and mass B at the antiresonance frequency f_A and at $2f_A$.
(c) Compare the phases of m_A and m_B just below and just above the second resonance frequency. (Which mass leads and by what phase angle?)

4.3. In the circuit shown in Fig. 4.9, $\omega_1 = 0.8\omega_a$. What is the ratio of mutual inductance M to self inductance L_a (assuming $L_a = L_b$)?

4.4. In the circuit in Fig. 4.12, let $C_a = 1000$ pF and $L_a = L_b = 100$ mH. Let C_b vary and make a plot similar to Fig. 4.13 ($\log \omega$ vs. $\log \omega_b$) for $C_c = 2C_a$ and $C_c = 4C_a$.

4.5. A freely supported drum has two identical heads with the same tension. Modal analysis indicates two modes with frequencies of 160 and 330 Hz, in which all parts of the batter head move in phase.
(a) Describe the motion of the entire drum at each of these frequencies.
(b) If the drum is placed in a bed of sand so that only the batter head can vibrate, what would the frequency of its lowest mode be?

4.6. A guitar string having a mass $m = 3.5$ g is coupled to a soundboard having an effective mass $M = 320$ g and $Q = 60$ at its lowest resonance ($f = 110$ Hz). Is this a case of strong coupling or weak coupling? Estimate the normal mode splitting from Fig. 4.18.

Chapter 5

5.1. Show, from a qualitative argument, that the output of a self-excited van der Pol oscillator contains no even-numbered harmonics.

5.2. Consider a single-mode system as described by Eq. (5.1) but with the spring-constant K a general weakly nonlinear function of y as in a real spring. Show that the nonlinearity generates harmonics of all orders, and that for small fundamental amplitude, the amplitude of the nth harmonic is proportional to the nth power of the amplitude of the fundamental. [Use $\cos^n \theta = 1/2^n(e^{j\theta} + e^{-j\theta})^n = 1/2^{n-1}(\cos n\theta + n\cos(n-2)\theta + \cdots)$]

5.3. Suppose that, in the system of Eq. (5.3), $g(y, \dot{y}, t) = -\omega_0^2 \beta y^3$, corresponding to a stiffening nonlinearity in the restoring force. Calculate the dependence of the natural frequency ω of the oscillator upon the amplitude a.

5.4. Consider a taut string with fundamental transverse frequency ω_0, damping factor k, and tension T_0, and suppose that T_0 is varied by adding to it a small time-varying tension $T \cos \omega t$. Find the initial rate of growth of the amplitude a of the mode ω_0 if $\omega = 2\omega_0$. This system is called a parametric oscillator.

5.5. This is a project, rather than a simple problem! Write a computer program to

simulate the behavior of the system described by Eq. (5.1) for the case in which the spring constant K is nonlinear, of the form $K = K_0(1 + \alpha y + \beta y^2)$, and the forcing function $F(t)$ is sinusoidal and of frequency ω. [Hint: Break the second-order DE into two first-order DEs by writing $dy/dt = z$. Integrate these two equations simultaneously.] Plot the output on the screen as
(a) an orbit in phase space (y, z) with t as a parameter;
(b) a Poincaré section in phase space, by plotting only one (y, z) point for each cycle of $F(t)$;
(c) a time series $y(t)$.
An interesting region to investigate has ω close to the resonance frequency $(K/m)^{1/2}$ or close to twice this frequency, $R/m \approx 0.02$, $\alpha = 0.3$, $\beta = 0.1$, and F/m in the range 10 to 25. Examples of simple orbits, multiple orbits, and chaotic behavior can be found. Display (b) will show the strange attractor for the chaotic case.

Chapter 6

Note: In these problems, assume the temperature to be 20°C unless told otherwise.

6.1. Calculate the amplitude of motion of the air near the threshold of human hearing for a sound of frequency 1000 Hz (taking this as 0 dB re 20 μPa rms). Repeat this near the threshold of discomfort at 120 dB.

6.2. Using Eqs. (6.33) and (6.20), find the intensity of a sound wave having an rms sound pressure of 20 μPa at 0°C. Compare this to the reference level I_0 for sound intensity. At what temperature does $L_p = L_I$ for a plane progressive wave?

6.3. The "ultimate" sound wave would be one that causes the total pressure to drop to zero during a rarefaction. Assuming the acoustic wave equations remain linear at such large amplitudes (they don't, of course), what would be the sound pressure level for such a wave?

6.4. The fact that the temperature in a sound wave is greater where the air is compressed than where it is expanded suggests that compressions propagate slightly faster than do rarefactions. Sketch pressure waveforms for waves of low, medium, and high amplitude to illustrate how this leads to steep pressure gradients at the leading edges of an intense sound wave (and eventually to the formation of shock waves).

6.5. (a) From the density (1000 kg m^{-3}) and bulk modulus (2.2×10^9 N m^{-2}) of water, calculate the velocity of sound in water and the wave impedance of water.
(b) Calculate, in decibels, the transmission coefficient for a sound wave normally incident from air on a water surface.

6.6. A loudspeaker is specified as producing 93 dB per watt on-axis at 1 meter. What is its efficiency, assuming that its radiation is confined to a solid angle of 3 steradians?

6.7. The sound from an amplified outdoor concert is annoying (SPL = 60 dB) at a distance of 1 km in still air at night. Calculate the approximate acoustic power being radiated, and estimate the power output of the amplifiers being used. (Neglect atmospheric absorption.)

6.8. A concert hall is 20 m wide, 60 m long, and 10 m high. The side walls and ceiling have an acoustic absorption coefficient of 0.2, the stage wall 0.1, and the rear wall 0.5. When empty, the carpeted floor has $\alpha = 0.3$. What is the empty reverberation time? If the hall seats 1500 people, each of whom contributes an effective absorbance of 0.2 m^2, what is the reverberation time when the hall is full? Comment on the results from a practical viewpoint.

6.9. A Helmholtz resonator consists of a spherical cavity of radius 2 cm with a neck of diameter 1 cm. How long must the neck be to give a resonance frequency of 500 Hz?

6.10. What is the impedance, in acoustic ohms at 1 kHz, of the following:
(a) an open cylindrical pipe of length 3 cm and diameter 5 mm;
(b) a cavity of volume 20 cm^3;
(c) a plug of cotton wool that allows a steady flow of 1 liter of air per minute when the over-pressure is 1 cm water gauge (100 Pa)?

Chapter 7

7.1. A simple source of sound radiates spherical waves with an acoustic power of 5 mW at 300 Hz. Calculate the intensity and the amplitudes of acoustic pressure, acoustic particle velocity, acoustic displacement, and acoustic condensation at a distance of 0.5 m from the source.

7.2. Calculate *ka* for the following cases, and comment on the relative magnitudes of acoustic resistance *R* and acoustic reactance *X*:
(a) a drumhead 66 cm in diameter vibrating at 140 Hz;
(b) a loudspeaker 20 cm in diameter vibrating at 100 Hz;
(c) a tone hole in a clarinet radiating A_4 ($f = 440$ Hz).

7.3. A small loudspeaker set in the wall of a closed box produces a sound pressure level of 70 dB at 100 Hz at a point on its axis at a distance of 10 m. What reduction in the sound level do you expect if the back is taken off the speaker box, the depth of the box being 15 cm?

7.4. A small constant-flow (high impedance) source produces a sound pressure level at 200 Hz of 70 dB at a distance of 5 m in free air. What level do you expect at this distance if the source is placed just in front of a large brick wall? What level do you expect if it is placed in a corner where three mutually perpendicular walls meet?

7.5. From Eq. (7.30) and a table of Bessel functions, find, in terms of *ka*, the angle at which the intensity radiated by a circular piston of radius *a* set in a plane baffle is less than the on-axis value by 3 dB.

7.6. The loudspeaker in a small radio has a diameter of 5 cm. How large must be its vibration amplitude if it is to produce an output power of 1 mW at 100 Hz, assuming that the radio is sealed at the back? Comment on the design of small radios.

7.7. A guitar string has a diameter of 1 mm and vibrates with an amplitude of 2 mm when the note A_4 (440 Hz) is played loudly. If the string length is 60 cm, what is the power radiated directly from the string? How does this compare with the power radiated by the guitar body, if the sound pressure level measured at 1 m is 80 dB?

Chapter 8

8.1. What is the characteristic impedance of an infinite cylindrical pipe of radius 1 cm?

8.2. From Eq. (8.7) calculate the phase velocity and group velocity of a higher mode, specified by its value of q_{mn}, in a duct. Sketch the behavior of these two velocities as the frequency approaches the mode cutoff frequency from above.

8.3. The probe tube of a microphone has diameter 1 mm and length 10 cm. Assuming the tube to be matched to the microphone so that there are no reflections, calculate the tube attenuation in decibels at 100 Hz, 1 kHz, and 10 kHz.

8.4. An organ pipe has a length of 2 m and a diameter of 20 cm. Using the results shown in Fig. 8.9, calculate the percentage inharmonicity of its first five resonances due to variation of the open end correction. [In practice the variation of the correction at the mouth end of the pipe is much more important.]

8.5. Show that the expressions (8.24) and (8.25) for rigidly stopped and ideally open pipes reduce to the expressions (9.5) and (9.3) respectively when the pipe is short compared with the wavelength of sound considered.

8.6. An exponential horn has a throat diameter of 2 cm, a mouth diameter of 60 cm, and a length of 1 m. Calculate its cutoff frequency. How can we lower this frequency and still achieve the same horn gain?

8.7. What is the cutoff frequency for the (1, 0) mode in:
(a) an automobile exhaust tailpipe with a diameter of 50 mm;
(b) a trumpet bore with a diameter of 11 mm;
(c) a heating duct with a diameter of 30 cm?
Comment on whether you would expect to find acoustic waves other than plane waves in these tubes.

Chapter 9

9.1. What is the acoustic impedance of an infinite air-filled pipe of diameter 30 mm and of a ventilation duct 50 cm × 30 cm?

9.2. What is the acoustic impedance at 1 kHz of a freely moving solid piston of thickness 1 mm and density 2000 kg m^{-3} placed in each of the ducts of example 1? Does the piston effectively block the tube or duct (a) at 1000 Hz, (b) at 100 Hz, (c) at 10 Hz?

9.3. A loudspeaker cone has diameter 200 mm, mass 50 g, and resonance frequency 100 Hz. What will be its resonance frequency when it is mounted in one side of an otherwise closed box of volume 0.05 m^3?

9.4. A loudspeaker having a cone with acoustic impedance Z_c is mounted in a bass-reflex enclose with cavity impedance Z_v and port impedance Z_p. Find an equation from which the two lowest bass resonance frequencies can be calculated.

9.5. Find an expression from which can be calculated the resonance frequencies for a pipe of length l and cross section S closed at one end by a slack diaphragm of mass m (equivalent to a free piston). Show that this goes to the expected limits for very light and for very heavy diaphragms.

9.6. [This is more a project than a simple problem.] Write a computer program to evaluate and plot as a function of frequency the input impedance p/U given by (9.17) for the Helmholtz resonator shown in Fig. 9.6, assuming reasonable values for the physical dimensions and neglecting losses in compression of the air in the cavity.

Chapter 11

11.1. An auditorium has dimensions 40 m × 20 m and a ceiling height of 15 m. The front and back walls are covered with plywood paneling; the side walls and ceiling are plaster. The floor is wood. There are 1100 wooden seats. Estimate the reverberation time (500 Hz) when
(a) The hall is empty;
(b) Half the seats are filled;
(c) All the seats are occupied.

11.2. Estimate the time delay t_1 of the first reflected sound for a person seated near the center of the auditorium described in Problem 11.1. Does the first reflection arrive from the side or from overhead?

11.3. If the ceiling in this auditorium were covered with acoustical tile, by how much would the reverberation time be decreased?

11.4. Find the reverberation time at 8000 Hz for a very live room having a volume of 1000 m^3 when the temperature is 20°C and the relative humidity is 30%. Assume that absorption by the walls in negligibly small. Would your answer be different if $V = 100\ m^3$ instead?

11.5. Calculate the frequencies of the first three resonances of a room with dimensions 5 m × 10 m × 2.5 m. Do they have any significance acoustically?

11.6. Design a perforated panel absorber that would have a resonance frequency at 500 Hz.

11.7. Estimate the number of modes between 0 and 100 Hz in a room 10 m × 8 m × 3 m.

11.8. Find the Schroeder cutoff frequency in the auditorium given in Problem 11.1 and also in the room in 11.7 if the reverberation time is 1 second.

Chapter 12

12.1. A noise source and a listener are both 1 m above a hard surface and 20 m apart. At what frequencies will the first two maxima and minima in sound level occur at the receiver?

12.2. What is the cutoff frequency at a source-to-receiver distance of 40 m over grass-covered ground?

12.3. By what percentage does the sound velocity fluctuate when the temperature fluctuates by 5°C?

12.4. Far more people are affected by truck noise than by airplane noise. Why are there fewer comlaints, citizens, protest groups, etc. directed against highway noise?

12.5. (a) If an automobile traveling at 60 mi/hr emits 0.01 W of acoustical power, estimate the average continuous power from 100 million automobiles, each of which travels 10,000 mi per year. (b) Make an estimate of the peak acoustical power. (How many automobiles might be traveling at a peak hour of the day?)

12.6. Calculate the total force on a wall of a typical house due to an overpressure of 20 N/m^2 in a sonic boom.

Chapter 13

13.1. A simple echo sounder receives an echo 20 s after the transmitted pulse. At what distance is the target?

13.2. Show by direct substitution that 12.4 is a solution of equation 12.3.

13.3. If the impedance of the sea bed is 3 times as great as that of sea water, what is the reflection coefficient for sound at the bottom?

13.4. (a) Estimate the frequency of the lowest mode of oscillation in the water in a coffee cup assuming longitudinal wave propagation in a column of water 7 cm long with one open and one closed end.
(b) Assume that adding instant coffee produces a foam that is 50% air, what frequency would the lowest mode of oscillation have?

Answers to Selected Problems

1.1. $t = 0$: $x = 1$ cm, $v = -21.8$ cm/s, $a = -157$ cm/s^2
$t = 2$: $x = 0.98$ cm, $v = -21.9$ cm/s, $a = -155$ cm/s^2.

1.3. (a) $\phi = \pi$; (b) $\phi = \pm 2\pi/3$.

1.7. $f_d = 9.99$ Hz; $x = Ae^{-3t}\cos(62.76\,t + \pi/2)$.

1.9. $D = 0.015$ m.

1.13. $4f$, $0.25f$, 0, $0.5f$, $0.75f$; 3.18 cycles.

2.1. 144 m/s.

2.4. 3.03×10^{-3} J, 1.99×10^{-3} J, 8.84×10^{-4} J, 1.90×10^{-4} J, 0, 8.46×10^{-5} J.

2.5. 3.48 Ns/m, 0.216 Ns/m.

2.7. $206 + 3.3 \times 10^{-3}$ Hz; 2.7×10^{-3}% greater.

2.9. (a) 395 Hz, 1089 Hz, 2135 Hz.
(b) 7202 Hz, 14,404 Hz, 21,606 Hz.
(c) 2208 Hz, 4416 Hz, 6624 Hz.

2.11. (a) $\beta = 9.26 \times 10^{-3}$, $f_2 = 55.5$ Hz.
(b) $\beta = 0.0104$, $f_2 = 889.7$ Hz.

2.13. $T = 1.53 \times 10^5$ N; far greater than the breaking force.

3.1. 144 Hz, 229 Hz, 307 Hz, 330 Hz.

3.3. 17 mm, 20.5 mm.

3.5. $v = 1.388 \dfrac{f_+^2 - f_-^2}{f_+^2 + f_-^2}$

3.7. About 0.02%.

4.1. (a) 0.352 Hz, 0.380 Hz.
(b) 17.9 s
(c) $x_A = 0.05(\cos\omega_1 t - \cos\omega_2 t)$; $x_B = 0.05(\cos\omega_1 t + \cos\omega_2 t)$.

4.3. 0.5625.

4.5. (b) 259 Hz.

5.3. $\omega_0\left(1 + \frac{3}{8}\beta a^2\right)$

5.4. $\dot{a} = \left[\frac{\omega_0 T}{4T_0} - k\right]a$

6.1. 1.1×10^{-11} m; 1.1×10^{-5} m.

6.3. 191 dB.

6.5. 1480 m/s; 1.48×10^6 rayl; -29 dB.

6.7. 6 W acoustic into a hemisphere; about 1 kW unless high-efficiency speakers are used.

6.9. 2.1 cm when 3 mm end correction is allowed at each end of the neck.

7.1. 1.6×10^{-3} W/m^2; 1.15 Pa; 2.8×10^{-3} m/s; 1.5×10^{-6} m; 1.1×10^{-5} (amplitudes are peak, not rms).

7.3. 5 dB

7.5. 1.62/ka

7.7. 12.7×10^{-7} W or 30 dB less than that radiated by the body.

8.1. 1.32×10^6 acoustic ohms.

8.3. 0.5 dB; 1.6 dB; 5.2 dB.

8.6. 185 Hz.

8.7. 4 kHz, no?; 18 kHz, no; 670 Hz, yes.

9.1. 5.9×10^5 and 2770 acoustic ohms (Pa s/m^3), respectively.

9.3. 107 Hz

9.5. $\omega \cot \frac{\omega l}{c} = -\frac{\rho c S}{m}$

Name Index

Subject Index